TJ840 MER

HYDRAULIC CONTROL SYSTEMS

Herbert E. Merritt

Section Head
Hydraulic Components Section
Product Development Department
Cincinnati Milling Machine Company

JOHN WILEY & SONS,

New York • Chichester • Brisbane • Toronto • Singapore

26 25 24

Library of Congress Catalog Card Number: 66-28759
Printed and bound in the United States of America
ISBN 0 471 59617 5

Preface

Although hydraulic control dates from the water regulating devices of ancient times, the branch of this field concerning the hydraulic control of machinery has made the greatest progress in this century, particularly since World War II. The growth of hydraulic control has paralleled developments in transportation, farm and earth moving equipment, industrial machinery, machine tools, ship control, fire control, aircraft, missiles, and numerous other applications. Government and industry supported research at several universities—the Dynamic Analysis and Controls Laboratory at the Massachussetts Institute of Technology is especially noteworthy—has accelerated hydraulic control technology. Increased usage of hydraulic control has brought demands for rational design techniques to replace effective but costly and time-consuming cut-and-try procedures and for a classification of the knowledge for instruction.

This book should be useful to both practicing engineers and students and is at a level attained after a basic college course in feedback control theory. Its purpose is to present a rational and well-balanced treatment of hydraulic control components and systems. A course in fluid mechanics would be helpful but not essential. The book is particularly well suited as a text for a college-level course in hydraulic control. Selected topics could be used to supplement feedback control theory courses with some instruction on components.

The analyses of many hydraulic components—electrohydraulic servo-valves in particular—are involved and tedious. However, in every case I have tried to wring conclusive design relations from these analyses rather than leave a mess of equations for the reader to untangle. This has sometimes necessitated making judgments and rules of thumb with which the reader may not agree.

The arrangement of the book follows in a fairly logical sequence. After some introductory remarks in Chapter 1, the physical and chemical properties of the working fluid are discussed in Chapter 2. Fluid flow through various passages and basic hydraulic equations are covered in Chapter 3. Hence these first chapters are basically a review of applicable topics in fluid mechanics.

The next four chapters are devoted to components encountered in hydraulic servo controlled systems. The characteristics of hydraulic actuators are discussed in Chapter 4. Hydraulic control valves, chiefly spool and flapper types, are covered in Chapter 5. The combination formed by a valve or pump controlling an actuator is the basic power element in hydraulic control servos, and the various combinations are discussed quite thoroughly in Chapter 6. Chapter 7 is devoted to the principal types of electrohydraulic servovalve and includes a static and dynamic analysis of torque motors.

The remaining five chapters treat systems oriented topics. Chapter 8 covers the major types of electrohydraulic servo. Hydromechanical servos are touched briefly in Chapter 9 because many comments in the previous chapter are applicable. Systems often perform somewhat differently than anticipated because of nonlinearities, and Chapter 10 discusses the effect of these on performance. Practical suggestions concerning testing and limit cycle oscillation problems are also given. Chapter 11 covers some common control valves useful in power generation, and Chapter 12 treats hydraulic power supplies and their interaction with the control.

Material for this book was taken from a set of notes used to teach a course in hydraulic control to engineers in industry. Much new information has been included, and I have tried to improve older treatments. Experience and the available literature also were sources. For the latter, I am indebted to the many original contributors, too numerous to mention.

I am particularly grateful to my good friend Mr. George L. Stocking of the General Electric Company for contributions to Sections 5-6 and 5-7.

Finally, I would like to express appreciation to my fellow associates at the "Mill," especially to Mr. James T. Gavin, for their help and encouragment.

HERBERT. E. MERRITT

Cincinnati, Ohio
December 1966

Contents

I

Introduction

The increasing amount of power available to man that requires control and the stringent demands of modern control systems have focused attention on the theory, design, and application of control systems. Hydraulics—the science of liquid flow—is a very old discipline which has commanded new interest in recent years, especially in the area of hydraulic control, and fills a substantial portion of the field of control. Hydraulic control components and systems are found in many mobile, airborne, and stationary applications.

1-1 ADVANTAGES AND DISADVANTAGES OF HYDRAULIC CONTROL

There are many unique features of hydraulic control compared to other types of control. These are fundamental and account for the wide use of hydraulic control. Some of the advantages are the following:

1. Heat generated by internal losses is a basic limitation of any machine. Lubricants deteriorate, mechanical parts seize, and insulation breaks down as temperature increases. Hydraulic components are superior to others in this respect since the fluid carries away the heat generated to a convenient heat exchanger. This feature permits smaller and lighter components. Hydraulic pumps and motors are currently available with horsepower to weight ratios greater than 2 hp/lb. Small compact systems are attractive in mobile and airborne installations.

2. The hydraulic fluid also acts as a lubricant and makes possible long component life.

3. There is no phenomenon in hydraulic components comparable to the saturation and losses in magnetic materials of electrical machines. The torque developed by an electric motor is proportional to current and is limited by magnetic saturation. The torque developed by hydraulic actuators (i.e., motors and pistons) is proportional to pressure difference

1

and is limited only by safe stress levels. Therefore hydraulic actuators develop relatively large torques for comparatively small devices.

4. Electrical motors are basically a simple lag device from applied voltage to speed. Hydraulic actuators are basically a quadratic resonance from flow to speed with a high natural frequency. Therefore hydraulic actuators have a higher speed of response with fast starts, stops, and speed reversals possible. Torque to inertia ratios are large with resulting high acceleration capability. On the whole, higher loop gains and bandwidths are possible with hydraulic actuators in servo loops.

5. Hydraulic actuators may be operated under continuous, intermittent, reversing, and stalled conditions without damage. With relief valve protection, hydraulic actuators may be used for dynamic breaking. Larger speed ranges are possible with hydraulic actuators. Both linear and rotary actuators are available and add to the flexibility of hydraulic power elements.

6. Hydraulic actuators have higher stiffness, that is, inverse of slope of speed-torque curves, compared to other drive devices since leakages are low. Hence there is little drop in speed as loads are applied. In closed loop systems this results in greater positional stiffness and less position error.

7. Open and closed loop control of hydraulic actuators is relatively simple using valves and pumps.

8. Other aspects compare less favorably with those of electromechanical control components but are not so serious that they deter wide use and acceptance of hydraulic control. The transmission of power is moderately easy with hydraulic lines. Energy storage is relatively simple with accumulators.

Although hydraulic controls offer many distinct advantages, several disadvantages tend to limit their use. Major disadvantages are the following:

1. Hydraulic power is not so readily available as that of electrical power. This is not a serious threat to mobile and airborne applications but most certainly affects stationary applications.

2. Small allowable tolerances results in high costs of hydraulic components.

3. The hydraulic fluid imposes an upper temperature limit. Fire and explosion hazards exist if a hydraulic system is used near a source of ignition. However, these situations have improved with the availability of high temperature and fire resistant fluids. Hydraulic systems are messy because it is difficult to maintain a system free from leaks, and there is

always the possibility of complete loss of fluid if a break in the system occurs.

4. It is impossible to maintain the fluid free of dirt and contamination. Contaminated oil can clog valves and actuators and, if the contaminant is abrasive, cause a permanent loss in performance and/or failure. Contaminated oil is the chief source of hydraulic control failures. Clean oil and reliability are synonymous terms in hydraulic control.

5. Basic design procedures are lacking and difficult to obtain because of the complexity of hydraulic control analysis. For example, the current flow through a resistor is described by a simple law—Ohm's law. In contrast, no single law exists which describes the hydraulic resistance of passages to flow. For this seemingly simple problem there are almost endless details of Reynolds number, laminar or turbulent flow, passage geometry, friction factors, and discharge coefficients to cope with. This factor limits the degree of sophistication of hydraulic control devices.

6. Hydraulics are not so flexible, linear, accurate, and inexpensive as electronic and/or electromechanical devices in the manipulation of low power signals for purposes of mathematical computation, error detection, amplification, instrumentation, and compensation. Therefore, hydraulic devices are generally not desirable in the low power portions of control systems.

The outstanding characteristics of hydraulic power elements have combined with their comparative inflexibility at low power levels to make hydraulic controls attractive primarily in power portions of circuits and systems. The low power portions of systems are usually accomplished by mechanical and/or electromechanical means.

1-2 GENERAL COMMENTS ON DESIGN

The term "design" has a broad meaning. It is often associated with the creativity required to produce sketches and rough layouts of possible mechanisms that will accomplish an objective. As a second meaning, it is sometimes associated with the engineering calculations and analyses necessary in the selection and sizing of hardware to form a component or system. Design is also associated with the many details of material selection, minor calculations, and making of complete engineering drawings. This book is directed toward the analysis and design (by paper and pencil) of control systems whose power elements are hydraulic. The term design is used in the sense of specifying proper size. Although considerations such as material, stress level, and seals are equally important to a finished device, they do not relate directly to the dynamic performance of a system and are treated with more authority elsewhere.

The differential equations that describe hydraulic components are non-linear and, in some cases, of high order. This has led control engineers toward analog and digital computer-aided design of servo systems using such components. Generally, the procedure is to write the equations that describe a system and then solve them with a computer. Coefficients are adjusted until the computed performance (stability, accuracy, and speed of response) is satisfactory. The system is then constructed, based on the computed results, with the hope that it will perform in a similar manner. More often than desirable, correlation with physical performance is poor. Lack of adequate correlation creates much concern that the basic assumptions used in the initial equations were not valid, that all "effects" had not been simulated, that some unsuspected nonlinearity had spoiled the expected result (usually the case), or that there was a gap in the theory.

Actually, a great deal of time and trouble can be saved if a paper and pencil analysis and design of the system is made before it is simulated on a computer for final refinements. If this is done carefully, with generous sprinklings of sound engineering judgments, then machine computation will not be necessary in most cases. In complicated cases in which judgments are most difficult, if not impossible, to make, machine computation is required; however, this requirement is exceptional. The development of digital computer programs in recent years to solve complex sets of nonlinear differential equations strengthens the argument for preliminary analysis, for now exact solutions are possible for comparison. In fact, preliminary analyses to determine approximate results are useful, and sometimes absolutely necessary, to obtain maximum benefits from machine computations. Availability of these programs allows more emphasis to be placed on the physics and mathematical formulation of problems and less on the solution techniques.

Preliminary dynamic analysis is necessarily restricted to linearized differential equations because only they may be solved without great difficulty. However, as far as dynamic performance is concerned, linearized analysis is an adequate tool considering the basic assumptions usually made to obtain initial equations, the preponderance of experimental correlation, and the fact that general performance indices have been developed only for linear systems. Furthermore, the algebraic or single-valued nonlinearities which occur in hydraulic equations are not usually the source of discrepancies between predicted and actual results. Discrepancies can be traced to two basic phenomena: multivalued nonlinearities such as backlash (which is notorious for causing limit cycle oscillations) and the types of quantity involved in hydraulic analysis.

Two basic types of physical quantity can be distinguished in hydraulic control analysis: hard and soft. A hard quantity is one that can be determined with fair precision and whose value remains relatively constant. In short, a hard quantity is easily identified, computed, and controlled. In contrast, a soft quantity is one whose value can, at best, be pinned down to a possible range of values. A soft quantity is unreliable, nebulous, and a function of variables not easily known or controlled. As an example, consider a simple spring-mass arrangement. The mass and spring constants are hard quantities and result in a hard, undamped natural frequency. However, the damping ratio, although it certainly has a value that can be measured, is difficult to compute and is soft quantity.

The most important asset of a servo system is stability, and therefore stability should be based on hard quantities. Indeed, the design of a system can be judged by the number of hard quantities on which its performance depends. If a certain performance index depends on soft quantities, correspondingly nebulous physical performance can be expected. For example, the stability of single stage relief valves depends on, among other things, the pressure sensitivity of the valve. This is a soft quantity because it depends on valve geometry at null, valve wear, and so on, and these valves are well known for their ability to oscillate. In contrast, the stability of an uncompensated electrohydraulic servo depends on hard quantities such as valve flow gain and piston area, and their stability is virtually assured. Therefore, an intent of this book is to instill a sense of judging the quality of quantities in addition to how quantities relate to performance. This sort of engineering judgment is absolutely necessary for the rational design of hydraulic controls. The designer should always ask whether the required performance depends on soft quantities. It is certainly safe to conclude that better systems can be built if more emphasis is placed on the quality of quantities and how this can be exploited to form a design rather than on a precise mathematical solution of a given set of equations which, supposedly, represents the system.

GENERAL REFERENCES ON HYDRAULIC CONTROL

[1] Blackburn, J. F., G. Reethof, and J. L. Shearer, *Fluid Power Control*. New York: Technology Press of M.I.T. and Wiley, 1960.
[2] Ernst, W., *Oil Hydraulic Power and Its Industrial Applications*, 2nd ed., New York: McGraw-Hill, 1960.
[3] Lewis, E. E., and H. Stern, *Design of Hydraulic Control Systems*. New York: McGraw-Hill, 1962.
[4] Morse, A. C., *Electrohydraulic Servomechanisms*. New York: McGraw-Hill, 1963.
[5] Pippenger, J. J., and T. G. Hicks, *Industrial Hydraulics*. New York: McGraw-Hill, 1962.
[6] Pippenger, J. J., and R. M. Koff, *Fluid-Power Controls*. New York: McGraw-Hill, 1959.
[7] Fitch, Jr., E. C., *Fluid Power and Control Systems*. New York: McGraw-Hill, 1966.

2

Hydraulic Fluids

Fluids, both liquids and gases, are characterized by their continuous deformation when a shear force, however small, is applied. Liquids and gases may be distinguished by their relative incompressibilities and the fact that a liquid may have a free surface while a gas expands to fill its confining container. In the field of controls, the term "hydraulic" is used to designate a system using a liquid and "pneumatic" applies to those systems using a gas. Because the liquid is the medium of transmission of power in a hydraulic system, knowledge of its characteristics is essential.

The purpose of this chapter is to define certain physical properties which will prove useful and to discuss properties related to the chemical nature of fluids, types of fluids available, and selection of fluids. The English system of units—that is, force in pounds, length in inches, time in seconds, and temperature in degrees Fahrenheit—is normally used in this country to measure performance characteristics of hydraulic systems. Although any system of units is applicable, the English system is used in this book to avoid confusion.

2-1 DENSITY AND RELATED QUANTITIES

Weight density is defined as the weight of a substance per unit of volume. The symbol γ is used for weight density and the units are lb/in.^{3}. For petroleum base fluids the approximate weight density is $\gamma = 0.03$ lb/in.^{3}. The weight density for a MIL-H-5606B hydraulic fluid is shown in Fig. 2-1.

Mass density is defined as mass per unit of volume. The symbol ρ is used to designate mass density with units of lb-sec^2/in.^{4}. The relation between weight density and mass density is

$$\rho = \frac{\gamma}{g} \qquad (2\text{-}1)$$

where g is the acceleration of gravity, $g = 386$ in./sec^2. For petroleum base fluids, the approximate mass density is $\rho = 0.78 \times 10^{-4}$ lb-sec^2/in.^{4}.

6

Specific gravity is the ratio of the mass (or weight) density of a substance at a certain temperature to the mass (or weight) density of water at the same temperature. Specific gravity is dimensionless, and the symbols σ and SG are often used. However, the temperature must be specified. The petroleum industry in the United States has selected 60°F as a standard temperature for the specific gravity of hydraulic fluids. Thus, the specific

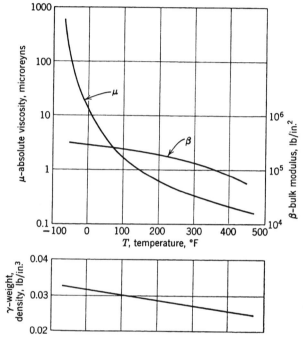

Figure 2-1. Absolute viscosity, bulk modulus, and weight density for a MIL-H-5606B hydraulic fluid.

gravity of oil at 60°F compared to water at 60°F is often designated $\sigma 60/60°F$. The specific gravity of hydraulic fluids ranges from about 0.8 for petroleum base fluids to as high as 1.5 for the chlorinated hydrocarbons.

2-2 EQUATION OF STATE FOR A LIQUID

The density of a liquid is a function of both pressure and temperature. A function relating density, pressure, and temperature of a fluid is, by definition, the equation of state. The equation of state for a liquid cannot be mathematically derived from physical principles. In contrast, the kinetic theory of gases yields an equation of state for gases. However, because changes in density as a function of pressure and temperature are small for a liquid, the first three terms of a Taylor's series for two variables may be used as an approximation. Therefore,

$$\rho = \rho_0 + \left(\frac{\partial \rho}{\partial P}\right)_T (P - P_0) + \left(\frac{\partial \rho}{\partial T}\right)_P (T - T_0) \qquad (2\text{-}2)$$

where ρ, P, and T are the mass density, pressure, and temperature, respectively, of the liquid about initial values of ρ_0, P_0, and T_0. A more convenient form for (2-2) is

$$\rho = \rho_0\left[1 + \frac{1}{\beta}(P - P_0) - \alpha(T - T_0)\right] \qquad (2\text{-}3)$$

where

$$\beta \equiv \rho_0\left(\frac{\partial P}{\partial \rho}\right)_T \quad \text{and} \quad \alpha \equiv -\frac{1}{\rho_0}\left(\frac{\partial \rho}{\partial T}\right)_P$$

Equation 2-3 is the linearized equation of state for a liquid. The mass density increases as pressure is increased and decreases with temperature increase. Because mass density is mass divided by volume, equivalent expressions for β and α are

$$\beta = -V_0\left(\frac{\partial P}{\partial V}\right)_T \qquad (2\text{-}4)$$

$$\alpha = \frac{1}{V_0}\left(\frac{\partial V}{\partial T}\right)_P \qquad (2\text{-}5)$$

where V is the total volume and V_0 is the initial total volume of the liquid.

The quantity β is the change in pressure divided by the fractional change in volume at a constant temperature and is called the *isothermal bulk modulus* or simply *bulk modulus* of the liquid. The bulk modulus is always a positive quantity, for $(\partial P/\partial V)_T$ is always negative, and has a value of about 220,000 lb/in.[2] for petroleum fluids. However, values this large are rarely achieved in practice because the bulk modulus decreases sharply with small amounts of air entrained in the liquid. As discussed in later chapters, the bulk modulus is the most important fluid property in determining the dynamic performance of hydraulic systems because it relates to the "stiffness" of the liquid. An *adiabatic bulk modulus*, β_a, may also be defined. However, it can be shown that the adiabatic and isothermal bulk moduluses are related by

$$\beta_a = \frac{C_p}{C_v}\beta \qquad (2\text{-}6)$$

where C_p/C_v is the ratio of specific heats. Because this ratio is only slightly in excess of unity for a liquid (see Section 2-4), it is difficult to justify the distinction between adiabatic and isothermal bulk moduluses and especially so in applications where entrained air and mechanical compliance are significant. Section 2-5 is devoted to the determination of practical values for the bulk modulus of a system. The reciprocal of β is often designated c and termed the *compressibility* of the liquid. The bulk modulus for a MIL-H-5606B hydraulic fluid is shown in Fig. 2-1.

The quantity α is the fractional change in volume due to a change in temperature and is called the *cubical expansion coefficient*. The cubical expansion coefficient for petroleum base fluids is about $\alpha = 0.5 \times 10^{-3}$ $(°F)^{-1}$, that is, there is about a 5% increase in volume for each 100°F of temperature increase.

2-3 VISCOSITY AND RELATED QUANTITIES

Viscosity is an important property of any fluid. It is absolutely necessary for hydrodynamic lubrication, and a suitable value is required for many

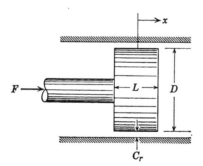

Figure 2-2. Piston concentric in cylinder.

other purposes. Close-fitting surfaces in relative motion occur in most hydraulic components. If the viscosity of the fluid is too low, leakage flows increase; if the viscosity is too large, component efficiencies decrease because of additional power loss in fluid friction. Viscosity is of such significance that it is common practice to designate the fluid by its viscosity at a certain temperature, for example, oil with 150 SSU at 130°F might be such a fluid designation.

Isaac Newton was the first to give a quantitative definition of viscosity. Referring to the piston and cylinder of Fig. 2-2, in which the radial clearance C_r is filled with a fluid, Newton observed that a force was necessary to cause relative motion. This force is a measure of the internal friction of the fluid or its resistance to shear and is proportional to the area in contact and to the velocity and is inversely proportional to the film thickness. Therefore,

$$F = \mu A \frac{\dot{x}}{C_r}$$

The constant of proportionality μ is known as the *absolute viscosity* (the terms "dynamic viscosity" and "coefficient of viscosity" are also used) of the fluid. For this case, since $A = \pi DL$, we obtain

$$F = \frac{\pi DL\mu}{C_r} \frac{dx}{dt} \qquad (2\text{-}7)$$

If the absolute viscosity at any given temperature is independent of shear rate, the fluid is called Newtonian; if it varies with shear rate, the fluid is termed non-Newtonian. Most hydraulic fluids, except for the water-in-oil emulsions, are Newtonian. In the English system of units absolute viscosity has units of lb-sec/in.^{2} and are called reyns; that is,

$$1 \text{ lb-sec/in.}^2 = 1 \text{ reyn}$$

A reyn is a very large unit and a smaller unit, or microreyn, is often used.

$$1 \text{ microreyn} = 10^{-6} \text{ reyn}$$

In the cgs system of units absolute viscosity has units of dyne-sec/cm^2 called poises; that is,

$$1 \text{ dyne-sec/cm}^2 = 1 \text{ poise}$$

Again, this is a large unit and a smaller unit, or centipoise (cp), is often used.

$$1 \text{ cp} = 10^{-2} \text{ poise}$$

Without difficulty it can be shown that conversion factors between the two systems of units are 1.45×10^{-7} reyn/cp and 1.45×10^{-5} reyn/poise. Thus, if absolute viscosity is given in cgs units, it may be converted to English units by multiplication of the appropriate factor.

The ratio of absolute viscosity to mass density occurs in many equations (Navier-Stokes, Reynolds number, etc.) and is easily measured by many viscometers. This ratio is, by definition, the *kinematic viscosity* ν of the fluid, that is,

$$\nu = \frac{\mu}{\rho} \qquad (2\text{-}8)$$

In the cgs system of units kinematic viscosity has units of cm^2/sec called stokes. Because this unit is large, the unit centistoke (cs) is often used: $1 \text{ cs} = 10^{-2}$ stoke. In English units kinematic viscosity has units of in.²/sec, but it has not been named. The conversion factor is 1.55×10^{-3} (in.²/sec)/cs.

Kinematic viscosity is easily measured using many instruments. The most well known of these in the United States is the Saybolt Universal Viscometer. Using this instrument, the time is measured for 60 cm³ of a sample to flow through a tube 0.176 cm in diameter and 1.225 cm long at a constant temperature. The resulting time in seconds is called *Saybolt Universal Seconds* and is abbreviated SSU or SUS. Similar instruments (Redwood in England and Engler in Germany) are used in Europe but the sample volumes are quite different, making conversion troublesome. SSU is commonly used to designate liquid viscosities in the petroleum industry. However, SSU does not have the appropriate units to be of use in engineering computations and equations and must be converted to other measures. The equivalent kinematic viscosity in centistokes v is closely approximated by

$$v = 0.216 \, \text{SSU} - \frac{166}{\text{SSU}}$$

for SSU greater than 32 sec. The tables in ASTM D 446 53 should be consulted if very accurate conversion is desired.

It is unfortunate that the name viscosity is attached to the quantity defined by (2-8), for it has confused the basic definition of viscosity and made conversion between absolute and kinematic, with the many measures of each, a trying experience. To this end a nomogram, Fig. 2-3, has been prepared to facilitate conversion of measures of viscosity which are used in this country. Because the English system of units is used in hydraulic control systems, the unit of the reyn is most convenient.

The viscosity of liquids decreases markedly with temperature increase and increases, but to a much lesser degree, with increased pressure. The viscosity variation with temperature is the more important and may be approximated by an equation of the form [1]

$$\mu = \mu_0 e^{-\lambda(T-T_0)} \qquad (2\text{-}9)$$

where μ = absolute viscosity at temperature T, lb-sec/in.²

μ_0 = viscosity at a reference temperature T_0, lb-sec/in.²

λ = a constant which depends on the liquid, (°F)⁻¹

T = temperature, °F

The most common method of presenting viscosity-temperature characteristics of fluids, particularly by those working extensively with fluids, is as in ASTM viscosity-temperature charts described in ASTM D 341.

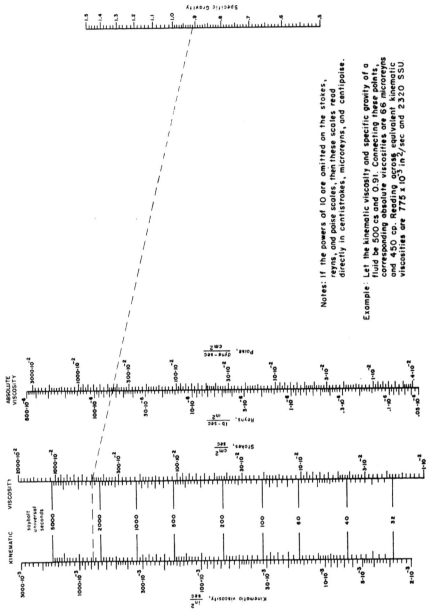

Figure 2-3. Viscosity conversion chart.

12

The virtue of these charts is that plotted viscosity-temperature characteristics are very nearly straight lines so that a reasonable curve can be obtained from few test points.

Many measures of the variation in viscosity of a fluid with temperature have been proposed. The simplest of these is the negative of the geometric slope of the viscosity-temperature curve plotted on an ASTM chart; however, this slope has little physical meaning. The most common measure of viscosity variation with temperature is the Dean and Davis Viscosity Index. The viscosity index is an empirical number and is computed according to tables given in ASTM D 567. The lower the viscosity index of a liquid, the greater the variation in viscosity with temperature, and vice versa. The viscosity index was originally conceived to range from 0 to 100. However, new fluid formulations and refining techniques have resulted in fluids which have a viscosity index greater than 100. Another measure of viscosity change with temperature is the Viscosity Temperature Coefficient, which is defined by

$$\text{VTC} = \frac{v_{100} - v_{210}}{v_{100}} = 1 - \frac{v_{210}}{v_{100}} \qquad (2\text{-}10)$$

where v_{210} and v_{100} are the kinematic viscosities of the liquid at 210°F and 100°F, respectively. None of these measures of the viscosity-temperature gradient are satisfactory for large temperature ranges, so the complete curve should be consulted.

Plots of weight density, absolute viscosity, and bulk modulus as a function of temperature for a MIL-H-5606B hydraulic fluid are shown in Fig. 2-1.

2-4 THERMAL PROPERTIES

Two thermal properties of liquids, specific heat and thermal conductivity, are of importance, especially in the design of hydraulic power supplies.

The *specific heat* of a liquid is the amount of heat required to raise the temperature of a unit mass by 1°. The symbol C_p is used to designate specific heat, and a typical value for petroleum base fluids is $C_p = 0.5$ Btu/lb-°F. The mechanical equivalent of heat, 1 Btu = 9339 in.-lbs, may be used for conversion of units. Specific heat at constant volume C_v and at constant pressure C_p must be distinguished for gases. However, liquids expand little with temperature so that the specific heats are nearly the same. For liquids it can be shown from general thermodynamic relations that the difference in specific heats is given by [2]

$$C_p - C_v = \frac{T\alpha^2\beta}{9339} \qquad (2\text{-}11)$$

Using values for petroleum base fluids of $\gamma = 0.03$ lb/in.³, $\beta = 220,000$ psi, $\alpha = 0.5 \times 10^{-3}$ (°F)⁻¹, $C_p = 0.5$ Btu/lb-°F, and assuming a fluid temperature of 100°F, we obtain a specific heat ratio of $C_p/C_v = 1.04$. At higher temperatures, the ratio of specific heats increase and may become significant in some computations. However, this increase is offset somewhat by the fact that the specific heat of liquids also increases with temperature.

Thermal conductivity is a measure of the rate of heat flow through an area for a temperature gradient in the direction of heat flow. For petroleum base oils the thermal conductivity is about 0.08 Btu/hr-ft²-(°F/ft).

2-5 EFFECTIVE BULK MODULUS

Interaction of the spring effect of a liquid and the mass of mechanical parts gives a resonance in nearly all hydraulic components. In most cases this resonance is the chief limitation to dynamic performance. The fluid spring is characterized by the value for the bulk modulus. The bulk modulus of a liquid can be substantially lowered by entrained air and/or mechanical compliance, and the purpose of this section is to develop equations and practical values for the resulting bulk modulus of a system.

Let us consider a flexible container filled with a fluid which is a mixture of liquid and a vapor or gas as shown in Fig. 2-4a. The gas is shown lumped, but in practice the gas could be entrained in the liquid in the form of bubbles and/or a pocket of gas could exist. Dissolved air in a liquid has little or no effect on the bulk modulus of a liquid. Initially, the total volume of the container V_t can be written

$$V_t = V_l + V_g \qquad (2\text{-}12)$$

where V_l and V_g are initial volumes of the liquid and gas, respectively. As the piston is moved to the left (Fig. 2-4b), a pressure increase ΔP is exerted on the fluid mixture and the container. As seen by the piston, there is a decrease in the initial volume of

$$\Delta V_t = -\Delta V_g - \Delta V_l + \Delta V_c \qquad (2\text{-}13)$$

where the subscripts g, l, and c refer to the gas, liquid, and container, respectively. Equation 2-13 is not to be confused with the physical volume change in the container. The effective or total bulk modulus β_e may be defined by

$$\frac{1}{\beta_e} = \frac{\Delta V_t}{V_t \Delta P} \qquad (2\text{-}14)$$

Combining (2-13) and (2-14) yields

$$\frac{1}{\beta_e} = \frac{V_g}{V_t}\left(-\frac{\Delta V_g}{V_g\,\Delta P}\right) + \frac{V_l}{V_t}\left(-\frac{\Delta V_l}{V_l\,\Delta P}\right) + \left(\frac{\Delta V_c}{V_t\,\Delta P}\right) \quad (2\text{-}15)$$

Now, the bulk modulus of a liquid has been shown to be (see Section 2-2)

$$\beta_l = -\frac{V_l\,\Delta P}{\Delta V_l} \quad (2\text{-}16)$$

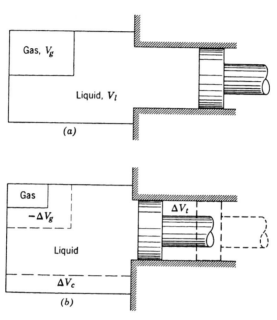

Figure 2-4. Flexible container filled with gas-liquid mixture under compression.

The bulk modulus of a gas may be defined by

$$\beta_g = -\frac{V_g\,\Delta P}{\Delta V_g} \quad (2\text{-}17)$$

The negative signs in (2-16) and (2-17) indicate a decrease in volume with pressure increase. The quantity

$$\beta_c = \frac{V_t\,\Delta P}{\Delta V_c} \quad (2\text{-}18)$$

may be defined as the bulk modulus of the container with respect to the total volume. Substituting (2-16), (2-17), and (2-18) into (2-15) gives the final result. Therefore

$$\frac{1}{\beta_e} = \frac{V_g}{V_t}\left(\frac{1}{\beta_g}\right) + \frac{V_l}{V_t}\left(\frac{1}{\beta_l}\right) + \frac{1}{\beta_c} \qquad (2\text{-}19)$$

This is a general equation which gives the equivalent bulk modulus for a liquid-gas mixture in a flexible container. Some approximations of (2-19) are useful. Solving (2-12) for V_l and substituting into (2-19),

$$\frac{1}{\beta_e} = \frac{1}{\beta_c} + \frac{1}{\beta_l} + \frac{V_g}{V_t}\left(\frac{1}{\beta_g} - \frac{1}{\beta_l}\right) \qquad (2\text{-}20)$$

Because $\beta_l \gg \beta_g$, (2-20) becomes

$$\frac{1}{\beta_e} = \frac{1}{\beta_c} + \frac{1}{\beta_l} + \frac{V_g}{V_t}\left(\frac{1}{\beta_g}\right) \qquad (2\text{-}21)$$

Because this equation involves reciprocals, the effective bulk modulus will be less than any one of the values β_c, β_l, or $(V_t/V_g)\beta_g$. This is analogous to the total resistance of parallel resistors. If no vapor or entrapped air is present, then

$$\frac{1}{\beta_e} = \frac{1}{\beta_c} + \frac{1}{\beta_l} \qquad (2\text{-}22)$$

The equations given are well and good but require the determination of many quantities. The total volume V_t is easily computed from geometric considerations. The bulk modulus of the liquid β_l is obtained from manufacturers' data. The isothermal bulk modulus of a gas is equal to the pressure level P, and the adiabatic bulk modulus of a gas is equal to $(C_p/C_v)P$. The adiabatic value should be used and for air, $\beta_g = 1.4P$. This leaves two quantities, V_g and β_c, to be determined.

Little work has been done on determining the bulk modulus of containers due to mechanical compliance. In some cases the elasticity of structural members, such as motor housings, can reduce the effective bulk modulus appreciably. Probably the major source of mechanical compliance is the hydraulic lines connecting valves and pumps to actuators. The radial displacement u at the inner surface of a thick-walled cylinder due to an internal pressure increase of ΔP is given by [3]

$$u = \frac{D\,\Delta P}{2E}\left(\frac{D_0^2 + D^2}{D_0^2 - D^2} + v\right) = \frac{D\,\Delta P}{2E}\left[\frac{(1+v)D_0^2 + (1-v)D^2}{2T(D_0 + D)}\right] \qquad (2\text{-}23)$$

where D is the inner diameter, D_0 is the outer diameter, T is the wall thickness $(2T = D_0 - D)$, E is the modulus of elasticity of the wall material, and v is Poisson's ratio for the material. By use of (2-18), the bulk modulus for a thick-walled cylindrical container then becomes

$$\frac{1}{\beta_c} = \frac{2}{E}\left[\frac{(1 + \nu)D_0{}^2 + (1 - \nu)D^2}{2T(D_0 + D)}\right] \tag{2-24}$$

For a thin-walled cylinder such that $D_0 \approx D$, and because $\nu \approx \frac{1}{4}$ for metals, (2-24) approximates to

$$\beta_c = \frac{TE}{D} \tag{2-25}$$

This formula is generally used for hydraulic tubing. For thick walled metal pipes in which $D_0 \gg D$, then (2-24) approximates to

$$\beta_c = \frac{E}{2(1 + \nu)} \approx \frac{E}{2.5} \tag{2-26}$$

If the wall thickness is equal to the inside radius, that is, $T = D/2$ and $D_0 = 2D$ then (2-24) becomes

$$\beta_c = \frac{3E}{2(5 + 3\nu)} \approx \frac{E}{3.83} \tag{2-27}$$

Thus there is not a great increase (about 50%) in the container bulk modulus as the wall thickness is increased from $D/2$ to infinite.

Many hydraulic lines are made of teflon or hard rubber with outside sheaths of a single or double braid of stainless steel. Such flexible hoses have a comparatively low bulk modulus with values in the range 10,000 to 50,000 psi common. The value for a particular hose can be readily computed from cubic expansion coefficients of the hose under pressure. These coefficients are available from hose manufacturers, and (2-18) is used to determine the bulk modulus.

With reference to (2-21), a small amount of entrapped air can drastically reduce the bulk modulus. For example, suppose that a fluid inside a steel pipe is at a pressure of 500 psi and contains 1 % (by volume) of entrapped air. Let the pipe diameter be six times the wall thickness so that the bulk modulus of the pipe becomes

$$\beta_c = \frac{1}{6} \times 30 \times 10^6 = 5 \times 10^6 \text{ psi}$$

An average value for the bulk modulus of petroleum base fluids is $\beta_l = 2.2 \times 10^5$ psi. The bulk modulus of the entrapped air is $\beta_g = 1.4 \times 500 = 700$ psi. Substituting these numbers into (2-21) yields

$$\frac{1}{\beta_e} = \frac{1}{5 \times 10^6} + \frac{1}{2.2 \times 10^5} + \frac{0.01}{700} = 1.904 \times 10^{-5}$$

Therefore $\beta_e = 52,600$ psi. In the absence of entrapped air the effective bulk modulus would be 210,000 psi. Thus a small percentage of air in a liquid can decrease the bulk modulus substantially. If the pressure level were 1000 psi, the effective bulk modulus would be 84,100 psi. This is one argument in favor of high pressure systems.

In any practical case it is difficult to determine the effective bulk modulus other than by direct measurement. Estimates of entrapped air in hydraulic systems runs as high as 20% when the fluid is at atmospheric pressure. As pressure is increased, much of this air dissolves into the liquid and does not affect the bulk modulus. In the author's experience, an effective bulk modulus of 100,000 psi has yielded reliable results. Blind use of the bulk modulus of the liquid alone without regard for entrapped air and structural elasticity can lead to gross errors in calculated resonances. Calculated resonances in hydraulic systems at best are approximate.

Because entrained air reduces the bulk modulus, the natural frequency of hydraulic actuators in servo systems may be lowered to such an extent that system instability results. This is especially noticeable when a hydraulic servo is first turned on after a period of shutdown has allowed air to collect in the system. Noisy, unstable performance occurs for a short time until the air is "washed out" of the system. Pumps often exhibit a noisy start until air is flushed out by the action of the flowing fluid. Antifoamants are usually added to hydraulic fluids to increase their ability to release air without forming emulsions. However, a major source of air entrainment is lack of adequate mechanical design of the fluid passages. Blind holes, pockets, and tortuous passages will allow air to collect which may not be flushed out by the moving fluid. Because of the lowered bulk modulus, the result can be a permanent degradation in performance in the form of lowered gains and bandwidths in servo loops and erratic actuator velocities.

2-6 CHEMICAL AND RELATED PROPERTIES

Fluids are subject to chemical reactions with their environment, and many properties have been defined which relate to their chemical behavior. Because of the complex nature of fluids, most of these properties are rather loosely defined.

Lubricity refers to the performance of a fluid as a boundary layer lubricant. Oil films should firmly attach to surfaces, commonly called oiliness, and have sufficient durability to resist the internal mechanical stress due to surfaces in relative motion so that a low coefficient of friction results. Lack of adequate lubricating properties promotes wear and shortens component life. The increased clearances between surfaces due to wear results in degraded performance in the form of increased leakages, loss in efficiency, and failure to build up pressures.

Thermal stability refers to the ability of fluids to resist chemical reactions and/or decomposition at high temperatures. Fluids react more vigorously as temperature is increased and may form solid reaction products which can clog filters, valves, pumps, and motors.

Oxidative stability refers to the ability of fluids to resist reaction with oxygen-containing materials, especially air. Solid reaction products, deposits, and acids may be formed which causes clogging, rusting, and corrosion of system hardware.

Hydrolytic stability refers to the ability of fluids to resist reaction with water. Undesirable formations of solids may result or a stable water-in-oil emulsion may be formed which degrades lubricating ability and promotes rusting and corrosion. Demulsifier additives are often used to inhibit emulsion formations.

Compatibility refers to the ability of fluids to resist reaction with materials commonly used in the systems. Some fluids tend to soften or liquify paints and sealing materials so that caution must be exercised. Recommendations of the fluid manufacturer should be followed in this regard.

Foaming refers to the ability of liquids to combine with gases, principally air, and form emulsions. Entrained air reduces the lubricating ability and bulk modulus of a liquid. A reduction in bulk modulus can severely limit dynamic performance (see Section 2-5). Fluids should have the ability to release air without forming emulsions, and antifoamant additives are used to encourage this ability.

Three quantitative measures have been defined which relate to the fire hazard of flammable fluids. The *flash point* is the oil temperature at which sufficient vapors are formed to cause a transient flame when a test flame is applied. The *fire point* is the oil temperature at which the transient flame is self-sustaining for a period of 5 sec and is usually about 50°F higher than the flash point for petroleum base fluids [4].

The spontaneous or *autogenous ignition temperature* is the temperature at which droplets of heated liquid will ignite when impinging on a hot surface in the presence of air; however, this temperature varies considerably with the exact conditions of the test and must be interpreted accordingly.

The *pour point* is the lowest temperature at which a fluid will flow when tested according to an ASTM procedure. This is a limiting temperature, and the lowest system operational temperature must be considerably higher.

Handling properties refer to the toxicity, odor, color, and storage characteristics of a fluid. Some fluids have highly toxic vapors and can cause skin irritations when in direct contact. Fluid odor should be pleasant or absent and the color should facilitate identification. Storage of the fluid for reasonable periods without alteration in its properties is obviously desirable.

2-7 TYPES OF HYDRAULIC FLUIDS

Two basic types of hydraulic fluids used in control systems can be distinguished: petroleum base fluids and synthetic fluids. The synthetic fluids may be subdivided into chemically compounded and water base fluids. Petroleum base fluids are obtained from refining crude oil. The major disadvantages of these fluids are their potential fire hazard and restricted operational temperature range. To overcome these difficulties, synthetic fluids have been formulated from compounds which are chemically resistant to burning or by the addition of "snuffer" agents, usually water, to flammable compounds to form water base fluids. Water, sometimes with soluble oil additives to increase lubricity and reduce rusting, is used in some industrial applications where large quantities of fluid are required and performance is not a premium. However, water is a poor hydraulic fluid because of its restrictive liquid range, low viscosity and lubricity, and rusting ability.

The properties of some commercially available fluids are listed in Table 2-1. However, hydraulic fluids and their formulations are continually changing, and the manufacturer should be consulted for latest data on available fluids and their properties. The comparative properties of hydraulic fluid base stocks are shown in Table 2-2.

Petroleum Base Fluids

Petroleum base oils are by far the most commonly used hydraulic fluid. Petroleum, a complex mixture of chiefly hydrocarbons, must be highly refined to produce a fluid with viscosity characteristics suitable for hydraulic control systems. Such mineral, turbine, or light oils, as they are often called, have a long history of satisfactory performance as a working fluid. Nearly all petroleum suppliers offer a wide variety of hydrocarbon fluids, ranging from straight refined petroleum to high formulated fluids containing additives to inhibit rust and oxidation, reduce foaming, and increase viscosity index and lubricity. A wide range of viscosity and viscosity-temperature characteristics are available from numerous manufacturers and should be consulted for specific properties. Military Specification MIL-H-5606B is the standard military specification for petroleum base hydraulic fluids.

Table 2-1 Properties of Some Commercial Hydraulic Fluids

	Univis J-43	Skydrol 500A	Pydraul F-9	Oronite 8515	Houghto-Safe 620
Designation					
Manufacturer	Esso Standard Oil Company	Monsanto Chemical Company	Monsanto Chemical Company	California Research Corporation	E. F. Houghton and Company
Base stock	Petroleum	Phosphate ester	Phosphate ester	Silicate ester	Water glycol
Use	Aircraft MIL-H-5606B	Aircraft	Industrial fire resistant	Aircraft	Industrial fire resistant
SG 60/60°F	0.848	1.07	1.28	0.93	1.055
Bulk Modulus, psi	270,000	308,000	387,000	248,000	285,000
Thermal expansion coefficient, 1/°F	0.0005	0.00045	0.00041	0.00044	0.00034
Viscosity, cs °F					
−65	2130	2300	—	2357	—
−40	500	480	—	600	—
0	100	90	4000/30°F	150	1079
100	14.3	11.5	47	24.3	43
210	5.1	3.9	5.5	8.1	16/155°F
400	1.9	—	—	2.6	—
Viscosity-temperature coefficient. (VTC)	0.645	0.695	0.883	0.67	—
Specific heat Btu/lb-°F	0.5	0.38	0.32	0.44	0.8
Thermal conductivity Btu/(hr)(ft²)(°F/ft)	0.08	0.078	0.067	0.083	0.25
Flash point, °F	225	360	430	395	none
Fire point, °F	275	425	675	450	none
Autogenous Ignition temperature, °F	700	1100	1100	755	—
Pour point, °F	−90	−85	−5	−100	−20

Table 2-2 Characteristics of Hydraulic Fluid Base Stocks[a]

Class	Viscosity Temperature	Volatility-Viscosity	Thermal Stability	Oxidative Stability	Hydrolytic Stability	Fire Resistance	Lubricating Ability	Additive Response
Petroleum	G	P	G	F	E	P	F	E
Phosphate esters	G	F	F	G	F	E	E	G
Silicate esters	E	G	E	G	F	F	G	G
Chlorinated hydrocarbons	P	F	E	E	G	E	F	G
Fluorinated hydrocarbons	P	P	E	E	G	E	P	P
Organic esters	G	G	F	F	G	F	G	G
Silicones	E	E	G	F	E	F	P	P
Water-glycol	G	P	F	G	E	E	P	G
Water-emulsion	G	P	F	G	E	E	P	G
Polyalkylene glycols	G	G	G	G	G	F	G	F
Polyphenyl ethers	F	G	E	G	E	F	G	G
Polyphenyls	P	F	E	P	E	F	P	F
Hydrocarbons	G	G	E	F	E	P	G	G
Silanes	F	F	E	G	E	F	F	F
No-beta hydrogen esters	G	G	G	F	G	F	G	G

[a] From Roger E. Hatton, *Introduction to Hydraulic Fluids*, New York, Reinhold, 1962.
E = excellent; G = good; F = fair; and P = poor.

22

Synthetic Hydraulic Fluids

Synthetic fluids on the whole have excellent fire resistant properties. Many of these fluids may be used at high temperatures, and some are quite expensive. Such fluids are named after their base stocks, that is, the predominant material, and their formulations are chemically involved.

Phosphate ester base fluids are used in both aircraft and industrial applications. Their thermal stability is rather poor for sustained operation at temperatures in excess of 300°F, but their lubricity is excellent [4]. These fluids are solvents for many types of paints and seals so that care must be used to ensure compatibility with system materials. Examples of commercially available fluids include Skydrol 500A, Pydraul F-9 and 150, Cellulube 220, Houghto-Safe 1000 series, and Nyvac 200.

Silicate ester base fluids have excellent thermal stability which permits their use as high temperature fluids, but they have poor hydrolytic stability. Commercial fluids include Monsanto OS-45 and Oronite 8515.

The halogens of chlorine and fluorine are united with hydrocarbons to form fluid base stocks of *chlorinated hydrocarbons* and *fluorinated hydrocarbons*. Such fluids have high thermal and oxidative stability required for high temperature applications, but relatively high freezing points limit their use at low temperatures. Commercial chlorinated hydrocarbons include Aroclor 1000 series and Pydraul A-200.

Silicone base fluids have excellent viscosity-temperature characteristics but are limited by their lubricating ability. Examples of commercial silicone fluids are Dow Corning F-60 and Versilube F-50.

The water base fluids are fire resistant and compatible with standard seal materials but have poor lubricating ability. *Water glycols* are a formulation of water and a glycol, which thickens the fluid to increase viscosity, with various additives to improve lubricity and corrosion resistance. Commercial water glycols include Ucon Hydrolube 100 series, Houghto-Safe 600 series, and Cellugard. *Water-in-oil emulsions* are formed by a stable suspension of water particles in a hydrocarbon oil. However, the water and oil does tend to separate and, if allowed to stand, agitation is required to maintain the dispersion of water in the oil. Checking the fluid while in use is desirable to ensure that the water content is at a satisfactory level. Commercial examples are Shell Irus 902, Sunsafe, and Houghto-Safe 5000 series fluids. Although their high temperature range is limited because of the water content, the water base fluids offer a satisfactory and economical industrial hydraulic fluid when properly used.

2-8 SELECTION OF THE HYDRAULIC FLUID

Many petroleum and synthetic fluids are available and more are being formulated. The highly technical formulations of the fluids with their various pros and cons makes the selection of such fluids difficult for those who are not thoroughly acquainted with the latest improvements and new formulations.

Generally, hydraulic fluids are chosen based on considerations of the environment of the application and chemical properties of the fluid. Physical properties such as viscosity, density, and bulk modulus are not usually basic considerations. Viscosity is very important, but usually a variety of viscosity characteristics are available in each fluid type. Bulk modulus should be large, but this requirement usually yields to the high temperature capability of the fluid. For example, the low bulk modulus of silicone fluids is more than offset by their high temperature range.

A basic judgment in fluid selection is required concerning the fire and explosion hazard posed by the application. If the environment and high temperature limit of the application are within the range of petroleum base fluids, then any number of suitable oils are available from numerous manufacturers. If the application requires a fire-resistant fluid, a choice must be made between the chemically compounded and water base synthetics. Factors to be considered are temperature range, cost, lubricity, compatibility, chemical, and handling characteristics of the fluid. Once a fluid type is selected, a number of viscosity and viscosity-temperature characteristics are usually made available, and a suitable matching must be made to the requirements of the system hardware. Consultation with representatives of hardware and fluid manufacturers is essential to ensure satisfactory compatibility and performance.

REFERENCES

[1] Blackburn, J. F., G. Reethof, and J. L. Shearer, *Fluid Power Control*. New York: Technology Press of M.I.T. and John Wiley, 1960.
[2] Van Wylen, G. J., *Thermodynamics*. New York: Wiley, 1959.
[3] Timoshenko, S., *Strength of Materials*, 2nd ed., Part II. New York: Van Nostrand.
[4] Hatton, R. E., *Introduction to Hydraulic Fluids*. New York: Reinhold, 1962.

3

Fluid Flow Fundamentals

Knowledge of the fundamental laws and equations which govern the flow of fluids is essential for the rational design of hydraulic control components and systems. This chapter will discuss the general equations of fluid motion, types of flow, and flow through conduits and orifices. The last section will summarize those relations normally used in hydraulic control analysis and design.

Fluids are made up of discrete particles—molecules. An accurate analysis would have to consider the motion of each particle, and this would be hopeless analytically. For example, the density at any geometrical point would depend on whether there exists a molecule at that point. Therefore, we must rely on "continuous" theory and consider the statistical properties of a fluid. This concept is in conflict with molecular theory, but it is sufficiently accurate for engineering purposes. However, at low pressures where there are large distances between molecules or when the distances between molecules are comparable in magnitude to the significant dimensions of the problem, the kinetic theory would be required.

3-1 GENERAL EQUATIONS

Analytic description of general fluid flow requires that the motion of a small cube of fluid be defined. If such a cube can be sufficiently defined, it would be possible to proceed to more complex situations. An infinitesimally small volume of fluid can be completely defined using eight parameters. These are the x, y, and z coordinates of the element and the pressure, temperature, density, and viscosity of the element and, of course, time. Therefore seven independent equations are required in order that they may be solved simultaneously to obtain any of the parameters as a function of another or, as is more usually the case, to find any parameter as a function of time.

The first three of these equations result when Newton's second law is applied to the three directions of motion and are known as the Navier-Stokes equations.* They are

$$\rho\left(\frac{\partial u}{\partial t} + u\frac{\partial u}{\partial x} + v\frac{\partial u}{\partial y} + w\frac{\partial u}{\partial z}\right) = \rho X - \frac{\partial P}{\partial x} + \mu\left(\frac{\partial^2 u}{\partial x^2} + \frac{\partial^2 u}{\partial y^2} + \frac{\partial^2 u}{\partial z^2}\right) \quad (3\text{-}1)$$

$$\rho\left(\frac{\partial v}{\partial t} + u\frac{\partial v}{\partial x} + v\frac{\partial v}{\partial y} + w\frac{\partial v}{\partial z}\right) = \rho Y - \frac{\partial P}{\partial y} + \mu\left(\frac{\partial^2 v}{\partial x^2} + \frac{\partial^2 v}{\partial y^2} + \frac{\partial^2 v}{\partial z^2}\right) \quad (3\text{-}2)$$

$$\rho\left(\frac{\partial w}{\partial t} + u\frac{\partial w}{\partial x} + v\frac{\partial w}{\partial y} + w\frac{\partial w}{\partial z}\right) = \rho Z - \frac{\partial P}{\partial z} + \mu\left(\frac{\partial^2 w}{\partial x^2} + \frac{\partial^2 w}{\partial x^2} + \frac{\partial^2 w}{\partial x^2}\right) \quad (3\text{-}3)$$

u, v, and w are velocity components in x, y, and z directions of Cartesian coordinates, and X, Y, and Z are body forces per unit volume in direction of coordinate axes, t is time, P is pressure per unit of area, ρ is the mass density, and μ is the absolute viscosity of the fluid. These equations are a result of the law of conservation of momentum. The terms on the left side of these equations are a result of fluid inertia. The last three terms on the right side result from viscous friction. If the inertia terms are neglected, the set of equations is called Stokes equations; if viscosity is neglected, the equations are called Euler's equations. The ratio of inertia force over viscous force is called Reynolds number and serves to weight the relative effects of viscosity and inertia terms of the Navier-Stokes equations. A large Reynolds number indicates that inertia terms are dominant, whereas a small number indicates the dominance of viscosity terms.

The fourth equation results from the law of conservation of mass. Consider a control volume (Fig. 3-1) in which there are weight flow rates W into and from the volume. Let the volume be V_0 and the accumulated or stored mass of fluid inside be m with a mass density of ρ. Since all fluid must be accounted for, as the medium is assumed continuous, the rate at which mass is stored must equal incoming mass flow rate minus outgoing mass flow rate. Therefore,

$$\sum W_{\text{in}} - \sum W_{\text{out}} = g\frac{dm}{dt} = g\frac{d(\rho V_0)}{dt} \quad (3\text{-}4)$$

Equation 3-4 is called the continuity equation because it is based on continuous theory, and the form given is convenient for the analysis of fluid components.

The fifth equation results from the law of conservation of energy and is called the first law of thermodynamics. Consider a volume (Fig. 3-2) in which weight flow rates in are W_{in} lb/sec and outflows are W_{out} lb/sec.

* The Navier-Stokes equations given here assume constant density and viscosity and are therefore a simplified form of the more general Navier-Stokes equations.

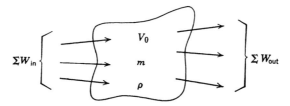

Figure 3-1. Flows entering and leaving a control volume.

The fluid inside the volume is doing external work (expansion, shaft, and shear) of dW_x/dt in-lb/sec, and heat is being transferred to the volume at a rate of dQ_h/dt in-lb/sec. The statement of the first law is that the energy flow in minus the energy flow out must equal the rate at which energy is stored inside the volume. Therefore

$$\frac{dQ_h}{dt} - \frac{dW_x}{dt} + \sum W_{in}h_{0in} - \sum W_{out}h_{0out} = \frac{dE}{dt} \qquad (3\text{-}5)$$

where $h_0 = h + V^2/2g + z$, total energy (internal, pressure, kinetic, and potential) per unit weight of fluid, in-lb/lb

E = total internal energy of fluid inside volume, in-lb

u = internal (intrinsic) energy, in-lb/lb

P = pressure, lb/in.²

γ = weight density, lb/in.³

V = velocity, in./sec

g = acceleration due to gravity, 386 in./sec²

z = elevation, in.

W = weight flow rate, lb/sec

$h = u + P/\gamma$, enthalpy of the fluid, in-lb/lb

This equation assumes the absence of capillary, electrical and magnetic forces, and that such a volume can be defined. For a liquid, the internal energy per pound is $u = 9339C_pT$, where T is the liquid temperature, °F,

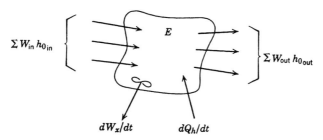

Figure 3-2. Energies entering and leaving a control volume.

C_p is the specific heat, Btu/lb-°F, and 9339 is the mechanical equivalent of heat, in-lb/Btu. Therefore for steady flow of an incompressible liquid (i.e., no energy stored in the volume, $dE/dt = 0$, and $\gamma_1 = \gamma_2 = \gamma$) which enters and leaves a control volume at only one place with negligible changes in elevation (Fig. 3-3), (3-5) becomes

$$\frac{dQ_h}{dt} - \frac{dW_x}{dt} + W_1\left(9339C_pT_1 + \frac{P_1}{\gamma} + \frac{V_1^2}{2g}\right)$$

$$- W_2\left(9339C_pT_2 + \frac{P_2}{\gamma} + \frac{V_2^2}{2g}\right) = 0 \quad (3\text{-}6)$$

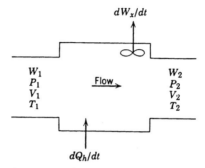

Figure 3-3   Flow entering and leaving a control volume with heat added and work being done.

The heat transferred to the volume would be determined by Fourier's law of heat conduction. This is not easy to define mathematically, and two extreme cases are usually considered which bracket all possibilities. Heat can be transferred at such a rate that the temperature remains constant. This condition is called isothermal and, since temperature is constant, the energy equation is not required. At the other extreme no heat is transferred, that is, $dQ_h/dt = 0$, and this condition is called adiabatic. In general, temperature changes have little effect on liquid flow because cubical expansion coefficients are small and cause negligible density change.

The sixth equation is the equation of state and may be written

$$\rho = f(P, T) \quad (3\text{-}7)$$

For a liquid (2-3) may be used. The seventh and last equation is that required to define viscosity as a function of pressure and temperature.

$$\mu = f_1(P, T) \quad (3\text{-}8)$$

Equation 2-9 could be used as this function for a liquid.

The equations which describe fluid flow are nonlinear partial differential equations with complex boundary conditions. Needless to say, no general solutions of these equations have been found. There is therefore no general theoretical treatment of fluid motion. The general equations do serve to define the scope of any problem involving fluids. In many instances certain approximations can be made which reduce the complexity of these equations and permit solutions accurate enough for most purposes.

3-2 TYPES OF FLUID FLOW

Since there is no general treatment of fluid flow, each particular situation must be considered a special case. Working formulas have evolved from a combination of experience and analysis for many cases of practical interest and considerable judgment is often required in their application. Flow in closed conduits is of particular interest and includes flow in pipes, sudden enlargements and contractions in pipe sections, flow-through fittings, and flow through restrictions in pipes such as orifices.

Some general comments on fluid flow can be made. The forces which affect fluid flow are due to body forces such as gravity and bouyancy, forces due to fluid inertia, forces arising from internal fluid friction (viscosity), and forces due to surface tension, electric and magnetic fields. In most cases, only those forces arising from fluid inertia and viscosity are significant. The Navier-Stokes equations are so formidable that either viscosity or inertia terms (not both) may be considered analytically. However, experience shows that flows in nature are generally dominated either by viscosity or inertia of the fluid. It is indeed fortunate that nature cooperates with our ability to analytically treat only simple flow cases. Therefore, it is useful to define a quantity which describes the relative significance of these two forces in a given flow situation. The dimensionless ratio of inertia force to viscous force is called Reynolds number and defined by

$$R = \frac{\rho \bar{u} a}{\mu} \tag{3-9}$$

where ρ is fluid mass density, μ is absolute viscosity, $\bar{u}$ is the average velocity of flow, and a is a characteristic dimension of the particular flow situation. For each flow case, the characteristic length is agreed upon and empirical values are obtained for the Reynolds number which describes transition from viscosity to inertia dominated flows.

Flow dominated by viscosity forces is referred to as *laminar* or *viscous* flow. Laminar flow is characterized by an orderly, smooth, parallel line motion of the fluid. Inertia dominated flow is generally *turbulent* and characterized by irregular, erratic, eddylike paths of the fluid particles. In some cases viscosity is important only in a layer, called the *boundary layer*, next to a solid boundary while the main body of flow outside of the boundary layer is inertia dominated and behaves in an orderly fashion similar to that of laminar flow. If the boundary layer forces can be neglected, the resulting flow is called *potential or streamline* flow, an example of which is flow through an orifice. Potential flow is nonturbulent, streamline, and frictionless, so that the Reynolds number is infinite. Thus inertia dominated flow may be either turbulent or potential; however, the term *turbulent* is generally used to designate flows at high Reynolds numbers.

Assuming one-dimensional, steady, incompressible, frictionless ($\mu = 0$) flow with no body forces, the Navier-Stokes equations reduce to

$$u \frac{\partial u}{\partial x} = - \frac{1}{\rho} \frac{\partial P}{\partial x}$$

which may be integrated to yield

$$\frac{P}{\gamma} + \frac{u^2}{2g} = \text{constant} \tag{3-10}$$

Equation 3-10 is Bernoulli's equation with negligible gravity forces and is applicable to a streamline of potential flow. If the velocity is uniform across a flow section, then the constant is the same for all streamlines and (3-10) becomes

$$\frac{P_1}{\gamma} + \frac{u_1^2}{2g} = \frac{P_2}{\gamma} + \frac{u_2^2}{2g} \tag{3-11}$$

Note that if the velocity u at a section increases, the pressure must decrease and vice versa, that is, the total head at any section is a constant.

Generally, laminar flows can be solved from the Navier-Stokes equations if the geometry of the flow is simple. Potential flows can be described by Bernoulli's equation. However, turbulent flow relationships are almost entirely empirical. Some specific cases of practical interest will be discussed in the following sections.

3-3 FLOW THROUGH CONDUITS

Flow in pipes may be laminar or turbulent. The characteristic length used for Reynolds number is inside pipe diameter D, and the average flow velocity is volumetric flow rate divided by pipe area, that is,

$$\bar{u} = \frac{Q}{A} = \frac{4Q}{\pi D^2} \tag{3-12}$$

Therefore the Reynolds number is given by

$$R = \frac{\rho \bar{u} D}{\mu} = \frac{4 \rho Q}{\pi \mu D} \tag{3-13}$$

Transition from laminar to turbulent flow has been experimentally observed to occur in the range $2000 < R < 4000$. Below $R = 2000$ the flow is always laminar; above $R = 4000$ the flow is usually, but not always, turbulent. It is possible to have laminar flow at Reynolds number considerably above 4000 if extreme care is taken to avoid disturbances which would lead to turbulence. However, these instances are exceptional, and the high limit of 4000 is a good rule.

Laminar Flow in Pipes

Let us first consider steady laminar flow in pipes. Such pipes are often termed capillary tubes because the small diameters usually result in laminar

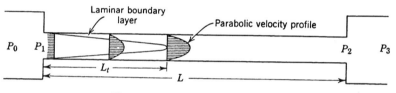

Figure 3-4   Laminar flow in a pipe.

flow. However, it should be recognized that low velocities or large viscosities can also result in laminar flow in pipes of larger diameter. As fluid enters a pipe (Fig. 3-4) the velocity profile is constant at a value $\bar{u}$ if there is rounding of the inlet. The fluid velocity at the pipe wall is zero, and this layer of fluid exerts considerable shear forces on the inner layers whose velocities must exceed $\bar{u}$ to satisfy the law of continuity. The boundary layer thus formed increases in thickness until the center of the pipe is reached. The velocity profile then becomes parabolic and remains parabolic throughout the length of pipe. Let u_0 denote the peak velocity of the entrance velocity profiles. For a parabolic profile, the peak velocity is $2\bar{u}$. The ratio $u_0/\bar{u}$ then varies from unity at the entrance to two at the length of pipe where a parabolic velocity profile is established. Langhaar [1] has studied this case in detail and the results are plotted in Fig. 3-5. P_0 is the pressure in the reservoir, and P_2 is the pressure after a length of pipe L (L is usually taken as the total length). The inlet length where the peak velocity is within 1 % of the final peak velocity of $2\bar{u}$ (i.e., $u_0/\bar{u} = 1.98$) is called the *transition length*. From Fig. 3-5 the transition length for laminar flow is

$$L_t = 0.0575 DR \qquad (3\text{-}14)$$

At the upper limit of laminar flow, that is, $R = 2000$, a transition length of 115 pipe diameters is required to establish fully developed laminar flow.

Both inertia and viscous forces affect the pressure drop in the transition length. However, only viscous forces are significant once a parabolic

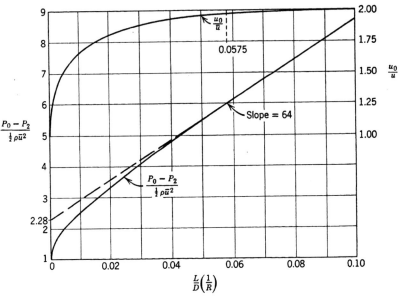

Figure 3-5   Pressure drop and peak velocity of steady laminar flow in a pipe.

velocity profile is achieved. For pipe lengths greater than the transition length, the pressure drop shown in Fig. 3-5 can be approximated by

$$\frac{P_0 - P_2}{\frac{1}{2}\rho \bar{u}^2} = 64 \frac{L}{D}\left(\frac{1}{R}\right) + 2.28 \qquad (3\text{-}15)$$

Combining (3-12), (3-13), and (3-15) yields

$$P_0 - P_2 = \frac{128\mu L Q}{\pi D^4}\left(1 + \frac{2.28}{64}\frac{DR}{L}\right) \qquad (3\text{-}16)$$

The first term in (3-16) is the well-known Hagen-Poiseuille law for fully developed laminar flow in pipes. The second term accounts for losses due to fluid inertia because the inner layers are being accelerated in the transition length. Entrance and exit losses are *not* accounted for in (3-16) and must be added (see Section 3-5). For example, an abrupt square-edged entrance from a large reservoir has a loss coefficient of 0.5. An abrupt exit into a large reservoir has a loss coefficient of 1, which means that all the kinetic energy of the issuing fluid is lost in the turbulent mixing with fluid in the reservoir; hence, $P_2 \approx P_3$. These losses can be factored into (3-16) by replacing 2.28 by 2.78. Thus, the pressure drop in a capillary with sharp-edged entrance and exit is

$$P_0 - P_3 = \frac{128\mu LQ}{\pi D^4}\left(1 + 0.0434 \frac{D}{L} R\right) \qquad (3\text{-}17)$$

If the length of pipe is less than the transition length, then the pressure drop is determined from the curve in Fig. 3-5. A trial and error solution is required if the pressure drop is known and flow is desired. If flow is known the abscissa may be directly computed and the corresponding ordinate selected to determine the pressure drop. Again, the entrance and exit losses must be added to the result.

Langhaar's theory seems to deviate from experimental values for $L/DR < 0.001$. An investigation of this range by Shapiro, Siegel, and Kline [2] gives

$$\frac{P_0 - P_2}{\frac{1}{2}\rho\bar{u}^2} = 1 + 13.74\sqrt{\frac{L}{DR}} \qquad (3\text{-}18)$$

for a short tube with well-rounded entry. Including the loss coefficient of 0.5 for a square-edged entry, assuming a square-edged exit such that $P_2 \approx P_3$, and because $\bar{u} = Q/A$, we obtain

$$P_0 - P_3 = \left(1.5 + 13.74\sqrt{\frac{L}{DR}}\right)\frac{\rho}{2}\left(\frac{Q}{A}\right)^2 \qquad (3\text{-}19)$$

This result is valid for short tubes with abrupt entrances and exits in the range $L/DR < 0.001$.

The linearity of pressure drop and flow, a characteristic of all laminar flows, is desirable in many circuits. Capillary tubes are often used to stabilize pressure control valves, as upstream restrictors in hydrostatic bearing, and placed across motor lines to improve damping. However, laminar flow is temperature sensitive, as it depends on viscosity, and this disadvantage must always be weighed.

Capillary tubes are usually formed by coiling small bore tubing or by screw thread passages. Although the formulas given are for straight tubes, computed results compare very well with test data for very tightly coiled capillaries because the coil radius is usually much larger than the capillary tube radius. A concern to the designer is the length of tube required for (*a*) Laminar flow through an elliptical tube

$$Q = \frac{\pi \, a^3 b^3}{4\mu L(a^2 + b^2)} (P_1 - P_2)$$

If $a = b = r$, then Hagen-Poiseuille law for a circular tube is obtained:

$$Q = \frac{\pi r^4}{8\mu L} (P_1 - P_2)$$

Passage cross sections

(*a*)

(*b*) Laminar flow through rectangular passages ($w \geq h$, i.e., w is selected to be the larger dimension)

$$Q = \frac{wh^3}{12\mu L}\left[1 - \frac{192h}{\pi^5 w} \tanh \frac{\pi w}{2h}\right](P_1 - P_2)$$

If $w = h$ (square cross section)

$$Q = \frac{w^4}{28.4\mu L} (P_1 - P_2)$$

If $w \gg h$, then

$$Q = \frac{wh^3}{12\mu L} (P_1 - P_2)$$

(*b*)

(*c*) Laminar flow through triangular passages
Equilateral triangle cross section

$$Q = \frac{s^4}{185\mu L} (P_1 - P_2)$$

Right angle triangle cross section

$$Q = \frac{s^4}{155.5\mu L} (P_1 - P_2)$$

(*c*)

(*d*) Laminar flow in annulus between circular shaft and cylinder ($c \ll r$)

$$Q = \frac{\pi r c^3}{6\mu L}\left[1 + \frac{3}{2}\left(\frac{e}{c}\right)^2\right](P_1 - P_2)$$

(*d*)

Nomenclature:

Q = volumetric flow rate, in.^{3}/sec
$P_1 - P_2$ = pressure drop in direction of flow, psi
L = passage length, in.
μ = fluid viscosity, lb-sec/in.^{2}
w = passage width, in.

h = passage height, in.
r = tube radius, in.
c = radial clearance, in.
e = eccentricity of shaft, in.
a, b = axes of ellipse, in.
s = side of a triangle, in.

Figure 3-6   Laminar flow through various passages with cross sections illustrated.

34

laminar flow to be dominant. Taking the extreme limit of laminar flow, $R = 2000$ and selecting a length to diameter ratio of 800, we find that the parenthesis in (3-17) has a value of 1.11. Therefore ratios of $L/D > 800$ give no more than 11% error in pressure drop when computed from the Hagen-Poiseuille law. If $R < 2000$, correspondingly shorter lengths can be used. Using 10% error in pressure computation as a criterion and referring to (3-17), we have design ratios given by

$$\frac{L}{D} \geq 0.434R \qquad (3\text{-}20)$$

that are satisfactory for engineering purposes. This relation insures that the capillary is the dominant resistance when pressures are measured in the end reservoirs. A conservative rule of thumb often used for capillary design is $L/D \geq 400$. However, much smaller ratios are satisfactory at lower Reynolds numbers.

Laminar flow in passages with other geometrical cross sections such as elliptical, rectangular, triangular, and annular can be derived [3], and the formulas are summarized in Fig. 3-6. The theoretical formulas for square and triangular cross sections are useful in sizing screw thread capillaries. However, experimental data is always preferred. It should be emphasized that laminar flow results only if the passage cross sections are comparatively small and/or have relatively long length. Otherwise, turbulent flow will exist. The Reynolds number, based on hydraulic diameter, should be computed to obtain a rough idea of the type of flow. Referring to Fig. 3-6, we note that flow through an annulus increases as the shaft becomes more eccentric. In the extreme case in which the shaft touches the cylinder wall, the eccentricity equals the radial clearance and the flow is 2.5 times that obtained with shaft and cylinder concentric. In practice the eccentricity is not known and an average flow between the two extremes might be used. This relation is useful in establishing the radial clearance in a seal so that leakage flow requirements are not exceeded.

Turbulent Flow in Pipes

Flow patterns and equations for turbulent flow in pipes are based largely on experimental observations. As flow enters the pipe (Fig. 3-7), the initial boundary layer is laminar but becomes turbulent (except for a very thin laminar sublayer) after a very short distance. This turbulent boundary layer increases in thickness to the center of the pipe in a transition length of about 25 to 40 pipe diameters [4]. A rather blunt velocity profile, with a peak velocity of about $1.2\bar{u}$, is then established and remains throughout the pipe length.

The empirical equation giving the pressure drop for fully developed turbulent flow is

$$P_1 - P_2 = f \frac{L}{D} \frac{\rho \bar{u}^2}{2} \qquad (3\text{-}21)$$

where f is the friction factor which depends on Reynolds number and pipe roughness. The additional pressure drop due to the transition length is about $0.09 \, \rho \bar{u}^2 / 2$ and is negligible in most computations. Pressure drops due to entrance and exit losses are also usually negligible.

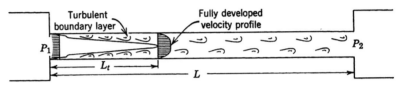

Figure 3-7   Turbulent flow in a pipe.

Equating (3-21) to the Hagen-Poiseuille law, the friction factor for laminar flow is

$$f = \frac{64}{R} \qquad (3\text{-}22)$$

In the turbulent flow range Blasius experimentally determined the friction factor to be

$$f = \frac{0.3164}{R^{0.25}} \qquad (3\text{-}23)$$

for smooth pipes and Reynolds numbers less than 100,000. Prandtl's universal law of friction for smooth pipes (3-24) is applicable

$$\frac{1}{\sqrt{f}} = 2 \log_{10} (R\sqrt{f}) - 0.8 \qquad (3\text{-}24)$$

for arbitrarily large Reynolds numbers but is somewhat difficult to manipulate mathematically. Fortunately (3-23) covers the cases in hydraulic control because the pipes are smooth and the flow velocities are normally kept below 15 ft/sec to avoid large pressure surges with sudden valve closures (see Section 3-7), and this results in Reynolds numbers being less than 10^5.

It is customary to plot friction factor versus Reynolds number for both flow regimes (Fig. 3-8) but iteration is required to determine flow for a given pressure difference since the Reynolds number is initially unknown.

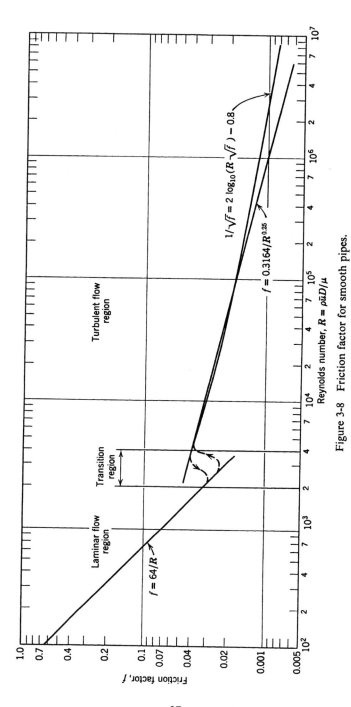

Figure 3-8 Friction factor for smooth pipes.

37

If the flow is known and the pressure drop is required, then Reynolds number can be directly computed and the friction factor selected from Fig. 3-8. The pressure drop is then obtained from (3-21). However, a much more convenient representation for laminar and turbulent flow can be obtained as follows [5]: The Hagen-Poiseuille law, applicable to fully developed laminar flow, can be written as

$$\frac{P_1 - P_2}{L} = \frac{128\mu}{\pi D^4} Q \tag{3-25}$$

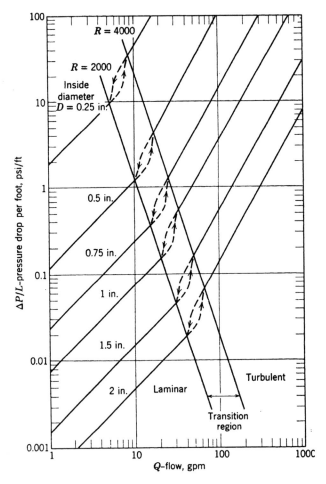

Figure 3-9   Pressure drop per foot for smooth pipe. Fluid density is 0.0307 lb/in^3 and viscosity is 4×10^{-6} lb-sec/in.^{2}

For fully developed turbulent flow (3-13), (3-21), and (3-23) may be combined to yield

$$\frac{P_1 - P_2}{L} = 0.242 \frac{\mu^{0.25} \rho^{0.75}}{D^{4.75}} Q^{1.75} \tag{3-26}$$

where $P_1 - P_2$ = pressure drop, psi
Q = volumetric flow rate, in.^{3}/sec
L = pipe length, in.
D = pipe inside diameter, in.
ρ = fluid mass density, lb-sec^2/in.^{4}
μ = fluid absolute viscosity, lb-sec/in.^{2}

For a given fluid and selected inside diameters (3-25) is plotted for $R <$ 2000, and (3-26) is plotted for $R > 4000$ as illustrated in Fig. 3-9. Such plots are very useful in design since flow, pressure drop, and pipe size are read directly without explicit computation of Reynolds number. Note the increase in the rate of pressure gradient along the pipe with flow in the turbulent region. For this reason laminar flow is desirable; however, the resulting pipe is usually unnecessarily large. Usually the flow is determined from load velocity requirements, and the pipe size is selected so that the pressure drop is moderate. Pipe selection criteria of 15 ft/sec maximum flow velocity and 1 psi/ft pressure drop are common.

Turbulent flow in closed conduits of noncircular cross section may be approximately computed from formulas given if the diameter is considered to be the hydraulic diameter. The hydraulic diameter is defined by

$$D_h = \frac{4A}{S} \tag{3-27}$$

where A is the flow section area and S is the flow section perimeter. For a circular section the hydraulic diameter becomes the inside pipe diameter. The concept of hydraulic diameter cannot be used for laminar flows because such flows are highly dependent on passage geometry. Transition Reynolds number may be approximately determined based on the hydraulic radius.

3-4 FLOW THROUGH ORIFICES

Orifices are a basic means for the control of fluid power. Flow characteristics of orifices plays a major role in the design of many hydraulic control devices. An orifice is a sudden restriction of short length (ideally zero length for a sharp-edged orifice) in a flow passage and may have a fixed or variable area. Two types of flow regime exist (Fig. 3-10), depending on whether inertia or viscous forces dominate. The flow velocity through an orifice must increase above that in the upstream region to satisfy the law of continuity. At high Reynolds numbers, the pressure drop across the orifice is caused by the acceleration of the fluid particles from the upstream velocity to the higher jet velocity. At low Reynolds numbers, the pressure drop is caused by the internal shear forces resulting from fluid viscosity.

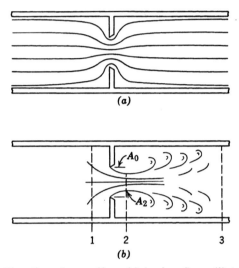

Figure 3-10   Flow through an orifice: (*a*) laminar flow; (*b*) turbulent flow.

Turbulent Orifice Flow

Since most orifice flows occur at high Reynolds numbers, this region is of greater importance. Such flows are often referred to as "turbulent", but the term does not have quite the same meaning as in pipe flow. Referring to Fig. 3-10*b*, the fluid particles are accelerated to the jet velocity between points 1 and 2. The flow between these points is streamline or potential flow and experience justifies the use of Bernoulli's equation in this region. The area of the issuing jet is smaller than the orifice area because the fluid particles have inertia and are moving in a curved path at the orifice opening. The point along the jet where the jet area becomes a minimum is called the *vena contracta*. The ratio of stream area at the vena contracta A_2 to the orifice area A_0 is called the contraction coefficient C_c.

$$A_2 = C_c A_0 \qquad (3\text{-}28)$$

For round orifices, the vena contracta occurs at approximately half an orifice diameter downstream and point 1 is about the same distance upstream (for a slit type orifice, these same distances are about $b/2$). Thus, the fluid is accelerated in a total distance of about one orifice diameter [6]. Between points 2 and 3 of Fig. 3-10b there is turbulence and violent mixing of the issuing jet with the fluid in the downstream region. The kinetic energy of the jet is converted into an increase in internal energy (temperature) of the fluid by the turbulence. Since the kinetic energy of the jet is not recovered, pressures P_2 and P_3 are approximately equal. This may be shown by analyzing the section between 2 and 3 as a sudden expansion (see Section 3-5).

The pressure difference required to accelerate the fluid particles from the lower upstream velocity u_1 to the higher jet velocity u_2 is found by applying Bernoulli's equation between points 1 and 2. Therefore

$$u_2{}^2 - u_1{}^2 = \frac{2}{\rho}(P_1 - P_2) \tag{3-29}$$

Applying the continuity equation for incompressible flow yields

$$A_1 u_1 = A_2 u_2 = A_3 u_3 \tag{3-30}$$

Combining (3-29) and (3-30) gives

$$u_2 = \left[1 - \left(\frac{A_2}{A_1}\right)^2\right]^{-\frac{1}{2}} \sqrt{\frac{2}{\rho}(P_1 - P_2)} \tag{3-31}$$

Because of viscous friction, the jet velocity is slightly less than that given by (3-31), and an empirical factor called the velocity coefficient C_v is introduced to account for this discrepancy. C_v is usually around 0.98 and is approximated by unity in most computations. Since $Q = A_2 u_2$, the volumetric flow rate at the vena contracta then becomes

$$Q = \frac{C_v A_2}{\sqrt{1 - (A_2/A_1)^2}} \sqrt{\frac{2}{\rho}(P_1 - P_2)} \tag{3-32}$$

Because it is more convenient to use orifice area rather than vena contracta area, (3-32) and (3-28) can be combined to yield

$$Q = C_d A_0 \sqrt{\frac{2}{\rho}(P_1 - P_2)} \tag{3-33}$$

where C_d, called the discharge coefficient, is given by

$$C_d = \frac{C_v C_c}{\sqrt{1 - C_c{}^2(A_0/A_1)^2}} \tag{3-34}$$

Since $C_v \approx 1$ and A_0 is usually much less than A_1, the discharge coefficient is approximately equal to the contraction coefficient. The contraction coefficient is difficult to compute but solutions have been made for round and slit-type sharp-edged orifices [6, 7] and are plotted in Fig. 3-11. Experience shows that the theoretical value of $C_c = \pi/(\pi + 2) = 0.611$ can be used for all sharp-edged orifices, regardless of the particular geometry, if the flow is turbulent and $A_0 \ll A_1$. For this reason a discharge coefficient of $C_d \approx 0.60$ is often assumed for all orifices and, since $C_d\sqrt{2/\rho} \approx 100$ in.²/$\sqrt{\text{lb}}$-sec, the orifice equation takes the familiar form

$$Q = 100A_0\sqrt{P_1 - P_2} \tag{3-35}$$

where pressures are in psi, orifice area is in in.², and volumetric flow rate is in in.³/sec.

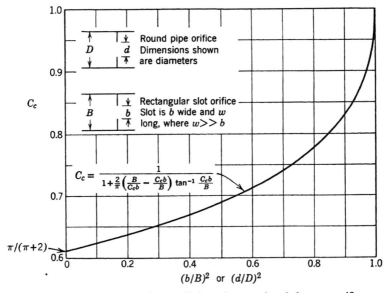

Figure 3-11   Contraction coefficients for round and slot type orifices.

Sharp-edged orifices are desirable for their predictable characteristics and insensitivity to temperature changes. However, cost frequently prohibits their use, especially as fixed restrictors, and orifices with length are often employed. An average discharge coefficient for such short tube orifices can be obtained as follows [2, 8]: Comparing (3-33) with (3-19) and (3-15), respectively, the discharge coefficient can be identified as

$$C_d = \left[1.5 + 13.74\left(\frac{L}{DR}\right)^{1/2}\right]^{-1/2} \quad \text{for} \quad \frac{DR}{L} > 50$$

$$C_d = \left(2.28 + 64\frac{L}{DR}\right)^{-1/2} \quad \text{for} \quad \frac{DR}{L} < 50$$

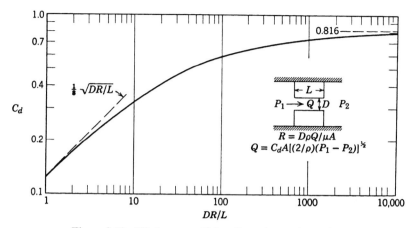

Figure 3-12   Discharge coefficient for a short tube orifice.

These relations are plotted in Fig. 3-12. Although this discharge coefficient is approximate, it is sufficient for design purposes. However, experimental data is always preferable.

Laminar Flow Through Orifices

At low temperatures, low orifice pressure drops, and/or small orifice openings, the Reynolds number may become sufficiently low to permit laminar flow. Reynolds number for an orifice is defined by

$$R = \frac{\rho(Q/A_0)D_h}{\mu} \qquad (3\text{-}36)$$

where Q/A_0 is the jet velocity at the orifice opening and D_h is the hydraulic diameter of the opening. For a circular orifice of diameter d the hydraulic diameter is $D_h = d$. For a rectangular slit orifice of width w and height b where $w \gg b$, the hydraulic diameter, defined by (3-27), becomes

$$D_h = \frac{4bw}{2(b + w)} \approx 2b \qquad (3\text{-}37)$$

Although the analysis leading to (3-33) is not valid at low Reynolds numbers, attempts have been made to extend this equation to the laminar region by plotting discharge coefficient as a function of Reynolds number. A typical plot of such data is shown in Fig. 3-13. For $R < 10$ many investigators have found the discharge coefficient to be directly proportional to the square root of Reynolds number; that is,

$$C_d = \delta\sqrt{R} \qquad (3\text{-}38)$$

The quantity δ depends on geometry and is called the laminar flow coefficient. Substituting (3-36) and (3-38) into (3-33) yields

$$Q = \frac{2\delta^2 D_h A_0}{\mu}(P_1 - P_2) \qquad (3\text{-}39)$$

for low Reynolds numbers. Note that flow is directly related to pressure difference and, since mass density is absent, dominated by fluid viscosity.

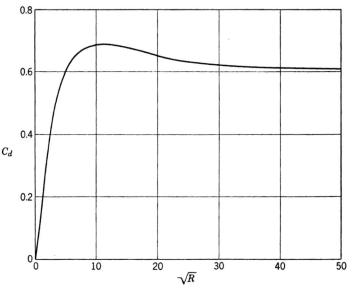

Figure 3-13   Typical plot of discharge coefficient versus Reynolds number for an orifice.

Wuest [9] has theoretically determined expressions for laminar flow through sharp-edged orifices. For a circular orifice in an infinite plane (i.e., $d \ll D$ in Fig. 3-11), the result is

$$Q = \frac{\pi \, d^3}{50.4\mu}(P_1 - P_2) \qquad (3\text{-}40)$$

For a rectangular slit of height b and width w in an infinite plane (i.e., $b \ll B$ in Fig. 3-11) with $w \gg b$, the result is

$$Q = \frac{\pi b^2 w}{32\mu}(P_1 - P_2) \qquad (3\text{-}41)$$

Equating (3-39) to (3-40) and to (3-41) gives $\delta = 0.2$ for a sharp-edged round orifice and $\delta = 0.157$ for a sharp-edged slit orifice. Viersma [10] represents the discharge coefficient by asymptotes defined by (3-38) in the laminar and $C_d = 0.611$ in the turbulent regions as shown in Fig. 3-14. The transition Reynolds number R_t is defined by the intersection point of the two asymptotes, that is

$$R_t = \left(\frac{0.611}{\delta}\right)^2 \tag{3-42}$$

For $\delta = 0.2$, the transition Reynolds number is $R_t = 9.3$ and increases as δ is increased.

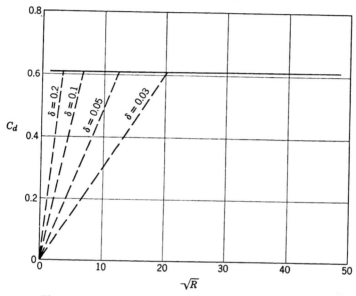

Figure 3-14   Asymptotic approximation of discharge coefficient.

In many instances it is desirable to evaluate the flow coefficient, $\Delta Q / \Delta(P_1 - P_2)$, of an orifice with no initial pressure drop. If turbulent flow is assumed, this coefficient is infinite as seen by differentiation of (3-33). However, the flow becomes laminar when the pressure drop is small and the flow coefficient has a finite value which can be estimated from (3-40) and (3-41).

In summary, orifice flow is laminar for $R < R_t$ with flow rates directly related to pressure drop as given by (3-39). In the vicinity of R_t, both inertia and viscosity are important. For $R > R_t$, the flow can be treated as turbulent and described by the orifice equation (3-33). The orifice equation is commonly used for all situations with a total disregard for the types of flow that can be encountered. This is justified in the majority of cases but can lead to gross errors in certain instances.

3-5 MINOR LOSSES

The term minor losses refers to those energy losses caused by bends, fittings, and sudden changes in flow cross section. These losses are empirically described by

$$H_L = K \frac{u^2}{2g} = \frac{K}{2g}\left(\frac{Q}{A}\right)^2 \tag{3-43}$$

where u is the fluid velocity, Q is the volumetric flow rate, and A is the passage area (if two areas are involved, the area of the *smaller* cross section

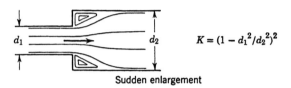

Sudall enlargement
$K = (1 - d_1^2/d_2^2)^2$

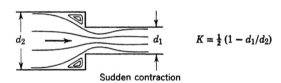

Sudden contraction
$K = \frac{1}{2}(1 - d_1/d_2)$

Resistance coefficient due to the geometry of pipe entrances and exits with large reservoirs, that is $d_1/d_2 \approx 0$

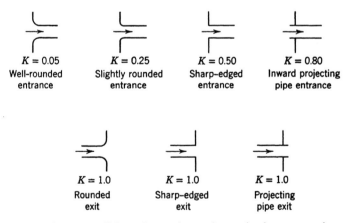

Figure 3-15   Resistance coefficients due to abrupt changes in pipe cross section and due to the geometry of pipe entrances and exits.

is used); K is called the resistance or loss coefficient, and values are given in Fig. 3-15 for many cases of interest.

Bernoulli's equation applied between points 2 and 3 of the abrupt exit in Fig. 3-15 would predict a pressure increase. However, there is a loss in energy which must be taken into account. Because $K = 1$ for all reasonably abrupt exits into a large reservoir (i.e., $d_2/d_1 \gg 1$), the head loss is $u^2/2g$. This means that the entire kinetic energy of the fluid entering the reservoir is converted into heat energy by the turbulent mixing which takes place. Hence, there is little or no recovery in pressure during the expansion so that $P_2 \approx P_3$. The exit must be smooth and diverge gradually to achieve significant pressure recovery. For most cases in hydraulic control it is sufficient to assume that $P_2 = P_3$ for an exit into a large chamber.

When fluid flow encounters a sudden contraction, a vena contracta is formed between approximately one half to one pipe diameters downstream. Because the conversion of pressure energy into kinetic energy at the inlet is very efficient, most of the energy loss occurs due to the expansion of the fluid stream from the vena contracta. Application of Bernoulli's equation to points 0 and 2 of the sudden contraction in Fig. 3-15 yields

$$\frac{P_0}{\gamma} + \frac{u_0^2}{2g} = \frac{P_2}{\gamma} + \frac{u_2^2}{2g} + K \frac{u_2^2}{2g}$$

where K is given in Fig. 3-15 for different inlet geometries. Because $u_0 \ll u_2$, u_0 can be neglected and we obtain

$$P_0 - P_2 = (1 + K)\frac{\rho}{2}u_2^2$$

Although point 2 is downstream of the vena contracta, the distinction between points 1 and 2 is often overlooked and $(P_0 - P_2)$ is referred to as the inlet pressure drop.

Centrifugal forces and secondary flow patterns result in a pressure drop in pipe bends. There is much variation in test data for resistance coefficients of bends and those values given in Fig. 3-16 should be considered approximate [4, 5]. Because the fluid velocities at points 1 and 2 in Fig. 3-16 are the same, application of Bernoulli's equation yields

$$P_1 - P_2 = K \frac{\rho}{2}\left(\frac{Q}{A}\right)^2$$

as the pressure drop due to the bend. The pressure drop for the length

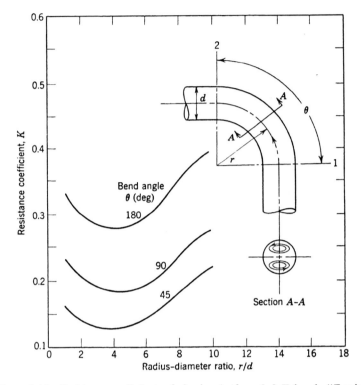

Figure 3-16   Resistance coefficients of pipe bends (from J. J. Taborek, "Fundamentals of Line Flow," *Machine Design Magazine*, April 16, 1959).

of the bend is determined from (3-21) and added to that due to the bend. Pressure drops in fittings and valves vary widely and data should be obtained from component manufacturers or by direct measurement.

3-6 POWER LOSS AND TEMPERATURE RISE

Hydraulic horsepower is the product of pressure drop and flow ΔPQ, and in many instances this power is consumed by fluid friction and increases the internal energy of the fluid. The power used by all hydraulic resistances such as orifices, valves, pipes, capillaries, and minor losses, is converted into a temperature increase of the fluid. Most of the horsepower produced by hydraulic motors is used as shaft work, but power used by internal and cross leakages is converted into heat.

The power converted into heat energy by a hydraulic resistance is $9339C_p\Delta TQ\gamma$. Equating this to the supplied hydraulic horsepower of

ΔPQ yields a temperature rise in the direction of flow of

$$\Delta T = \frac{\Delta P}{9339 C_p \gamma} \tag{3-44}$$

Thus there is about a $1°F$ rise in temperature across any hydraulic resistance (orifice, pipe, etc.) for each 140 psi drop if the fluid has a petroleum base for which $\gamma = 0.03$ lb/in.^{3}.

Equation 3-44 may be derived in a more rigorous manner from (3-6). If there is no heat added or work done and the flow is steady (i.e., $W_1 = W_2$) then (3-6) becomes

$$9339 C_p T_1 + \frac{P_1}{\gamma} + \frac{V_1^2}{2g} = 9339 C_p T_2 + \frac{P_2}{\gamma} + \frac{V_2^2}{2g} \tag{3-45}$$

In most cases the inlet and outlet is about the same size so that $V_1 \approx V_2$ and (3-45) reduces to (3-44).

The temperature rise given by (3-44) is useful in determining the heat generated in hydraulic systems (see Section 12-6).

3-7 PRESSURE TRANSIENTS IN HYDRAULIC CONDUITS

When fluid flowing in a conduit is suddenly stopped due to a rapid valve closure at the end of the conduit, a very large pressure transient may result. This phenomenon is called *waterhammer* because it is usually accompanied by considerable noise. The fluid adjacent to the valve is stopped initially and a pressure wave, which heads the increasing amount of fluid being brought to a standstill, travels back to the fluid source at velocity c given by

$$c = \left(\frac{\beta_e}{\rho}\right)^{1/2} \tag{3-46}$$

where $c =$ velocity of sound in the fluid, in./sec
   $\beta_e =$ effective bulk modulus (includes fluid and mechanical compliance), lb/in.^{2}
   $\rho =$ mass density of fluid, lb-sec^2/in.^{4}

Values of c in the range 35,000 to 50,000 in./sec are common. When the pressure wave arrives at the source end of the conduit (in L/c seconds where L is the conduit length), then the kinetic energy of the moving mass of fluid has been completely stored as potential energy in the elasticity of conduit and fluid and the pressure of the compressed fluid, P_{IC}, is a maximum. At this time a decompression wave forms and travels back to the valve. These waves continue to travel back and forth with the associated interchanges of kinetic and potential energies until friction expends the energy involved.

At the instant of valve closure the kinetic energy of the moving fluid is

$$KE = \tfrac{1}{2}M_f v_0{}^2 = \tfrac{1}{2}\rho L A v_0{}^2 \qquad (3\text{-}47)$$

where A is the conduit area and v_0 is the initial velocity of the fluid. The potential energy stored in the compressed fluid is

$$PE = \frac{1}{2}\frac{LA}{\beta_e} P_{IC}{}^2 \qquad (3\text{-}48)$$

where P_{IC} is the pressure rise due to the instant valve closure. Equating these energies yields an expression for pressure rise.

$$P_{IC} = \rho c v_0 \qquad (3\text{-}49)$$

Using typical values for ρ and c, this expression becomes $P_{IC} \approx 50v_0$ psi where v_0 has units of ft/sec.

It is apparent from (3-49) that the most effective and only way, since ρ and c are fixed, to reduce this pressure surge is to design pipe systems to have low original fluid velocities by keeping pipe areas large. If fluid velocities are limited to a maximum of 15 ft/sec, then the instant closure pressure rise (above the steady state level) is about 750 psi, which is generally considered a safe design value and is a criterion for conduit selection. It is interesting to note that P_{IC} is independent of line length.

Equation 3-49 is valid and the closure is considered instantaneous if the valve closure time T is less than that required for one round trip of the pressure wave, that is,

$$T \le T_c = \frac{2L}{c} \qquad (3\text{-}50)$$

where T_c is commonly called the critical closure time. For short lines this inequality is generally not satisfied. In this event the pressure rise will depend on line length, steady state pressure level, and valve closure time in addition to P_{IC}. The mathematical equations which describe this situation are unwieldy but solutions can be made and expressed in graphical form. Quick's chart [11] assumes uniform valve closure and is the most convenient graphical technique (Fig. 3-17). To use Quick's chart first

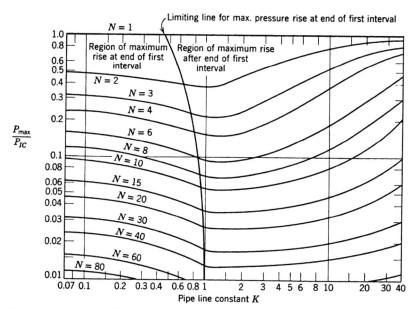

Figure 3-17   Quick's Chart showing maximum pressure rise with uniform valve closure (from *Applied Fluid Mechanics* by M. P. O'Brien and G. H. Hickox, Copyright ©, 1937 and used by permission of the McGraw-Hill Book Co.).

compute the quantities K and N in which

$$K = \frac{P_{IC}}{2P_0} = \text{pipe line constant, dimensionless}$$

$$N = \frac{T}{T_c} = \text{number of critical closure time intervals, dimensionless}$$

P_0 = steady state pressure in conduit, psi

$$T_c = \frac{2L}{c} = \text{critical closure time, sec}$$

T = time required for uniform valve closure, sec

$P_{\max}$ = pressure rise in conduit above P_0 due to uniform valve closure (for instant closure, $P_{\max} = P_{IC}$), psi

Now, from the chart obtain the value of $P_{\max}/P_{IC}$ from which, since P_{IC} is computed from (3-49), $P_{\max}$ is computed. Note that $P_{\max}$ will always be less than P_{IC}.

In summary, the most severe pressure surge in a single pipe is that caused by an instantaneous valve closure (i.e., $T \leq T_c$) and is given by (3-49). If the valve closure is not instantaneous (i.e., $T > T_c$) then uniform valve closure should be assumed and P_{max} obtained from Quick's chart. This latter case is typical for short lines (less than 10 ft if $T > 0.005$ sec). However, because P_{IC} is the maximum surge pressure, instant valve closure is usually assumed for design purposes.

3-8 SUMMARY

There is always the possibility of confusion when many equations are presented. Therefore, a summary of those basic equations ordinarily used in the analysis of hydraulic controls is in order.

At the outset it was determined that seven equations are required to define a situation involving fluids. These seven equations reduce considerably if the fluid is a liquid. The first three equations, the Navier-Stokes equations, are reduced to the application of certain formulas which were discussed in Sections 3-3, 3-4 and 3-5. As a general rule, only those equations that describe intentionally inserted hydraulic resistances are used in a dynamic analysis because these are usually the dominant restrictors. Resistances of flow passages such as pipes, bends, and fittings, are often neglected. Therefore the formulas most often used are the orifice equation, (3-33) and those given in Fig. 3-6.

Because cubical expansion coefficients are small for liquids, the direct effect of temperature on fluid density and, consequently, on fluid flow is often negligible. This is not to say that thermal gradients never exist. It is simply that these gradients have little influence on flow conditions. Therefore it is usually sufficient to include temperature by evaluating fluid properties at the operating temperature. It is generally assumed that isothermal conditions exist in liquid flow. The assumption of constant temperature eliminates the need for the energy equation and reduces the equation of state to the simple form

$$\rho = \rho_i + \frac{\rho_i}{\beta} P \tag{3-51}$$

where ρ_i and β are the mass density and bulk modulus at zero pressure. The continuity equation (3-4) can be written

$$\sum W_{in} - \sum W_{out} = g \frac{d(\rho V_0)}{dt} = g\rho \frac{dV_0}{dt} + gV \frac{d\rho}{dt} \tag{3-52}$$

Noting that weight flow rate can be written $W = g\rho Q$, we can combine

(3-51) and (3-52) to yield

$$\sum Q_{in} - \sum Q_{out} = \frac{dV_0}{dt} + \frac{V_0}{\beta}\frac{dP}{dt} \tag{3-53}$$

Thus the continuity equation and the equation of state are combined into the more useful form given by (3-53). The first term on the right side is the flow consumed by expansion of the control volume; if the volume is fixed, this term is zero. The second term is the compressibility flow and describes the flow resulting from pressure changes.

Need for the seventh equation is eliminated by assuming that viscosity is constant. Therefore all seven of the initial equations have been accounted for. In pneumatic systems the temperature may vary and a slightly different reduction of the initial equations is required.

REFERENCES

[1] Langhaar, H. L., "Steady Flow in the Transition Length of a Straight Tube," *J. Appl. Mech.*, June 1942, A55-A58.

[2] Shapiro, A. H., R. Siegel, and S. J. Kline, "Friction Factor in the Laminar Entry of a Smooth Tube," Proc. *2nd* U.S. *Natl Congr. Appl. Mech.*, June 1954, 733-741.

[3] Dryden, H. L., F. D. Murnaghan, and H. Bateman, *Hydrodynamics.* New York: Dover, 1956, p. 197.

[4] Schlichting, H., *Boundary Layer Theory.* New York: McGraw-Hill, 1960.

[5] Taborek, J. J., "An Engineering Approach to Hydraulic Lines," *Machine Design*, April 16, 30; May 14, 28; June 11; July 9, 1959.

[6] Rouse, H., and A. Abul-Fetouh, "Characteristics of Irrotational Flow Through Axially Symmetric Orifices," *J. Appl. Mech.*, December 1950, 421-426.

[7] Cisotti, U., *Idromeccanica Piana*, vol. 2. Milan: Librerio Editrice Politecnica, 1922, pp. 237-383.

[8] Kreith, F. and R. Eisenstadt, "Pressure Drop and Flow Characteristics of Short Capillary Tubes at Low Reynolds Numbers," *ASME Trans.*, July, 1957, 1070-1078.

[9] Wuest, W., "Stromung durch Schlitz-und Lochblenden bei kleinen Reynolds-Zahlen," *Ingenieur Archiv*, No. 22, 1954, 357-367.

[10] Viersma, T. J., "Designing Load-Compensated Fast Response Hydraulic Servos," *Control Eng.*, May 1962.

[11] Quick, Ray S., "Comparison and Limitations of Various Waterhammer Theories," *Mech. Eng.*, **49**, No. 5a, May 1927, 524-530.

4

Hydraulic Pumps and Motors

Hydraulic pumps and motors are used to convert mechanical energy into hydraulic energy and vice versa, respectively. These machines may be broadly classified as *hydrodynamic* or *positive displacement*. In hydrodynamic machines, such as turbine, centrifugal, and jet pumps, the flow is continuous from inlet to outlet and results from energy being directly imparted to the fluid stream. These machines are basically low pressure with high volume output. They are quite inefficient and not suited to control purposes but are used for auxiliary functions such as coolant pumps.

In the positive displacement machine fluid passes through the inlet into a chamber which expands in volume and fills with fluid. The volume expansion causes shaft rotation in a motor in contrast to a pump where the volume expansion is caused by shaft rotation. The volume of trapped fluid is then sealed from the inlet by some mechanical means and then transported to the outlet side where it is discharged. A succession of small volumes of fluid transported in this manner gives a fairly uniform flow. Thus a positive or definite amount of fluid is displaced or transported through the machine per unit of shaft revolution. Positive displacement machines are quite efficient and find extensive use in control systems.

In this chapter we discuss types of positive displacement machines with particular emphasis on steady-state performance parameters and characteristics. A meaningful dynamic analysis cannot be made without including the control elements associated with the pump or motor. Chapter 5 treats the steady-state performance of valves. Valves and pumps and motors are then combined to form basic hydraulic power elements in Chapter 6 and dynamic analyses are made.

4-1 BASIC TYPES AND CONSTRUCTIONS

Positive displacement machines may be broadly classified as either *limited* or *continuous travel* devices. Examples of limited travel actuators are the common piston (Fig. 4-1) and the limited rotation motor (Fig. 4-2).

These machines are simple in construction and are available in a wide selection of rod or shaft configurations and types of mounting.

Any continuous travel (i.e., rotating) positive displacement machine must have a mechanical element (such as a gear tooth, vane, or piston head) on which pressure acts, a lever arm mechanism (gear radius, eccentric rotor, bent piston axis, wobble plate, or cam) to convert this force to shaft torque (or convert shaft torque to a force in the case of a pump), some

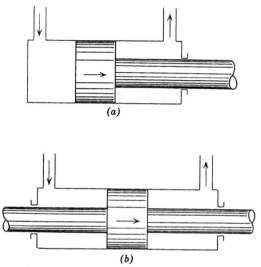

(a)

(b)

Figure 4-1   Schematic diagrams of double-acting piston actuators: (a) single rod; (b) double rod.

method of sealing inlet from outlet, and some method of porting fluid to the mechanical elements on which the pressure acts. Continuous travel positive displacement machines are usually classified by mechanical construction and at least five basic types can be distinguished: *screw*, *gear*, *vane*, *rotary abutment*, and *piston*. With certain restrictions, all types can be operated as either pumps or motors.

Screw-type machines consist of two or three screws intermeshed and suitably housed. Fluid is propelled axially in the sealed enclosures formed by the meshing of the screws. These units are quiet in operation and deliver a smooth flow. The volumetric displacement cannot be varied and pressure range is limited to under 1000 psi.

Gear machines may be subdivided into *external* and *internal gear* configurations. External gear machines are further subdivided by the type of gear teeth employed: *spur*, *helical*, and *herringbone*. A gear pump or motor consists of two meshed gears (Fig. 4-3), suitably housed with a shaft to one of the gears. Close tolerances are held between the housing and the gear sides and periphery to prevent excess leakage. Fluid is transported in the spaces between the teeth in the periphery of both gears, and the meshed segment of the gears serve as a seal between inlet and outlet. The

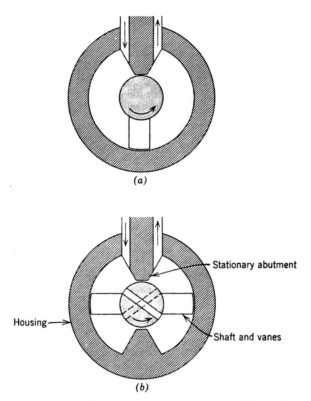

Figure 4-2   Limited rotation actuators: (*a*) single vane; (*b*) double vane.

basic hydrostatic unbalance of these devices results in considerable friction and wear which limits performance to the lower pressures (up to 1000 psi). However, hydrostatically balanced units can be operated at higher pressures and is a desirable feature in units which must function as motors.

Two types of internal gear machines can be distinguished: *crescent seal* and *gerotor*. The crescent seal unit (Fig. 4-4) consists of an inner and outer gear separated by a crescent or moon-shaped piece which acts as a seal. Both gears rotate in the same direction, but the inner gear speed is faster than that of the outer gear. However, the relative speed of the gears

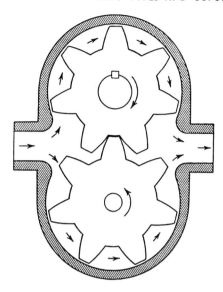

Figure 4-3 Schematic of a spur gear pump.

is low. Fluid enters where the gears begin to unmesh and is trapped in the spaces between the teeth of both gears and carried past both sides of the crescent to the outlet port where the gears begin to mesh. The gerotor pump or motor (Fig. 4-5) consists of a pair of gear shaped elements ingeniously mated so that each tooth of the inner gear is always in sliding contact with the outer gear to form sealed pockets of fluid. Both gears rotate in the same direction at low relative speeds with the inner gear being faster. Fluid enters the chamber with increasing volume, is trapped in the spaces between the teeth, and is transported to the outlet. Frictional torques due to hydraulic unbalance restrict internal gear machines to the lower pressures. It is not possible to vary the displacement of gear type pumps and motors.

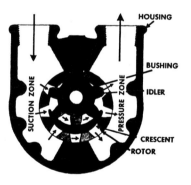

Figure 4-4 Crescent seal internal gear pump (courtesy of Tuthill Pump Company, Chicago, Illinois).

Vane pumps and motors, illustrated in Fig. 4-6, are usually classified by the type of vane design: *sliding* vane, *swinging* vane, *rolling* vane, and so on. The sliding vane machine, by far the most common design, has vanes which fit into slots in the rotor periphery and are held in contact with the housing by springs, centrifugal force, and/or pressure. The swinging vane design has vane members hinged in the rotor and rotated by centrifugal force to contact the housing. In both units, vane wear is automatically compensated. The vanes of the roller vane pump are semi-cylindrical in shape and are rotated by gearing to the shaft to keep them

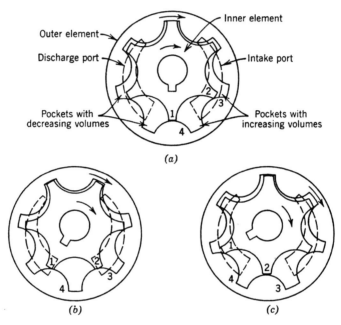

Figure 4-5   Operating principle of the generated-rotor pump or motor. The inner and outer elements travel in the same direction with the inner element moving slightly faster. Note how the pocket formed by teeth 1, 2, 3, and 4 is filled with fluid and transported from intake to discharge as the shaft rotates.

parallel as the rotor turns. In all of these machines the rotor is mounted eccentrically in the housing. Fluid enters the chamber with increasing volume and is trapped between the vane elements and transported to the chamber with decreasing volume. The displacement of these devices can be varied by changing the rotor eccentricity, and operation is generally limited to low to medium working pressures (up to 1500 psi) because of the basic hydrostatic unbalance. Hydraulically balanced units are preferred for operation at higher pressures or as motors. In the balanced design (Fig. 4-7) the vanes run in an oval-shaped bore with the rotor centered. Two regions of changing volumes are thus formed around the rotor periphery through which fluid is transported. The displacement of hydraulically balanced machines is not variable.

Rotary abutment pumps and motors are basically limited rotation actuators, such as those shown in Fig. 4-2, but with the abutments properly shaped and free to rotate so that they can move around the rotor vanes and

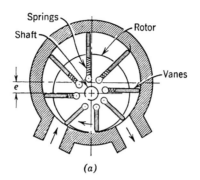

(a)

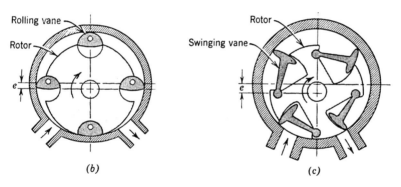

(b)   (c)

Figure 4-6   Vane type pumps and motors: (a) sliding vane; (b) rolling vane; (c) swinging vane.

permit continuous rotation. Single and double vane designs can be distinguished. The unit in Fig. 4-8 is basically a single vane device because rotary pistons P_1 and P_2 alternate as the active vane with each being effective for one-half revolution. The center member V, called a rotary abutment, provides a rolling contact on the active piston to seal between inlet and outlet. All three members are geared to rotate in a 1:1 ratio, and power is taken from a shaft on the rotary abutment. No sealing contact is required as the inactive vane passes in the recess of the rotary abutment

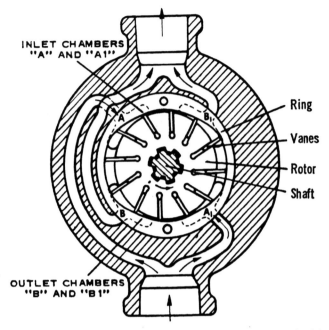

Figure 4-7   Vickers balanced-vane type pump or motor (courtesy of Vickers, Inc., Machinery Hydraulics Division, Ferndale, Michigan).

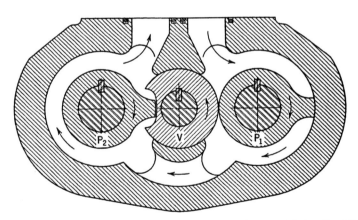

Figure 4-8   Rotary abutment pump or motor (courtesy of Tyrone Hydraulics, Division of Oliver Tyrone Corp., Corinth, Mississippi).

because pressure is equalized around the vane. Rotary abutment motors have constant volumetric displacement and operate smoothly at lower speeds.

Piston type pumps and motors are classified by the motion of the pistons relative to the shaft axis as *axial*, *bent-axis*, and *radial*, as illustrated in Figs. 4-9 to 4-12. These units can be distinguished also by the type of

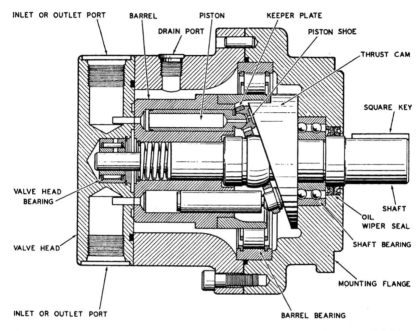

Figure 4-9   Axial piston pump or motor (courtesy of the Dynex Company, Division of Applied Power Industries, Inc., Pewaukee, Wisconsin).

lever arm mechanism employed to convert force to torque and by the basic mechanism used for porting flow to the pistons. Common lever mechanisms are wobble plate, bent axis, cam, and eccentric rotor. Three basic porting arrangements are in use: valve plate, pintle, and piston, as illustrated in Figs. 4-10, 4-11, and 4-12, respectively. Piston-type machines can be used at working pressures in excess of 3000 psi and the displacement can be varied. These devices are quite rugged, versatile, and efficient and are used extensively for hydraulic control in both aircraft and industrial systems.

Figure 4-9 shows an axial piston pump or motor having a stationary wobble plate and using valve plate porting. The valve plate ports the inlet fluid to half of the cylinder barrel, and the pistons receiving this

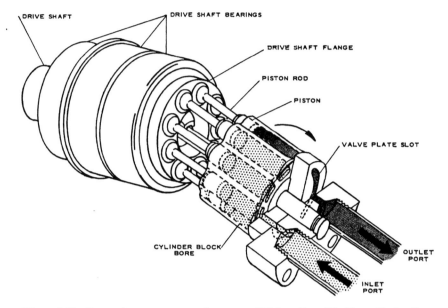

Figure 4-10   Bent-axis pump or motor (courtesy of Vickers, Inc., Machinery Hydraulics Division, Ferndale, Michigan).

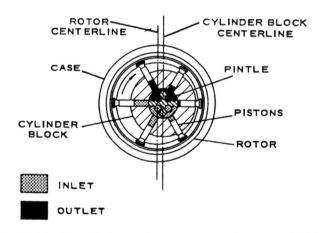

Figure 4-11   Radial piston, pintle ported pump or motor (courtesy of Vickers, Inc., Machinery Hydraulics Division, Ferndale, Michigan).

fluid are forced against the inclined wobble plate. Because the plate is stationary, the cylinder barrel must rotate and is connected with the drive shaft. Variable displacement can be achieved by varying the angle of the inclined plate. Some axial piston motors have a stationary cylinder barrel and a rotating wobble plate that turns the drive shaft as a result of piston motion.

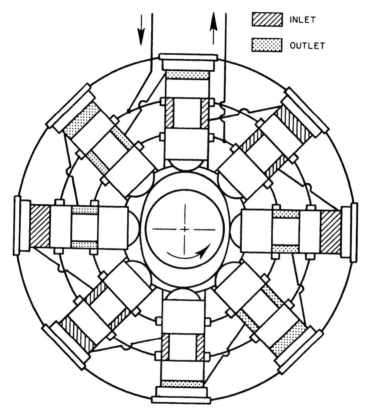

Figure 4-12   Radial piston cam-roller hydraulic motor using piston porting (courtesy of the Cincinnati Milling Machine Company, Cincinnati, Ohio).

The bent-axis machine (Fig. 4-10) also uses valve plate porting. Fluid is ported by the valve plate to one half of the cylinder barrel and forces the pistons in that portion of the barrel away from the valve plate. These pistons rotate the drive shaft as they move to the maximum distance from the valve plate because the piston axis is bent from that of the drive shaft. The cylinder barrel is driven through a universal joint shaft from the drive shaft.

A radial piston machine is illustrated in Fig. 4-11. The cylinder block is eccentric in the housing, and fluid from the pintle enters half of the cylinder block bores and forces the pistons outward radially. However, the pistons can move radially only by revolving the cylinder block which then rotates the output shaft. The pistons ported by the pintle to the outlet move in a manner to decrease the chamber volume, thereby expelling fluid to the outlet.

The radial piston motor in Fig. 4-12 uses an oval-shaped cam to convert the pressure force on the pistons to shaft torque. The cam is shaped so that the volumetric displacement of the motor is absolutely constant. The displacement of most piston motors varies periodically a few percent at a frequency, depending on the number of pistons and shaft speed. This displacement variation causes undesirable velocity variations. Each piston in Fig. 4-12 is ported to inlet or outlet by the adjacent piston. This motor is basically a low speed and high torque device.

Of all the positive displacement machines discussed, only the vane and piston units are capable of variable displacement. This feature is quite desirable in pumps because it is possible to adjust the delivery to meet the circuit needs and thus improves efficiency. However, this feature is of limited usefulness in units used as motors. In both vane and piston units the displacement is varied by altering the lever arm. When operated as a motor the reduction in lever arm is limited by internal friction because binding will occur. As a practical limit, motor displacement can be reduced to only about one quarter of the maximum displacement.

4-2 IDEAL PUMP AND MOTOR ANALYSIS

An ideal pump or motor is defined as having no power losses due to friction and leakages and, consequently, has an efficiency of 100%. Although this is certainly not true in practice, hydraulic machines are quite efficient, and system design is often based on ideal machines. Consider an ideal hydraulic motor. The mechanical power output is

$$hp \mid_{\text{out}} = T_g \dot{\theta}_m \tag{4-1}$$

where T_g = torque generated by motor and delivered to load, in.-lb
  $\dot{\theta}_m$ = shaft speed of motor, rad/sec
The hydraulic power supplied to the motor is

$$hp \mid_{\text{in}} = P_L Q_L \tag{4-2}$$

where P_L = pressure difference across motor lines, psi
  Q_L = flow through the motor, in.³/sec
Because the motor is assumed to be 100% efficient, these expressions can

be equated to yield

$$T_g = \frac{Q_L}{\dot{\theta}_m} P_L \qquad (4\text{-}3)$$

Now, by definition,

$$D_m \equiv \frac{Q_L}{\dot{\theta}_m} \qquad (4\text{-}4)$$

where D_m = volumetric displacement (or simply displacement) of the motor, in.^{3}/rad. Combining equations gives

$$T_g = D_m P_L \qquad (4\text{-}5)$$

Equations 4-4 and 4-5 are the fundamental relations for an ideal motor (or pump). Only one parameter (D_m) is required to define the ideal machine, and this quantity is also the single most important parameter for practical machines. In fact, hydraulic motor sizes are designated by the ideal theoretical displacement. This analysis also holds for the ideal pump, except that the power flow is reversed, that is, mechanical power is transformed into hydraulic power. Pump sizes are customarily designated by the flow obtained at a certain shaft speed. The ratio of these quantities is the volumetric displacement.

A similar analysis can be made for an ideal piston type actuating device. The piston area is then the parameter analogous to the displacement of a rotary device.

4-3   PRACTICAL PUMP AND MOTOR ANALYSIS

Leakage flows and friction are the sources of losses in hydraulic machines. Let us now examine these losses and include them in an analysis of steady state performance. Consider the axial piston motor illustrated in Fig. 4-13. This motor has a stationary wobble plate to convert piston motion into rotary motion and uses valve plate porting. Only two pistons, one each from the two chambers, are drawn to simplify the illustration. The leakage and friction losses of all pistons are lumped at these pistons.

From Fig. 4-13 it is apparent that two types of leakage flow can exist: *internal* or *cross-port leakage* between the lines and *external leakage* from each motor chamber past the pistons to case drain. Because all mating clearances in a motor are intentionally made small to reduce losses, these leakage flows are laminar and, therefore, proportional to the first power of pressure. The internal leakage is proportional to motor pressure difference and may be written

$$Q_{im} = C_{im} P_L \qquad (4\text{-}6)$$

where C_{im} = internal or cross-port leakage coefficient, in.^{3}/sec/psi
$P_L = P_1 - P_2$ = pressure difference across motor, psi
The external leakage in each chamber is proportional to the particular chamber pressure (assuming negligible drain pressure) and may be written

$$Q_{em1} = C_{em}P_1 \tag{4-7}$$

$$Q_{em2} = C_{em}P_2 \tag{4-8}$$

where C_{em} = external leakage coefficient (assumed to be the same for each chamber), in.^{3}/sec/psi
P_1 = pressure in forward chamber, psi
P_2 = pressure in return chamber, psi

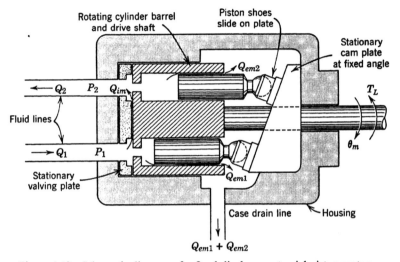

Figure 4-13 Schematic diagram of a fixed displacement axial piston motor.

The steady-state continuity equation (i.e., without compressibility flow) for the motor chambers are

$$Q_1 - C_{em}P_1 - C_{im}(P_1 - P_2) - D_m\dot{\theta}_m = 0 \tag{4-9}$$

$$D_m\dot{\theta}_m + C_{im}(P_1 - P_2) - C_{em}P_2 - Q_2 = 0 \tag{4-10}$$

where D_m = ideal volumetric displacement of motor, in.^{3}/rad
$\dot{\theta}_m$ = motor shaft speed, rad/sec
Q_1 = forward flow to motor, in.^{3}/sec
Q_2 = return flow from motor, in.^{3}/sec
These two equations completely describe the flows in the motor. If leakage coefficients are zero then $Q_1 = Q_2 = D_m\dot{\theta}_m$, which was the result for the ideal motor.

Subtracting (4-10) from (4-9) yields

$$Q_L = D_m \dot{\theta}_m + \left(C_{im} + \frac{C_{em}}{2} \right) P_L \qquad (4\text{-}11)$$

where by definition

$$Q_L = \frac{Q_1 + Q_2}{2} \qquad (4\text{-}12)$$

The quantity Q_L, commonly called the *load flow*, represents the average of the flows in the two motor lines; Q_L equals the flow in each line only if external leakage is zero. The concept of load flow is useful because it permits reduction of the two flow equations to a single equation which relates load flow to motor pressure difference and speed only. Very often (4-11) is written directly and Q_L is interpreted as the flow in each line; however, this is valid only if $C_{em} = 0$. Defining Q_L as the average flow avoids this restriction and properly shows that external leakage acts like internal leakage as far as pressure difference is concerned. We shall see in Chapter 6, in which the dynamic case is treated, that external leakage also contributes to motor damping.

Let us now consider the torques which act in a motor. The ideal generated torque is

$$T_g = D_m(P_1 - P_2) \qquad (4\text{-}13)$$

However, there are at least three sources of torque losses which detract from the generated torque:

1. A torque proportional to motor speed exists because torque is required to shear the fluid in the small clearances between mechanical elements in relative motion. This damping torque can be written as

$$T_d = B_m \dot{\theta}_m = C_d D_m \mu \dot{\theta}_m \qquad (4\text{-}14)$$

where $B_m = C_d D_m \mu$ = viscous damping coefficient, in-lb-sec
C_d = dimensionless damping coefficient
μ = absolute viscosity of fluid, lb-sec/in.^{2}
Pressure drop in internal passages and rotor windage also result in torque losses which depend on speed, but these are less significant.

2. In piston motors some sort of lever mechanism is required to translate piston motion into rotary shaft motion. An examination of the forces on each piston will show a friction force opposing motion of the piston in its bore that is proportional to the pressure acting on the piston area. Other motor elements, such as bearings, are also loaded proportional to the motor pressures and cause friction torques. Therefore, there results an opposing friction torque proportional to the motor displacement and to

the sum of the motor pressures, as illustrated in Fig. 4-14. For steady state performance the stiction portion of the curve is usually ignored and the torque loss is written as

$$T_f = \frac{\dot{\theta}_m}{|\dot{\theta}_m|} C_f D_m (P_1 + P_2) \qquad (4\text{-}15)$$

where C_f = internal friction coefficient, dimensionless
       C_{fs} = internal static friction coefficient, dimensionless

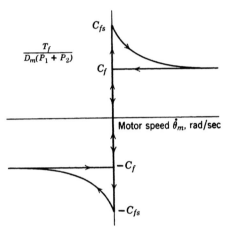

Figure 4-14   Starting and running friction torque due to pressures in motor chambers.

3. There is also a small torque required to overcome seal friction that is constant but reverses direction with speed. This torque will be denoted $(\dot{\theta}_m / |\dot{\theta}_m|)T_c$ but can be neglected altogether in most instances.

These torques can now be summed so that the net torque delivered to the load is

$$T_L = D_m(P_1 - P_2) - C_d D_m \mu \dot{\theta}_m - \frac{\dot{\theta}_m}{|\dot{\theta}_m|} C_f D_m (P_1 + P_2) - \frac{\dot{\theta}_m}{|\dot{\theta}_m|} T_c \quad (4\text{-}16)$$

where T_L = external load torque on motor shaft, in.-lbs. For a pump, the negative signs become positive and T_L would represent the torque applied to the shaft.

Equations 4-9, 4-10, and 4-16 define the static performance of the motor. However, there are six variables (Q_1, Q_2, P_1, P_2, T_L, and $\dot{\theta}_m$) but only three equations. It is apparent then that three quantities must be specified in order to find the other three by simultaneous solution. The forward flow Q_1 and the load torque T_L are arbitrary quantities which should properly be regarded as inputs and as such they must be specified. One

other quantity must also be specified or another equation must be found.*
Although a given load torque requires a certain pressure difference across
the motor, there are an infinite number of individual pressures which can
have that difference. Therefore the final bit of information required is
obtained by specifying P_2 or P_1 or by obtaining a relation between these
quantities. It is quite important to realize that the individual chamber
pressures are not established solely by conditions within the motor and
load but also depend on the circuit in which the motor is placed. Thus
one of the motor pressures can be correctly considered an input and must
be specified.

Two common techniques of controlling a motor involve the use of a
pump or a servovalve. With pump control, one line is at a very low pressure
(replenish pressure) while the other line pressure modulates to handle the
load. With this type of control the return pressure may be neglected;
that is, $P_2 \approx 0$. When servovalve control is used, both motor pressures
vary equal but opposite amounts above and below one half of supply
pressure to match the load. With this type of control the sum of the line
pressures is approximately equal to the supply pressure. Therefore

$$P_1 + P_2 = P_s \qquad (4\text{-}17)$$

Because $P_L = P_1 - P_2$ these relations can be solved simultaneously to
yield

$$P_1 = \frac{P_s + P_L}{2} \qquad (4\text{-}18)$$

$$P_2 = \frac{P_s - P_L}{2} \qquad (4\text{-}19)$$

Thus with servovalve control each pressure may be defined in terms of P_L,
and the pressure variables have been reduced from two to one. It should
be emphasized that these two control techniques are by no means the only
possibilities. In some cases it is desirable to place a rate valve in the return
line of the motor to control speed. This control technique would establish
a constant return pressure which may not be negligible.

However, it is possible to completely specify motor performance with
$P_2 = 0$ and this condition is generally assumed in the interests of uniform-
ity of data presentation and simplicity of analysis. But it should be
recognized that once motor parameters are established from tests, then

* This situation is avoided in some treatments by letting P_2 be zero or by erroneously
letting the internal motor friction be proportional to pressure difference rather than
sum so that the motor pressures are completely characterized by their difference.

performance can be computed for any combination of line pressures. With $P_2 = 0$, (4-9), (4-10), and (4-16) become

$$Q_1 - (C_{em} + C_{im})P_1 - D_m\dot\theta_m = 0 \qquad (4\text{-}20)$$

$$D_m\dot\theta_m + C_{im}P_1 - Q_2 = 0 \qquad (4\text{-}21)$$

$$T_L = D_mP_1 - C_dD_m\mu\dot\theta_m - C_fD_mP_1 - T_c \qquad (4\text{-}22)$$

The quantity $\dot\theta_m/|\dot\theta_m|$, which determines the sign of the friction terms, has a value of $+1$ because the rotation would be in one direction.

It is possible to obtain some well-defined expressions for slip and efficiency with the condition that $P_2 = 0$. Referring to (4-20), the *slip flow* may be defined as

$$Q_s = (C_{em} + C_{im})P_1 \qquad (4\text{-}23)$$

and represents the flow supplied to the motor that is not converted into shaft rotation. Some authors assume no external leakage and define slip flow as the internal leakage. However, the definition given here is logical and well defined since $P_2 = 0$. If $P_2 \neq 0$, then slip could be defined as $Q_{im} + Q_{em1}$ (see Fig. 4-13), but this is a function of both forward pressure and pressure difference and would be difficult to compute and identify physically.

The slip flow is usually laminar and, therefore, inversely proportional to viscosity. Test results also indicate that the slip flow is related to motor displacement. Therefore

$$Q_s = C_s\frac{D_m}{\mu}P \qquad (4\text{-}24)$$

where $C_s = \dfrac{\mu}{D_m}(C_{em} + C_{im})$ = coefficient of slip, dimensionless

μ = absolute viscosity of fluid, lb-sec/in.^{2}

The *volumetric efficiency* is defined as the ratio of flow which results in motor speed (the ideal flow) to the flow supplied to the motor. Therefore by definition

$$\eta_v \equiv \frac{D_m\dot\theta_m}{Q_1} \qquad (4\text{-}25)$$

Obtaining Q_1 from Eq. (4-20), the volumetric efficiency becomes

$$\eta_v = \left(1 + \frac{C_sP_1}{\mu\dot\theta_m}\right)^{-1} \qquad (4\text{-}26)$$

The *torque* or *mechanical efficiency* is defined as the ratio of actual to ideal torque delivered by the motor.

$$\eta_t \equiv \frac{T_L}{D_m P_1} \tag{4-27}$$

By obtaining T_L from (4-22) and neglecting T_c the torque efficiency becomes

$$\eta_t = 1 - \frac{C_d \mu \dot\theta_m}{P_1} - C_f \tag{4-28}$$

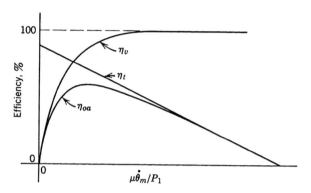

Figure 4-15   Typical efficiency curves for a motor.

The *over-all efficiency* is defined as the ratio of actual horsepower output to the hydraulic horsepower supplied.

$$\eta_{oa} \equiv \frac{\text{hp}\big|_{\text{out}}}{\text{hp}\big|_{\text{in}}} = \frac{T_L \dot\theta_m}{Q_1 P_1} = \frac{T_L}{D_m P_1} \times \frac{D_m \dot\theta_m}{Q_1} = \eta_v \eta_t \tag{4-29}$$

Thus the over-all efficiency is simply the product of the volumetric and torque efficiencies. Combining with (4-26) and (4-28), the over-all efficiency becomes

$$\eta_{oa} = \eta_v \eta_t = \frac{1 - (C_d \mu \dot\theta_m/P_1) - C_f}{1 + (C_s P_1/\mu \dot\theta_m)} \tag{4-30}$$

Therefore, static performance of a motor with zero return pressure can be defined by the parameters C_s, C_d, C_f, and the dimensionless quantity $\mu \dot\theta_m/P_1$. Typical efficiency curves for a motor are illustrated in Fig. 4-15.

Although the discussion has been concerned with motors, it should be apparent that an analogous discussion can be made for pumps. There will be only one continuity equation (for the high pressure side), and the power source driving the pump must overcome the internal torque and leakage losses. The pressure at the inlet side can be neglected so that only one pressure variable would be involved. The resulting over-all efficiency would be

$$\eta_{oa} = \frac{1 - (C_s P_1/\mu N_p)}{1 + (C_d \mu N_p/P_1) + C_f} \tag{4-31}$$

where N_p = pump speed, rad/sec
P_1 = output pressure, psi

4-4 PERFORMANCE CURVES AND PARAMETERS

So far we have analyzed the steady-state behavior of motors and pumps and defined certain useful performance parameters such as C_{im}, C_{em}, C_{fs}, C_f, and C_d. The problem now at hand is to define the series of tests necessary to establish these parameters. Because efficiencies are often high, it is apparent that accurate instrumentation would be required if the motor is externally loaded because a few percent error in measurement of input or output horsepower may be larger than the losses involved. Hence, it is far better to measure the losses and then compute motor performance with external loads.

The motor leakage characteristics are easily determined by locking the motor shaft and letting the return line be vented to atmosphere (i.e., $P_2 = 0$). Pressure P_1 is then applied to the forward chamber and the flows in the return and drain lines are measured, taking care to avoid silting. Under these conditions the return line flow is the internal leakage flow Q_{im} and the drain line flow is the external leakage flow Q_{em1} of the forward chamber. The external leakage of the return chamber could also be obtained but is usually not necessary because the two chambers are nearly identical. Typical leakage curves are illustrated in Fig. 4-16. The leakage flows at any two motor pressures can be determined from these curves by interpreting P_1 as pressure difference to obtain cross-port leakage and by interpreting P_1 as the particular chamber pressure to obtain each external leakage flow. The slopes of the curves in Fig. 4-16 give the leakage coefficients C_{im} and C_{em}.

The coefficients C_{fs} and C_f which define the friction torque caused by pressure (Fig. 4-14) may be determined by measuring the pressure difference required to initiate rotation ΔP_s of the unloaded motor at various return pressure levels. Once rotation is achieved, the pressure difference is gradually decreased until rotation ceases to determine the running friction level. Thus, the curves illustrated in Fig. 4-17 will result. Because $\dot{\theta}_m = 0$, $T_L = 0$, and $\dot{\theta}_m/|\dot{\theta}_m| = 1$ for the test conditions, (4-16) becomes

$$D_m \Delta P_c = C_f D_m (P_1 + P_2) + T_c \tag{4-32}$$

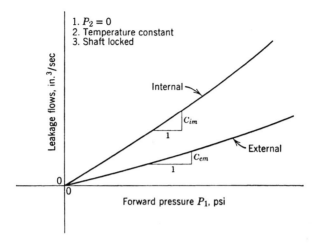

Figure 4-16 Leakage flows with shaft locked.

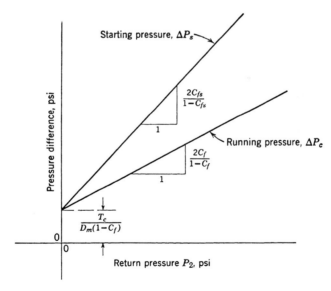

Figure 4-17 Pressure differences required to overcome starting and running friction at various return pressure levels.

73

where ΔP_c = pressure difference required to overcome running friction (i.e., internal coulomb friction), psi

ΔP_s = pressure difference required to overcome starting friction (i.e., internal static friction), psi

Because $P_1 = \Delta P_c + P_2$ by definition, (4-32) becomes

$$\Delta P_c = \frac{2C_f}{1 - C_f} P_2 + \frac{T_c}{D_m(1 - C_f)} \qquad (4\text{-}33)$$

which is the equation for the running friction curve in Fig. 4-17. The slope value gives C_f and the intercept value gives T_c.* Since C_f is usually small, the slope is approximately $2C_f$. Using similar reasoning, the starting pressure difference is (neglecting T_c)

$$\Delta P_s = \frac{2C_{fs}}{1 - C_{fs}} P_2 \qquad (4\text{-}34)$$

and C_{fs} is determined from the slope of the starting friction curve.

The torque losses which depend on speed are found by measuring the forward pressure required to run the unloaded motor at various speeds with no (i.e., negligible) return pressure. A typical curve is illustrated in Fig. 4-18. Because (4-22) applies and $T_L = 0$, then the pressure required to run is

$$P_1 = \frac{C_d \mu}{1 - C_f} \dot{\theta}_m + \frac{T_c}{D_m(1 - C_f)} \qquad (4\text{-}35)$$

Thus, the parameter C_d is obtained from the slope of the curve in Fig. 4-18 because C_f and μ are known. This curve may sweep upward for some motors indicating that the pressure depends on higher powers of $\dot{\theta}_m$. This type of characteristic is due to the flow resistance of internal passages.

The curves in Figs. 4-16, 4-17, and 4-18 completely define the performance parameters of a motor. These parameters can be used in (4-9), (4-10), and (4-16) to compute the steady-state running characteristics, such as all

---

* An alternate method of measuring internal friction is to connect a line between both motor ports and apply a test pressure P_t. In this manner the same pressure is applied to both motor ports (i.e., $P_1 = P_2 = P_t$) and no pressure drop exists across the motor. A load torque is then applied with a spring scale to rotate the shaft slowly. Because $P_1 - P_2 = 0$ and $\theta \approx 0$, Eq. (4-16) reduces to

$$-T_L = 2C_f D_m P_t + T_c$$

A series of readings of $-T_L$ and P_t will give a plot from which C_f and T_c can be obtained from the slope and intercept. If the static breakaway torque is measured, then C_{fs} can be found.

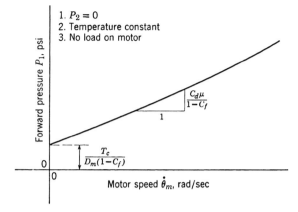

1. $P_2 = 0$
2. Temperature constant
3. No load on motor

$\dfrac{C_d \mu}{1 - C_f}$

1

$\dfrac{T_c}{D_m(1 - C_f)}$

Forward pressure P_1, psi

Motor speed $\dot{\theta}_m$, rad/sec

Figure 4-18   Pressure difference required to run motor with no load or return pressure.

efficiencies, flow, and pressure required to yield a desired speed and torque output, and internal and external leakage flows for any given application. The breakaway or starting pressure is often of interest in the static accuracy of servo systems and can be obtained from Fig. 4-17.

REFERENCES

[1] Blackburn, J. F., G. Reethof, and J. L. Shearer, *Fluid Power Control*. New York: Technology Press of M.I.T. and Wiley, 1960, Chapter 4.

[2] Ernst, W., *Oil Hydraulic Power And Its Industrial Applications*, 2nd ed. New York: McGraw-Hill, 1960.

[3] Norrie, D. H., *An Introduction to Incompressible Flow Machines*. New York: American Elsevier, 1963.

[4] Wilson, W. E., *Positive-Displacement Pumps and Fluid Motors*. New York: Pitman Publishing, 1950.

[5] Thoma, Jean U., *Hydrostatic Power Transmission*. Trade and Technical Press, Surrey, England, 1964.

[6] Hadekel, R., *Displacement Pumps and Motors*. London: Sir Isaac Pitman, 1951.

5

Hydraulic Control Valves

Hydraulic control valves are devices that use mechanical motion to control a source of fluid power. They vary in arrangement and complexity, depending upon their function. Because control valves are the mechanical (or electrical) to fluid interface in hydraulic systems, their performance is under scrutiny, especially when system difficulties occur. Therefore knowledge of the performance characteristics of valves is essential. The purpose of this chapter is to discuss the characteristics and design criteria for the principal types of hydraulic control valves. Although emphasis is placed on valves for servo control, the principles involved apply equally well to valves used in other applications, such as solenoid valves, pressure reducing valves, and flow control valves.

5-1 VALVE CONFIGURATIONS

Valves may be divided into three classifications: sliding, seating, and flow dividing. Examples of these are spool, flapper, and jet pipe valves, respectively, which are shown schematically in Fig. 5-1.

The most widely used valve is the sliding valve employing spool type construction. Typical spool valve configurations are shown in Fig. 5-1a, b, and c. Spool valves are classified by (a) the number of "ways" flow can enter and leave the valve, (b) the number of lands, and (c) the type of center when the valve spool is in neutral position. Because all valves require a supply, a return, and at least one line to the load, valves are either three-way or four-way.* A three-way valve (Fig. 5-1d) requires a bias pressure acting on one side of an unequal area piston for direction reversal. Usually the head-side area is twice the rod-side area, and supply pressure acts on the smaller area to provide the bias force for reversal. A four-way valve would have two lines to the load. The number of lands on a spool vary from one in a primitive valve to the usual three or four,

---

* There are, of course, two-way valves. However, two-way valves cannot provide a reversal in the direction of flow.

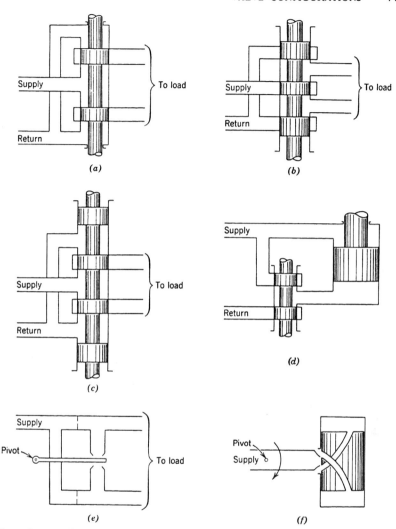

Figure 5-1 Typical hydraulic servovalves: (a) two-land–four-way spool valve; (b) three-land–four-way spool valve; (c) four-land–four-way spool valve; (d) two-land–three-way spool valve with piston load; (e) two-jet flapper valve; (f) jet pipe valve.

and special valves may have as many as six lands. If the width of the land is smaller than the port in the valve sleeve, the valve is said to have an *open center* or to be *underlapped*. A *critical center* or *zero lapped* valve has a land width identical to the port width and is a condition approached by practical machining. *Closed center* or *overlapped* valves have a land width greater than the port width when the spool is at neutral.

Certain characteristics of a valve may be directly related to the type of valve center. The most important of these characteristics is the flow gain which has the shape shown in Fig. 5-2 for the three types of center. In fact, it is better to define the type of valve center from the shape of the flow gain near neutral rather than from geometrical considerations. A critical center valve may be defined as the geometrical fit required to achieve a linear flow gain in the vicinity of neutral and usually necessitates a slight overlap to offset the effect of radial clearance.

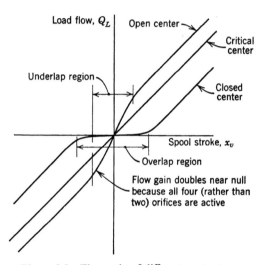

Figure 5-2   Flow gain of different center types.

The vast majority of four-way valves are manufactured with a critical center because of the emphasis on the linear flow gain. Closed center valves are not desirable because of the deadband characteristic in the flow gain. Deadband results in steady state error and, in some cases, can cause backlash which may lead to stability problems. Open center valves are used in applications which require a continuous flow to maintain reasonable fluid temperatures and also in constant flow systems. However, the large power loss at neutral position, the decrease in flow gain outside the underlap region, and the decreased pressure sensitivity of open center valves restrict their use to special applications. The gain of systems using open center valves must be adjusted with the valve at neutral because of the increased flow gain near null. Therefore the system error and bandwidth are adversely affected when the valve is away from neutral because of the decreased flow gain. This aspect of open center valves is most undesirable.

Spool valve manufacture requires that close and matching tolerances be held so that such valves are relatively expensive and sensitive to fluid contamination. Tolerances required for flapper valves are not as stringent, which makes them attractive with regard to these two aspects. However, the relatively large leakage flows limit their application to low power levels. Flapper valves are used extensively as the first stage valve in two-stage electrohydraulic and hydromechanical servovalves. Poppet-type valves are basically two way valves and therefore are restricted to applications such as check and relief valves, in which reversal in flow direction is not required.

The jet pipe valve is not as widely used as the flapper valve because of large null flow, characteristics not easily predicted, and slower response. The main advantage of jet pipe valves is their insensitivity to dirty fluids. However, the more predictable flapper valve has similar performance characteristics and is usually preferred.

5-2 GENERAL VALVE ANALYSIS

In this section we define some general performance characteristics, such as pressure-flow curves and valve coefficients, which are applicable to all types of valves. Although the analysis is illustrated with a spool type of valve, the principles involved are quite general. The general relations derived in this section are applied to particular valve configurations in future sections.

General Flow Equations

Consider the four-way spool valve shown in Fig. 5-3. The four orifices are completely analogous to the four arms of a wheatstone bridge, and this analogy is often helpful in visualizing valve operation. Arrows at the ports indicate the assumed directions of flows, and the numbers at ports refer to the subscripts of the flow and the area at the ports. Let the spool be given a positive displacement from the null or neutral position, that is, the position $x_v = 0$, which is chosen to be the symmetrical position of the spool in its sleeve.

Because we are interested only in the steady-state characteristics, the compressibility flows are zero and the continuity equations for the two valve chambers are

$$Q_L = Q_1 - Q_4 \tag{5-1}$$

$$Q_L = Q_3 - Q_2 \tag{5-2}$$

A dynamic analysis would require inclusion of the compressibility flows which depend on the valve chamber volumes. However, this is best

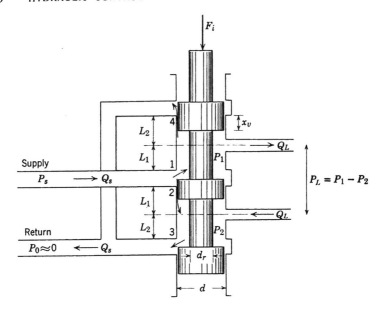

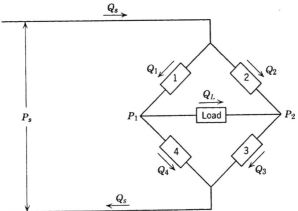

Figure 5-3 Three-land–four-way spool valve.

achieved by considering the combination of valve and motor, because the motor and lines contribute appreciable volumes; this is the subject of Chapter 6. Now, by definition,

$$P_L = P_1 - P_2 \qquad (5\text{-}3)$$

Q_L is the flow through the load and P_L is the pressure drop across the load. Flows through the valving orifices are described by the orifice equation (3-33). Therefore

$$Q_1 = C_d A_1 \sqrt{\frac{2}{\rho}(P_s - P_1)} \tag{5-4}$$

$$Q_2 = C_d A_2 \sqrt{\frac{2}{\rho}(P_s - P_2)} \tag{5-5}$$

$$Q_3 = C_d A_3 \sqrt{\frac{2}{\rho} P_2} \tag{5-6}$$

$$Q_4 = C_d A_4 \sqrt{\frac{2}{\rho} P_1} \tag{5-7}$$

The return line pressure P_0 is neglected because it is usually much smaller than the other pressures involved. If the return pressure is appreciable, then P_s can be interpreted as the pressure difference, that is, supply pressure minus return pressure, across the valve. The orifice areas depend on valve geometry and four equations are required to define the areas A_1, A_2, A_3, and A_4 as a function of valve displacement. Therefore,

$$A_1 = A_1(x_v) \qquad A_2 = A_2(-x_v) \qquad A_3 = A_3(x_v) \qquad A_4 = A_4(-x_v) \tag{5-8}$$

Thus a total of 11 equations is required to define the pressure-flow behavior of a general four-way valve. These 11 equations can be solved simultaneously to yield load flow as a function of valve position and load pressure;* that is,

$$Q_L = Q_L(x_v, P_L) \tag{5-9}$$

The plot of (5-9) is known as the *pressure-flow* curves for the valve and is a complete description of steady-state valve performance. All of the performance parameters, such as valve coefficients, can be obtained from these curves.

Although (5-9) can be found theoretically, simultaneous solution for the general case is tedious because the algebraic equations involved are nonlinear. However, this is no serious obstacle because valves are never so complex as to have all four orifice areas differently described. In the vast majority of cases the valving orifices are *matched* and *symmetrical*. Matched orifices require that

$$A_1 = A_3 \tag{5-10}$$

$$A_2 = A_4 \tag{5-11}$$

---

* For example, selecting a value for x_v will numerically determine all the orifice areas. Then choose a series of values for Q_L and solve (5-1) and (5-2) for P_1 and P_2, respectively, for each Q_L value. Thus it is possible to tabulate x_v, Q_L, P_1, P_2, and also P_L.

and symmetrical orifices require that

$$A_1(x_v) = A_2(-x_v) \tag{5-12}$$

$$A_3(x_v) = A_4(-x_v) \tag{5-13}$$

Therefore, at the neutral position of the spool, all four orifice areas are equal.

$$A_1(0) = A_2(0) \equiv A_0 \tag{5-14}$$

With these restrictions on the orifice areas only one orifice area need be defined because the other areas follow from it. In fact, if the orifice areas are linear with valve stroke, as is usually the case, only one defining parameter is required: the width of the slot in the valve sleeve w. The rate of change of orifice area with stroke is called the *opening* or *area gradient* of the valve. If the valve is linear, then the area gradient of each orifice (and so the whole valve) is w in.²/in. and is the single most important parameter for such a valve.

It should be emphasized that great care is taken in manufacture to ensure that orifices are matched and symmetrical, otherwise the valve will exhibit peculiar valve coefficients near neutral. However, it is possible that certain applications may require special valve fits, but these instances are exceptional. If the orifices are both matched and symmetrical, the flows in diagonally opposite arms of the bridge in Fig. 5-3 are equal; that is,

$$Q_1 = Q_3 \tag{5-15}$$

$$Q_2 = Q_4 \tag{5-16}$$

Substituting (5-4), (5-6), and (5-10) into (5-15) yields

$$P_s = P_1 + P_2 \tag{5-17}$$

Equation 5-16 may be similarly treated to give the same result. Equations 5-3 and 5-17 may be solved simultaneously to obtain

$$P_1 = \frac{P_s + P_L}{2} \tag{5-18}$$

$$P_2 = \frac{P_s - P_L}{2} \tag{5-19}$$

Thus for a matched and symmetrical valve with no load (i.e., $P_L = 0$), the pressure in each motor line is $\frac{1}{2}P_s$. As load is applied, the pressure in one line increases as the pressure in the other line decreases by the same amount. Thus the pressure drop across orifices 1 and 3 in Fig. 5-3 are identical and, because the areas are the same, (5-15) is verified. A similar argument can be used for the validity of (5-16).

One other general relation will prove useful in the study of valves. Referring to Fig. 5-3, note that the total supply flow can be written

$$Q_s = Q_1 + Q_2 \qquad (5\text{-}20)$$

$$Q_s = Q_3 + Q_4 \qquad (5\text{-}21)$$

In summary, for a matched and symmetrical valve (5-15), (5-16), (5-18), and (5-19) are applicable, and (5-1) and (5-2) both become

$$Q_L = C_d A_1 \sqrt{\frac{1}{\rho}(P_s - P_L)} - C_d A_2 \sqrt{\frac{1}{\rho}(P_s + P_L)} \qquad (5\text{-}22)$$

A similar treatment yields

$$Q_s = C_d A_1 \sqrt{\frac{1}{\rho}(P_s - P_L)} + C_d A_2 \sqrt{\frac{1}{\rho}(P_s + P_L)} \qquad (5\text{-}23)$$

for (5-20) and (5-21).

Linearized Analysis of Valves—Valve Coefficients

In making a dynamic analysis it is necessary that the nonlinear algebraic equations which describe the pressure-flow curves be linearized. Equation 5-9 is a general expression for the load flow, and we can express this function as a Taylor's series about a particular operating point $Q_L = Q_{L1}$. Therefore

$$Q_L = Q_{L1} + \frac{\partial Q_L}{\partial x_v}\bigg|_1 \Delta x_v + \frac{\partial Q_L}{\partial P_L}\bigg|_1 \Delta P_L + \cdots$$

If we confine ourselves to the vicinity of the operating point, the higher order infinitesimals are negligibly small, and we may write

$$Q_L - Q_{L1} \equiv \Delta Q_L = \frac{\partial Q_L}{\partial x_v}\bigg|_1 \Delta x_v + \frac{\partial Q_L}{\partial P_L}\bigg|_1 \Delta P_L \qquad (5\text{-}24)$$

The partial derivatives required are obtained by differentiation of the equation for the pressure-flow curves or graphically from a plot of the curves. These partials define the two most important parameters for a valve. The *flow gain* is defined by

$$K_q \equiv \frac{\partial Q_L}{\partial x_v} \qquad (5\text{-}25)$$

The *flow-pressure coefficient* is defined by

$$K_c \equiv -\frac{\partial Q_L}{\partial P_L} \qquad (5\text{-}26)$$

It can be shown (see page 90) that $\partial Q_L/\partial P_L$ is negative for any valve configuration which makes the flow-pressure coefficient always a positive number. Another useful quantity is the *pressure sensitivity* defined by

$$K_p = \frac{\partial P_L}{\partial x_v} \qquad (5\text{-}27)$$

and which is related to the other quantities by the well-known relation from Calculus

$$\frac{\partial P_L}{\partial x_v} = -\frac{\partial Q_L/\partial x_v}{\partial Q_L/\partial P_L} \quad \text{or} \quad K_p = \frac{K_q}{K_c} \qquad (5\text{-}28)$$

With these definitions, the linearized equation of the pressure-flow curves becomes

$$\Delta Q_L = K_q\,\Delta x_v - K_c\,\Delta P_L \qquad (5\text{-}29)$$

and is applicable to all valves whether spool, flapper, or otherwise. The coefficients K_q, K_c, and K_p are called *valve coefficients* and are extremely important in determining stability, frequency response, and other dynamic characteristics. The flow gain directly affects the open loop gain constant in a system and therefore has a direct influence on system stability. The flow-pressure coefficient directly affects the damping ratio of valve-motor combinations. The pressure sensitivity of valves is quite large, which accounts for the ability of valve-motor combinations to breakaway large friction loads with little error.

The values of the valve coefficients vary with the operating point. The most important operating point is the origin of the pressure-flow curves (i.e., where $Q_L = P_L = x_v = 0$) because system operation usually occurs near this region, the valve flow gain is largest, giving high system gain, and the flow-pressure coefficient is smallest, giving a low damping ratio. Hence this operating point is the most critical from a stability viewpoint, and a system stable at this point is usually stable at all operating points. The valve coefficients evaluated at the operating point are called the *null valve coefficients.*

5-3   CRITICAL CENTER SPOOL VALVE ANALYSIS

In this section we derive and/or define the pressure-flow curves, valve coefficients, leakage flow curves, and stroking forces for critical center spool valves.

It is useful to classify the critical center valve as having either ideal or practical geometry. Ideal geometry implies that the orifice edges are perfectly square with no rounding and that there is no radial clearance between the spool and sleeve. Although these geometrical perfections are not possible in practice, it is possible to construct a valve with a relatively linear flow gain near null position, as described in Section 5-1. Such a critical center valve with practical geometry, that is, with radial clearance, is in many respects an optimum valve because leakage flows are minimum, flow gain is linear, and design procedures are usually based on such a valve. Because the vast majority of four-way spool valves are of this type, a detailed discussion of its characteristics is warranted.

Pressure-Flow Curves

Referring to Fig. 5-3, let us use the general equations developed in Section 5-2 to derive the pressure-flow curves for an ideal critical center valve with matched and symmetrical orifices. The leakage flows (Q_2 and Q_4 when x_v is positive; Q_1 and Q_3 when x_v is negative) for such a valve are zero because the geometry is assumed ideal. Therefore by substituting (5-18) and (5-4) into (5-1) we obtain

$$Q_L = C_d A_1 \sqrt{\frac{2}{\rho}\left(\frac{P_s - P_L}{2}\right)} \quad \text{for} \quad x_v > 0 \qquad (5\text{-}30)$$

Equation 5-2 may be similarly treated to give this result. For negative valve displacements $Q_L = -Q_4$, and (5-18) and (5-7) may be used to give

$$Q_L = -C_d A_2 \sqrt{\frac{2}{\rho}\left(\frac{P_s + P_L}{2}\right)} \quad \text{for} \quad x_v < 0 \qquad (5\text{-}31)$$

As before, (5-2) will give the same result. Because the valve is assumed symmetrical, (5-12) is applicable and (5-30) and (5-31) can be combined into a single relation:

$$Q_L = C_d |A_1| \frac{x_v}{|x_v|} \sqrt{\frac{1}{\rho}\left(P_s - \frac{x_v}{|x_v|} P_L\right)} \qquad (5\text{-}32)$$

This is the general equation for the pressure-flow curves of an ideal critical center valve with matched and symmetrical orifices. If rectangular ports are used with an area gradient of w for each port, (5-32) becomes

$$Q_L = C_d w x_v \sqrt{\frac{1}{\rho}\left(P_s - \frac{x_v}{|x_v|} P_L\right)} \qquad (5\text{-}33)$$

and is plotted in a normalized manner in Fig. 5-4. The quantity x_{vm} is the maximum valve stroke. It is possible to be in quadrants II and IV of

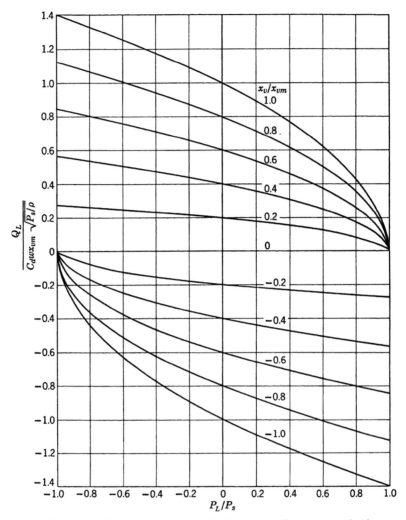

Figure 5-4   Pressure-flow curves of critical center four-way spool valve.

the pressure-flow curves only during transient conditions. For instance, a sudden change in x_v could reverse the pressure in the lines to the load but, because of the inertia of the fluid and load, the load could remain in the same direction; that is, the flow instantaneously remains in the same direction.

For petroleum base fluids, $\rho \approx 0.78 \times 10^{-4}$ lb-sec^2/in.^{4}, and, because $C_d \approx 0.61$ (see Section 3-4), $C_d\sqrt{1/\rho} \approx 70$ in.$^2/\sqrt{\text{lb-sec}}$ and a common form of (5-33) is

$$Q_L = 70wx_v\sqrt{P_s - \frac{x_v}{|x_v|}P_L} \qquad (5\text{-}34)$$

where English units are used. This equation is the basis for most design procedures involving hydraulic servovalves.

Valve Coefficients

The valve coefficients for the important case of an ideal critical center valve can be obtained by differentiation of (5-33). The flow gain is

$$K_q = C_d w\sqrt{\frac{1}{\rho}(P_s - P_L)} \qquad (5\text{-}35)$$

The flow-pressure coefficient is

$$K_c = \frac{C_d w x_v\sqrt{(1/\rho)(P_s - P_L)}}{2(P_s - P_L)} \qquad (5\text{-}36)$$

and by (5-33) the pressure sensitivity is

$$K_p = \frac{2(P_s - P_L)}{x_v} \qquad (5\text{-}37)$$

As discussed in Section 5-2, the null operating point is most important. Evaluation of these coefficients at the point $Q_L = P_L = x_v = 0$ gives the null coefficients for the ideal critical center valve. Therefore

$$K_{q0} = C_d w\sqrt{\frac{P_s}{\rho}} \qquad (5\text{-}38)$$

$$K_{c0} = 0 \qquad (5\text{-}39)$$

$$K_{p0} = \infty \qquad (5\text{-}40)$$

In English units with usual values for ρ and C_d the null flow gain is commonly written as

$$K_{q0} = 70w\sqrt{P_s} \quad \text{in.}^3/\text{sec/in.} \qquad (5\text{-}41)$$

and is a simple function of two well-known and easily measured quantities: the valve area gradient and the system supply pressure. The computed null flow gain has been amply verified by tests of practical critical center valves and may be used with confidence. We are fortunate indeed that system stability depends on this quantity which is easily computed and controlled. This is one of the major reasons why hydraulic servos enjoy a reputation for dependable stability. However, the computed values for K_{c0} and K_{p0} are far from that obtained in tests of practical center valves.

It is possible to compute more realistic values for these two null coefficients once the leakage characteristics for such valves have been investigated. We shall return to this problem after such an investigation.

Leakage Characteristics of Practical Critical Center Spool Valves

It is the leakage characteristics which actually differentiate a practical from an ideal critical center valve. The ideal valve has perfect geometry so that leakage flows are zero. The practical valve has radial clearance and

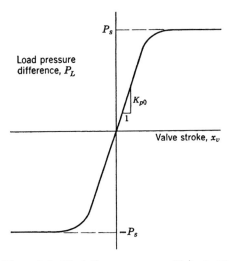

Figure 5-5   Block-line pressure sensitivity curve.

perhaps minute under or overlap of well under 0.001 in. The leakage performance of such valves dominates their behavior and the associated pressure-flow curves for these small valve openings (say $|x_v| < 0.001$ in.). Outside this region the theoretical equation (5-33) fits very well.

Consider the bridge circuit of a valve, assumed to have matched and symmetrical orifices, shown in Fig. 5-3. Let us further assume that the motor lines are blocked, perhaps with pressure gauges for measurement purposes, so that the load flow Q_L is zero. We can now define and measure three characteristic curves for this valve. By stroking the valve and recording load pressure P_L and total supply flow Q_s (which is actually a leakage flow since load flow is zero) for a given supply pressure, we can plot the *blocked-line pressure sensitivity* and the *leakage flow* curves shown in Figs. 5-5 and 5-6, respectively. The load pressure difference P_L quickly increases to full supply pressure after a very small spool displacement. An experimental value for the null pressure sensitivity can be obtained from Fig.

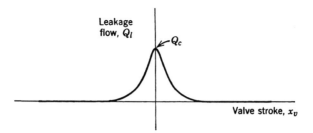

Figure 5-6 Typical leakage curve.

5-5 and is the reason for the importance of this curve. The leakage flow is maximum at valve neutral and decreases rapidly with valve stroke because the spool lands overlap the return valve orifices. This curve is a measure of hydraulic power loss and has little significance otherwise.

A third characteristic is obtained by measuring the total flow through the valve with the spool centered (i.e., all four orifice areas equal) as supply pressure is varied. This flow is called the *center flow* Q_c and the resulting curve is called the *center flow curve*. To avoid confusion of symbols, the center flow Q_c is simply the particular supply flow Q_s when the spool is centered. The center flow curve for a valve is illustrated in Fig. 5-7 and reveals the flow is laminar for a new valve and becomes orifice flow after some length of service because abrasive materials in the fluid erode the orifice edges increasing their areas. The center flow for a particular supply pressure can be selected from this curve and is identical to the maximum leakage flow in Fig. 5-6 for the same supply pressure. The center flow curve is useful because the shape of the curve (whether linear or square root) indicates the quality of valve fit, a single value for center

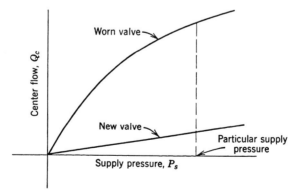

Figure 5-7 Typical center flow curves.

flow at a selected pressure can be used as a manufacturing tolerance, and the null flow-pressure coefficient for the valve can be obtained from this curve. This latter aspect will now be investigated.

Assuming that the valve orifices are matched and symmetrical, (5-22) and (5-23) give the load and supply flows, respectively, for the valve. Differentiation of (5-22) yields*

$$\frac{\partial Q_L}{\partial P_L} = - \frac{C_d A_1}{2\sqrt{(1/\rho)(P_s - P_L)}} - \frac{C_d A_2}{2\sqrt{(1/\rho)(P_s + P_L)}} \tag{5-42}$$

Now, differentiation of (5-23) with respect to P_s gives the negative of (5-42). Therefore we conclude that

$$\frac{\partial Q_s}{\partial P_s} = - \frac{\partial Q_L}{\partial P_L} \equiv K_c \tag{5-43}$$

This result is valid for any matched and symmetrical valve and whether the flow is laminar or orifice type.

Now, we are interested in obtaining the null flow-pressure coefficient. Because the center flow curve is made with $Q_L = P_L = x_v = 0$, (5-43) justifies taking the slope of this curve at the particular operating pressure as the null flow-pressure coefficient. This is a simple step to perform and avoids a difficult direct measurement of $-\partial Q_L/\partial P_L$. Comparing this coefficient for a new and a worn valve (Fig. 5-7), we note that although the center flow at a particular pressure may increase drastically, the slope increase is considerably less. Hence the null flow-pressure coefficient and, consequently, the null pressure sensitivity do not vary greatly (perhaps by a factor of 2 to 3) as the valve wears. This is fortunate; otherwise relatively sluggish system performance would result in comparatively short order with contaminated fluids.

Because the center flow is laminar for a new valve, we can use (3-41) to describe the flow through the·sharp-edged orifices due to radial clearance when the valve spool is centered. The pressure drop and flow

---

* Because the areas are always positive, we note that $\partial Q_L/\partial P_L$ is always a negative quantity. This may also be concluded from an examination of the flows in Fig. 5-3. If P_1 is increased, letting orifice areas be fixed, the flow through orifice 1 is decreased and the flow through orifice 4, increased. Under these circumstances the continuity equation at this node is satisfied only if the load flow is in a direction opposite to that indicated. Hence the load flow direction is toward the point of *higher* pressure and $\partial Q_L/\partial P_L$ is not a leakage coefficient, as it is sometimes considered, because leakage flows seek a point of lower pressure. The name flow-pressure (rather than leakage) coefficient is used to emphasize this special distinction. It should be noted that as P_1 is increased, P_2 is decreased, and compatible behavior takes place at the other node.

associated with one orifice are $P_s/2$ and $Q_c/2$, respectively. Therefore (3-41) becomes

$$Q_c = \frac{\pi w r_c^2}{32\mu} P_s \qquad (5\text{-}44)$$

where w = area gradient of valve, in.²/in.
   r_c = radial clearance between spool and sleeve, in.
   μ = absolute viscosity, lb-sec/in.²
   P_s = supply pressure to valve, psi
   Q_c = center flow (i.e., flow through valve with blocked load ports and centered spool), in.³/sec
   K_{c0} = null flow-pressure coefficient, in.³/sec/psi
Using (5-43) and (5-26), we obtain

$$K_{c0} = \frac{\pi w r_c^2}{32\mu} \qquad (5\text{-}45)$$

This expression gives an approximate value for the null flow-pressure coefficient of a practical critical center valve and is considerably more correct than the theoretical value of $K_{c0} = 0$. A more or less standard radial clearance is $r_c = 0.0002$ in. and this value may be used in preliminary calculations of K_{c0}. It is important to note that K_{c0} varies directly with valve area gradient and, therefore, with valve size.

An approximate expression for the null pressure sensitivity of a practical critical center valve can be obtained by dividing (5-38) by (5-45). Therefore

$$K_{p0} = \frac{32\mu C_d \sqrt{P_s/\rho}}{\pi r_c^2} \qquad (5\text{-}46)$$

Using typical values of $\mu = 2 \times 10^{-6}$ lb-sec/in.², $C_d\sqrt{1/\rho} = 70$ in.²/$\sqrt{\text{lb}}$-sec and $r_c = 2 \times 10^{-4}$ in., we can get a feel for this quantity. Therefore

$$K_{p0} = 35,700\sqrt{P_s} \text{ psi/in.} \qquad (5\text{-}47)$$

Thus with $P_s = 1000$ psi, one might expect a null pressure sensitivity of 1.13×10^6 psi/in. Indeed, practice confirms that values of 10^6 psi/in. are easily achieved with supply pressures of 1000 psi, and this value is commonly used as a rule-of-thumb for all critical center spool valves. This is somewhat justified, for the null pressure sensitivity indicated by (5-46) is independent of valve area gradient and therefore independent of valve size. It is perhaps worth emphasizing that (5-45) and (5-46) are not rigorous relations and must be considered accordingly. However, the resulting values have compared well with test data in the author's experience.

Stroking Forces

So far we have been concerned with the pressure-flow behavior of critical center valves. Let us now examine the forces on the spool of such a valve. The most important of these forces are the flow forces discussed in Section 5-6.

With reference to Fig. 5-3, only two orifices (1 and 3 for $+x_v$; 2 and 4 for $-x_v$) are active at one time in an ideal critical center valve. If we assume a positive spool displacement, (5-90) and (5-93) can be applied to orifices 1 and 3 (both have the same area) to give the total flow force, steady-state and transient, opposing the spool motion. Therefore

$$F_R = 2C_dC_v(\cos\theta)wx_v(P_s - P_1) - L_1\rho\frac{dQ_1}{dt}$$

$$+ 2C_dC_v(\cos\theta)wx_vP_2 + L_2\rho\frac{dQ_3}{dt} \quad (5\text{-}48)$$

Assuming that the valve is matched and symmetrical, (5-18) and (5-19) can be used and the derivative of (5-4) and (5-6) can be substituted into (5-48) to yield

$$F_R = 2C_dC_vw(\cos\theta)(P_s - P_L)x_v + (L_2 - L_1)C_dw\sqrt{\rho(P_s - P_L)}\frac{dx_v}{dt}$$

$$(5\text{-}49)$$

The pressure derivative dP_L/dt terms are neglected because there is little evidence to indicate they play any substantial role in valve performance. The equation of motion of the valve spool in Fig. 5-3 can now be written

$$F_i = M_s\frac{d^2x_v}{dt^2} + B_f\frac{dx_v}{dt} + K_fx_v \quad (5\text{-}50)$$

where M_s = spool mass, lb-sec²/in.

$B_f = (L_2 - L_1)C_dW\sqrt{\rho(P_s - P_L)}$ = damping coefficient due to transient flow force, lb-sec/in.

$K_f = 2C_dC_vw(\cos\theta)(P_s - P_L) \approx 0.43w(P_s - P_L)$ = flow force spring rate, lb/in.

Thus we see that the steady-state flow force acts as a centering spring on the valve, and the transient flow force behaves like a viscous damping. Both quantities are somewhat nonlinear because of changes in P_L. If $L_2 < L_1$, the transient flow force is negative and may cause valve instability. Therefore such a valve should be designed so that $L_2 \geq L_1$ to prevent this possibility. This is easily done because it is simply a matter of location of the two load lines.

We can obtain an idea of the significance of the transient flow force coefficient B_f by computing the damping ratio in (5-50). The damping ratio for the quadratic involved is

$$\delta_s = \frac{B_f}{2\sqrt{K_f M_s}} = \frac{(L_2 - L_1)C_d\sqrt{\rho}}{2\sqrt{0.43}}\sqrt{\frac{w}{M_s}} \tag{5-51}$$

and is independent of the pressure $P_s - P_L$. Assuming some geometrical proportions for the spool in Fig. 5-3, we can determine values for δ_s. The spool length is usually six times the diameter, that is,

$$L \approx 6d \tag{5-52}$$

The total damping length is about one third of the spool length

$$L_1 + L_2 \approx \frac{L}{3} = 2d \tag{5-53}$$

and the spool rod diameter is usually

$$d_r \approx \frac{d}{2} \tag{5-54}$$

By subtracting the two annular valve chamber volume from the spool cylinder volume, the spool volume is found to be

$$\frac{\pi}{4}d^2L - 2\left(\frac{\pi}{4}d^2 - \frac{\pi}{4}d_r^2\right)(L_1 + L_2) = \left(\frac{1}{2}\right)\frac{\pi}{4}d^2L = \frac{3\pi}{4}d^3 \tag{5-55}$$

If the spool material is steel (weight density of 0.3 lb/in.³), then the spool mass is

$$M_s = \frac{(0.3)(3\pi/4)d^3}{386} = 1.83 \times 10^{-3} d^3 \text{ lb-sec}^2/\text{in.} \tag{5-56}$$

Assuming that $C_d = 0.61$ and $\rho = 0.78 \times 10^{-4}$ lb-sec²/in.⁴, (5-51) becomes

$$\delta_s = 0.34\left(\frac{L_2 - L_1}{L_2 + L_1}\right)\sqrt{\frac{w}{\pi d}} \tag{5-57}$$

using the assumed geometrical proportions. Now the maximum value for w is

$$w = \pi d \tag{5-58}$$

and occurs when the ports are the full periphery of the spool. Thus, the maximum value for $\sqrt{w/\pi d}$ is unity.

With full periphery ports and one damping length 50% greater than the other (say $L_2 = 1.5L_1$), a damping ratio of $\delta_s = 0.068$ would result; if $L_1 = 1.5L_2$, δ_s would be negative but with the same value. The nature of (5-57) is such that δ_s can never be greater than $0.34\sqrt{w/\pi d}$ but obvious physical constraints on L_1 and L_2 would limit δ_s to a much lesser amount of perhaps no more than $0.1\sqrt{w/\pi d}$. In practice, valve designs are such that $L_2 \approx L_1$, and there is little attempt to exploit the transient flow force as a source of damping because of its very limited potential, especially when the valve area gradient is considerably less than πd.

Both the viscous damping coefficient and the spring rate due to the flow forces are maximum when $P_L = 0$. Thus, null values for these quantities can be defined, and the null flow force spring rate is

$$K_{f0} = 0.43 w P_s \qquad (5\text{-}59)$$

Therefore, a $\frac{1}{2}$ in. diameter spool with full periphery ports and operating with 1000 psi supply pressure would have a spring rate of $0.43\pi(\frac{1}{2})1000 = 675$ lb/in. which is certainly a sizable value. Thus, a valve stroke of 0.020 in. would require a theoretical force of 13 lb to overcome the flow force. In addition, as explained in Section 5-6, the measured flow force spring rate near null is often nearly twice that computed by (5-59). This fact causes the flow force to be somewhat higher than predicted when the valve is off from null, even though the flow force spring approaches that of theory for larger valve strokes. In practice, the source stroking the valve should have a force capability much in excess of the flow force so that there is force potential to shear dirt particles which might lodge in the orifices.

Flow forces are normally quite large and are the major contributor to the total force required to stroke spool valves. Several compensation schemes to reduce or eliminate such flow forces have been investigated (see Section 5-6) but none have met with general acceptance because of manufacturing cost and the nonlinear flow force which results from imperfect compensation. The practical solution to this problem has been the two-stage servovalve in which a hydraulic first stage, usually a flapper valve, provides a quite adequate force to stroke the second-stage spool valve.

It should be borne in mind throughout this discussion of (5-50) that the mass, damping, and spring constants of the source driving the spool (e.g., an electric torque motor) must be added to that of the spool itself and the coefficients M_s, B_f, and K_f must be interpreted accordingly. In many instances the source coefficients may be larger than those of the valve.

5-4   OPEN CENTER SPOOL VALVE ANALYSIS

In Section 5-2 performance parameters and characteristics were defined for valves in general. The results were then applied in Section 5-3 to the particular case of a critical center spool valve which was considered in detail. In this section a similar discussion will be given for the open center spool valve. Although this type valve is more limited in application, knowledge of its characteristics will broaden our understanding of valves. Also, the critical center valve behaves somewhat like an open center valve for small valve strokes.

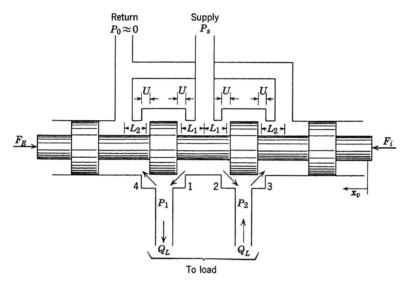

Figure 5-8   Four-land–four-way open center spool valve.

Consider the four-way spool valve shown in Fig. 5-8. When the valve is centered, the underlap of the supply and return ports are identical with a value of U. The case where the supply and return ports have different underlap is treated elsewhere [1]. Arrows at the ports indicate flow directions and the numbers at the ports refer to the subscripts of the flow and the area at the port. It is assumed that the valve is matched and symmetrical and that its operation remains within the underlap region; that is, $|x_v| \leq U$. The orifice areas therefore are

$$A_1 = w(U + x_v) = A_3 \qquad (5\text{-}60)$$

$$A_2 = w(U - x_v) = A_4 \qquad (5\text{-}61)$$

where w is the area gradient of each port and thus of the whole valve. Substituting (5-60) and (5-61) into the general equation (5-22), we obtain

$$\frac{Q_L}{C_d w U \sqrt{P_s/\rho}} = \left(1 + \frac{x_v}{U}\right)\left(1 - \frac{P_L}{P_s}\right)^{1\!/\!2} - \left(1 - \frac{x_v}{U}\right)\left(1 + \frac{P_L}{P_s}\right)^{1\!/\!2} \qquad (5\text{-}62)$$

This is the equation for the pressure-flow curves of an open center four-way spool valve for operation within the underlap region and may be plotted

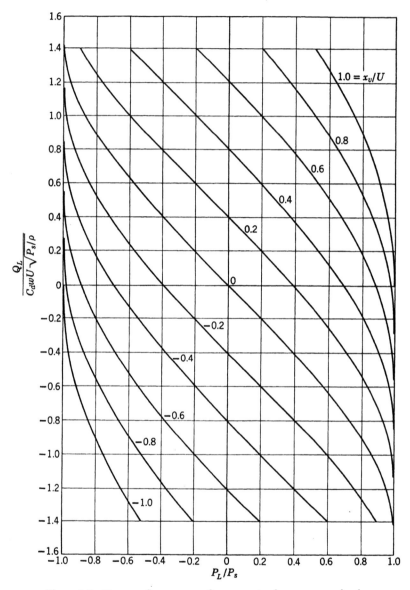

Figure 5-9 Pressure-flow curves of open center four-way spool valve.

in a normalized manner as shown in Fig. 5-9. These curves are quite linear compared with those for a critical center valve. Outside the underlap region the valve acts as a critical center valve, because only two orifices are active at a time, with pressure-flow curves shaped like those in Fig. 5-4.

The null valve coefficients can be obtained by differentiating (5-62) and evaluating the derivatives at $Q_L = P_L = x_v = 0$. Therefore

$$K_{q0} = 2C_d w \sqrt{\frac{P_s}{\rho}} \tag{5-63}$$

$$K_{c0} = \frac{C_d w U \sqrt{P_s/\rho}}{P_s} \tag{5-64}$$

$$K_{p0} = \frac{2P_s}{U} \tag{5-65}$$

As discussed in Section 5-2, the null operating point is most important. The null coefficients are useful in system dynamic analysis and valve design. The flow gain is most important, and note that it is twice that for a critical center valve. Note further that K_{c0} depends on w, and K_{p0} is independent of the valve area gradient w. This strengthens similar conclusions made for the critical center valve from a study of its leakage curves.

Leakage flow curves similar to those defined for the critical center valve can be made for the open center valve. The total center flow through the valve is useful because it gives the power loss at the null operating point. At this point, $P_L = x_v = 0$ so that $A_1 = A_2 = wU$, and (5-23) gives the total center flow as

$$Q_c = 2C_d w U \sqrt{\frac{P_s}{\rho}} \tag{5-66}$$

The final item of interest is the force equation for the open center valve, and this requires a determination of the total flow force which reacts on the spool to oppose the driving force F_i. The steady-state flow force is found by applying (5-90) to each of the valve ports in Fig. 5-8 and adding the result. Signs are determined from the fact that the steady-state flow force at a particular port acts on the valve in such a direction as to close the particular port. Thus, the steady-state flow force is

$$F_{Rss} = 2C_d C_v(\cos \theta)[A_1(P_s - P_1) + A_3 P_2 - A_2(P_s - P_2) - A_4 P_1] \tag{5-67}$$

Combining with Eqs. (5-60), (5-61), and (5-3) yields

$$F_{Rss} = 4C_d C_v(\cos \theta)(P_s - P_L)w x_v = 0.86w(P_s - P_L)x_v \tag{5-68}$$

Using (5-93) and following the rule given in section 5-6 to determine signs of the terms, we can write the transient flow force

$$F_{R(trans)} = \rho L_1 \frac{dQ_1}{dt} - \rho L_1 \frac{dQ_2}{dt} - \rho L_2 \frac{dQ_3}{dt} + \rho L_2 \frac{dQ_4}{dt} \quad (5\text{-}69)$$

Using (5-15) and (5-16) and combining with (5-18) and (5-19), we can write the flows through the orifices

$$Q_1 = Q_3 = C_d A_1 \left[\frac{2}{\rho}(P_s - P_1)\right]^{1/2} = C_d w(U + x_v)\left[\frac{1}{\rho}(P_s - P_L)\right]^{1/2} \quad (5\text{-}70)$$

$$Q_2 = Q_4 = C_d A_2 \left[\frac{2}{\rho}(P_s - P_2)\right]^{1/2} = C_d w(U - x_v)\left[\frac{1}{\rho}(P_s + P_L)\right]^{1/2} \quad (5\text{-}71)$$

The transient flow force can now be written

$$F_{R(trans)} = \rho(L_1 - L_2)\frac{dQ_1}{dt} - \rho(L_1 - L_2)\frac{dQ_2}{dt} \quad (5\text{-}72)$$

The required flow derivatives are found by differentiation of (5-70) and (5-71). Neglecting the pressure derivative (dP_L/dt) terms, the final expression for the transient flow force becomes

$$F_{R(trans)} = (L_1 - L_2)C_d w \sqrt{\rho}\left[\sqrt{P_s + P_L} + \sqrt{P_s - P_L}\right]\frac{dx_v}{dt} \quad (5\text{-}73)$$

In most dynamic analysis it is sufficient to consider $P_L \approx 0$. This value also gives the largest flow forces. Therefore, by combining the flow forces with the spool inertia force the force balance equation for the spool in Fig. 5-8 becomes

$$F_i = M_s \frac{d^2 x_v}{dt^2} + B_f \frac{dx_v}{dt} + K_f x_v \quad (5\text{-}74)$$

where

$$K_f = 4C_d C_v w(\cos\theta)P_s = 0.86wP_s$$
$$B_f = 2(L_1 - L_2)C_d w \sqrt{\rho P_s}$$

and symbol definition is the same as that in (5-50). As with the critical center valve, the steady-state flow force acts as a centering spring, and the transient flow force acts as a viscous damping. However, it is interesting to note that both the spring rate and viscous damping coefficient are twice that of a critical center valve. The open center valve should be designed so that $L_1 \geq L_2$ in Fig. 5-8 to prevent valve instability due to transient flow forces. A discussion of this phenomenon similar to that given in Section 5-3 can be made.

5-5 THREE-WAY SPOOL VALVE ANALYSIS

The two most common types of four-way spool valves were analyzed in Sections 5-3 and 5-4. The present section will give a brief discussion of the two types, critical center and open center, of three-way spool valves. Since many of the comments in the last two sections are applicable, we

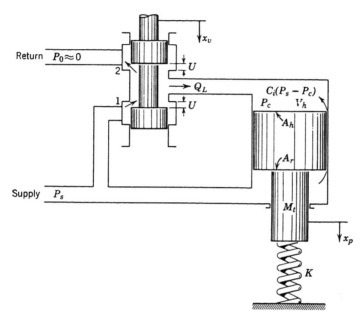

Figure 5-10   Three-way spool valve with piston load.

shall merely outline the principal features and equations of three-way valves. The combination of three-way valve and a piston will be treated in Section 6-3.

Three-way spool valves must be used with an unequal area piston, Fig. 5-10, to provide direction reversal. The requirement prevents the use of three-way valves with rotary type motors. The rod and head side areas of the piston should be such that a steady-state control pressure of about $P_s/2$: that is

$$P_{c0} = \frac{P_s}{2} \tag{5-75}$$

acts on the head area. This design relation allows the control pressure P_c to rise or fall and to provide equal acceleration and deceleration capability. With no loads on the piston, (5-75) is satisfied by making the head area twice the rod area; that is,

$$A_h = 2A_r \qquad (5\text{-}76)$$

and this rule is generally used for piston sizing even with load forces. However, the areas should be sized to satisfy (5-75) if there are unidirectional force load components.

As explained in Section 6-3, three-way valves find greatest use in hydromechanical position servos, and the critical center type is usually preferred. Referring to Fig. 5-10, let us first consider the critical center case in which either orifice 1 or 2 (not both) is active at one time. Letting $U = 0$ to achieve a critical center, the equations for the pressure-flow curves can be written directly. Therefore

$$Q_L = C_d A_1 \left[\frac{2}{\rho} (P_s - P_c) \right]^{1/2} = C_d w x_v \left[\frac{2}{\rho} (P_s - P_c) \right]^{1/2} \quad \text{for} \quad x_v \geq 0$$

$$Q_L = -C_d A_2 \left[\frac{2}{\rho} P_c \right]^{1/2} = C_d w x_v \left[\frac{2}{\rho} P_c \right]^{1/2} \quad \text{for} \quad x_v \leq 0 \qquad (5\text{-}77)$$

A normalized plot of these equations is identical to Fig. 5-4 except that the abscissa scale must be altered so that $P_c/P_s = 0$ at $P_L/P_s = -1$ and $P_c/P_s = 1$ at $P_L/P_s = 1$.

The null operating point for a three-way valve is defined by $Q_L = x_v = 0$ and $P_c = P_s/2$. Evaluating the derivatives of (5-77) at this point, we find that the null coefficients for the critical center three-way valve become

$$K_{q0} = \left. \frac{\partial Q_L}{\partial x_v} \right|_0 = C_d w \sqrt{\frac{P_s}{\rho}} \qquad (5\text{-}78)$$

$$K_{c0} = -\left. \frac{\partial Q_L}{\partial P_c} \right|_0 = \left. \frac{2 C_d w x_v \sqrt{P_s/\rho}}{P_s} \right|_{x_v=0} = 0 \qquad (5\text{-}79)$$

$$K_{p0} = \left. \frac{\partial P_c}{\partial x_v} \right|_0 = \left. \frac{P_s}{x_v} \right|_{x_v=0} = \infty \qquad (5\text{-}80)$$

Comparing the null coefficients, we note that the flow gain is the same but the pressure sensitivity is half that of a four-way critical center valve. Therefore constant and friction load forces will cause twice the static error in systems with three-way valves. In Section 6-3 it is shown that dynamic errors are also about doubled. This is the most serious objection to the use of three-way valves and offsets to a large extent their manufacturing simplicity. Thus three-way valves are best suited to hydromechanical servos which have little or no loads or can tolerate the error.

Leakage flows and practical values for null coefficients for the critical center three-way valve may be treated similar to that in Section 5-3.

For operation within the underlap region (open center three-way valve), the pressure flow curves given by

$$\frac{Q_L}{C_d w U \sqrt{2P_s/\rho}} = \left(1 + \frac{x_v}{U}\right)\left(1 - \frac{P_c}{P_s}\right)^{\frac{1}{2}} - \left(1 - \frac{x_v}{U}\right)\left(\frac{P_c}{P_s}\right)^{\frac{1}{2}} \quad (5\text{-}81)$$

and the null coefficients are

$$K_{q0} = \frac{\partial Q_L}{\partial x_v}\bigg|_0 = 2C_d w \left(\frac{P_s}{\rho}\right)^{\frac{1}{2}} \quad (5\text{-}82)$$

$$K_{c0} = -\frac{\partial Q_L}{\partial P_c}\bigg|_0 = \frac{2C_d w U \sqrt{P_s/\rho}}{P_s} \quad (5\text{-}83)$$

$$K_{p0} = \frac{\partial P_c}{\partial x_v}\bigg|_0 = \frac{P_s}{U} \quad (5\text{-}84)$$

The leakage or center flow at null is given by

$$Q_c = C_d w U \left(\frac{P_s}{\rho}\right)^{\frac{1}{2}} \quad (5\text{-}85)$$

Since three-way valves are usually used in hydromechanical servos, the source positioning the spool is mechanical and quite stiff compared with the force loads imposed by the spool. For this reason the force equation describing the spool motion is usually unimportant and flow forces are not of interest.

5-6 FLOW FORCES ON SPOOL VALVES

These forces are also referred to as *flow induced forces*, *Bernoulli forces*, or *hydraulic reaction forces*. These names refer to those forces acting on a valve as a result of fluid flowing in the valve chambers and through the valve orifices.

Let us first investigate the *steady-state flow force*. The inherent fluid accelerating property of an orifice results in a jet force of

$$F_j = \overset{\text{mass}}{\overbrace{\rho V}}\overset{\text{acc}}{\overbrace{\frac{Q_2^{\,2}}{A_2 V}}} = \frac{\rho Q_2^{\,2}}{A_2} = \frac{\rho Q_2^{\,2}}{C_c A_0} \quad (5\text{-}86)$$

acting normal to the plane of fluid at the vena contracta in Fig. 5-11.

The following symbols are defined:

Q_2 = volumetric flow rate through orifice, in.^{3}/sec

V = volume of fluid being accelerated, in.^{3}

$A_0 = wx_v$ = orifice area, in.^{2}

C_c = contraction coefficient, dimensionless

ρ = mass density of fluid, lb-sec^2/in.^{4}

w = area gradient of orifice, in.^{2}/in.

A_v = area of the valve land, in.^{2}

θ = jet angle, degrees

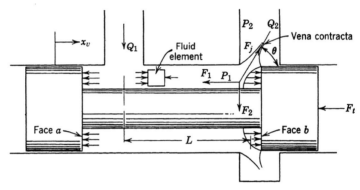

Figure 5-11   Flow forces on a spool valve due to flow leaving a valve chamber.

By Newton's third law this jet force has an equal and opposite reaction force which may be resolved into two components:

$$F_1 = -F_j \cos \theta \quad \text{(axial component)} \tag{5-87}$$

$$F_2 = -F_j \sin \theta \quad \text{(lateral component)} \tag{5-88}$$

The lateral component tends to push the valve spool sideways against the sleeve and cause sticking. However, it is compensated in practice by locating the valve ports symmetrically around the spool. The axial force is not compensated and acts in a direction to close the valve port, as may be seen by a comparison of the pressure distribution on faces a and b in Fig. 5-11. Neglecting compressibility in the small valve chambers (i.e., assuming the fluid to be incompressible), continuity requires that $Q_1 = Q_2$ and the orifice equation (3-33) can be used to describe the flow. Therefore

$$Q_1 = Q_2 = C_d A_0 \left[\frac{2}{\rho} (P_1 - P_2) \right]^{1/2} = C_c C_v A_0 \left[\frac{2}{\rho} (P_1 - P_2) \right]^{1/2} \tag{5-89}$$

which may be combined with Eq. (5-86) to yield the steady-state axial flow force acting on the valve spool in Fig. 5-11. Hence

$$F_1 = 2C_dC_vA_0(P_1 - P_2)\cos\theta \tag{5-90}$$

If the orifice is rectangular and the peripheral width is large compared with its axial length, then the flow can be considered two-dimensional and LaPlace's equation can be solved to determine the jet angle θ if the flow is assumed irrotational, nonviscous, and incompressible. This solution was performed by Von Mises and θ turns out to be 69° if there is no

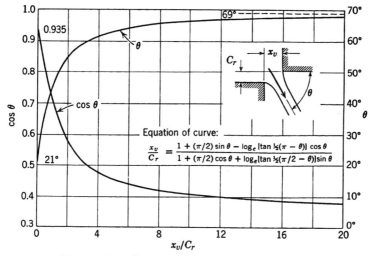

Figure 5-12 Effect of radical clearance on the jet angle.

radial clearance between the valve spool and sleeve. Using $C_d = 0.61$, $C_v = 0.98$, $\cos 69° = 0.358$, and defining $\Delta P = P_1 - P_2$, (5-90) becomes

$$F_1 = 0.43w\,\Delta Px_v = K_f x_v \tag{5-91}$$

which is the usual form of the steady-state flow force equation. Thus, the steady-state flow force is directly proportional to orifice area gradient, pressure drop, and spool displacement and always acts in a direction to close the orifice. Because this force depends on valve position, it is completely analogous to a centering spring on the valve. At small orifice openings the jet angle becomes 21° because of radial clearance C_r, and the steady-state flow force is given by [2]

$$F_1 = 2C_dC_vw(\Delta P)\sqrt{x_v^2 + C_r^2}\cos\theta \tag{5-92}$$

with $\cos\theta$ obtained from Fig. 5-12. Note that the factor $\cos\theta$ varies by $0.933/0.358 = 2.6$ so that large deviations from (5-91), due to radial clearance, are possible at short valve strokes.

Experimentally, the flow force is somewhat larger than that computed from (5-91). Typically, the curve of force versus valve displacement has a higher slope at null, perhaps $1.6K_f$ to $2K_f$, and the curve bends over with a slope approaching K_f as the valve is opened a few thousandths. There are two reasons for the higher slope at null: (1) when the valve is near null the radial clearance becomes more significant and the jet angle is reduced; therefore, a multiplicative factor of $\cos 21° = 0.933$ should be used in (5-90) instead of 0.358, and (2) critical center (zero lapped) four-way spool valves usually have a slight underlap; therefore, since open center valves have twice the flow force of critical center valves—compare (5-68) with the first term in (5-49)—a higher flow force spring rate should be expected near null.

Thus far the discussion has considered only the steady-state flow force. If the slug of fluid in the valve chamber is accelerated, then a force is produced which reacts on the face of the spool valve lands. The direction of this force can be seen by considering the acceleration of a small element of fluid (Fig. 5-11) in the direction of flow. If the fluid element is being accelerated, the pressure on the left side of the element must be greater than the pressure on the right side. Therefore, the pressure on face a must be greater than the pressure on face b. Thus, the transient flow force is due to *acceleration of the fluid in the annular valve chamber*. The direction of this force for the case in Fig. 5-11 is such that it tends to close the valve port; however, this is not the general rule.

The magnitude of the transient flow force is given by Newton's second law as

$$F_3 = Ma = \rho L A_v \frac{d(Q_1/A_r)}{dt} = \rho L \frac{dQ_1}{dt} \tag{5-93}$$

Obtaining dQ_1/dt from (5-89), the transient flow force becomes

$$F_3 = LC_d w \sqrt{2\rho(P_1 - P_2)} \frac{dx_v}{dt} + \frac{LC_d w x_v}{\sqrt{(2/\rho)(P_1 - P_2)}} \frac{d(P_1 - P_2)}{dt} \tag{5-94}$$

Thus we note that the transient flow force is proportional to spool velocity and pressure changes. The velocity term is the more significant because it represents a damping force. There is little direct evidence to indicate that the pressure rate term contributes substantially to valve dynamics, and therefore it is usually neglected. The quantity L is the axial length between incoming and outgoing flows and is called the *damping length*.* The

* All damping lengths are considered positive quantities; however, the associated flow forces can be positive or negative. In the literature, however, the signs of the forces are usually assigned to the damping lengths so that positive or negative damping lengths are possible. The concept of a negative length is somewhat confusing and clouds the physics of the forces. Therefore this practice is not followed in this book.

damping length comes about because the fluid mass in the annular valve chamber is proportional to L. For the case in Fig. 5-11 the transient flow force is stabilizing because it opposes valve motion. For the reverse flow arrangement where fluid enters the valve chamber through an orifice (Fig. 5-13) the transient flow force is destabilizing.

In general, a valve consists of a series-parallel combination of several orifices and establishing the sign of the various transient flow force terms can become confusing. A recommended method is that used in this discussion and consists of analyzing the reaction force on the valve lands due to the fluid being accelerated in the annular valve chambers whose

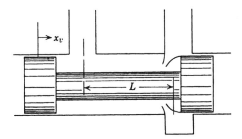

Figure 5-13   Flow entering a valve chamber.

lengths are the damping lengths. The transient flow force is stabilizing if the reaction force acts on the valve lands to oppose the force stroking the valve. For example, orifices 3 and 4 in Fig. 5-3 and 1 and 4 in Fig. 5-8 are associated with stabilizing forces while those of the remaining orifices are destabilizing. The total steady-state and transient flow forces for more complex valve spool configurations are discussed in Sections 5-3 and 5-4.

Steady State Flow Force Compensation

As discussed with regard to (5-59) ,the steady-state flow force is a significant contributor to the force required to stroke spool valves. For larger valves (four-way, 1 in. diameter spool, 0.020 in. stroke, and $P_s = 1000$ psi), this force can exceed 20 lb. There is no particular difficulty if such forces are available from the stroking source. This is usually the case in hydromechanical servovalves because the valve is stroked from a mechanical source with ample force. However, the trouble arises when a valve is stroked from an electromagnetic device, such as a torque motor or solenoid which has distinct force limitations (usually on the order of 10 to 20 lb). Such force limitations in turn limit the largest size possible for single-stage electrohydraulic servovalves, as discussed in Section 7-3. To obtain larger valve sizes, one must resort to a two-stage configuration or use some technique to reduce or compensate the steady-state flow force. The two-stage solution to this problem is almost universally used in spite of the additional complexity and cost.

Methods of compensating steady-state flow forces are costly in manufacture, not very effective, or result in a nonlinear flow force versus stroke characteristic which is undesirable. Although no method of compensation has found wide acceptance, a brief description of four investigated techniques is warranted [2, 3].

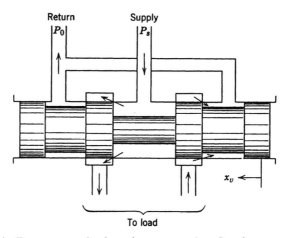

Figure 5-14   Four-way spool valve using pressure drop flow force compensation.

The steady-state flow force equation (5-90) states that the force is zero if the jet angle θ is 90°. Such a condition can be approximated by using many small holes symmetrically placed around the sleeve to make the orifice area. For each hole the jet angle is 69° for small openings and reaches 90° when the hole is completely uncovered. The combined results of many *radial hole orifices* gives a substantially reduced flow force but at the expense of valve linearity at small openings because of the round holes.

A second method of compensation is to increase the shank diameter of the spool at the spool ends (Fig. 5-14). Because the annular passage is reduced, a pressure drop is required for large flows which reacts on the valve lands to oppose the centering action of the steady-state flow force. Although *pressure drop compensation* is quite simple, it is effective only at large flows. It is not effective with partial periphery ports because the fluid jets tend to spread after leaving the orifices and it becomes difficult to restrict the flows.

If the fluid that leaves a valve chamber can be directed back on the

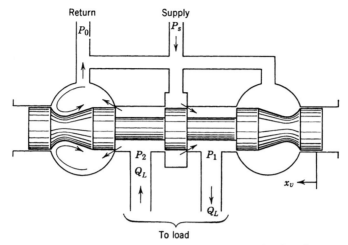

Figure 5-15 Four-way spool valve using recirculation lands for flow force compensation.

spool, then an opening force will exist which compensates the steady-state flow force. This is the principle of the *recirculation lands* in Fig. 5-15. This method of compensation is more effective at higher flows because considerable fluid is recirculated, making the net force versus stroke characteristic quite nonlinear.

A method of compensation similar to that of the recirculation lands is that due to *negative force ports* (Fig. 5-16). The returning flow in a motor

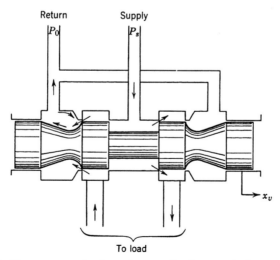

Figure 5-16 Four-way spool valve using negative force ports for compensation.

line impinges directly on the valve spool, creating an opening force. Use of negative force ports provide nearly perfect compensation at manufacturing expense.

The design of a particular compensation scheme should be worked out empirically following the suggestions given by the original investigators. Results to be expected are illustrated in Fig. 5-17. The saturation type characteristic of the compensated curve is typical because most schemes increase in effectiveness at higher flows. This makes the compensated curve

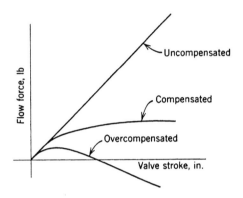

Figure 5-17   Typical flow force curves.

nonlinear and, consequently, undesirable in many systems, especially single stage servovalves. It is possible to overcompensate with any of the schemes to the point where the slope of the curve is negative. This may be undesirable because of the adverse affect on stability in some systems. With recirculation lands or negative force ports it is possible to overcompensate even further to the point at which the curve actually crosses the abscissa (Fig. 5-17) and the force becomes an opening rather than closing force. This feature is also undesirable because of stability considerations.

5-7 LATERAL FORCES ON SPOOL VALVES

In the last section we were concerned with steady-state flow forces on spool valves. These forces could be resolved into axial and lateral components. The lateral component of the steady-state flow force is compensated by placing the ports symmetrically around the spool. However, another source of lateral forces results from leakage flow (usually laminar) across the valve lands. These forces cause excessive friction and can result in the spool being held securely against the sleeve—a condition referred to as *hydraulic lock*. These forces are the subject of this section.

There is no net lateral force on a perfectly cylindrical piston in a perfectly cylindrical bore (the axes of piston and bore need not be coincident) because of leakage flow past the piston. However, it is not possible to achieve such a precision fit in practice. Usually there is a small clearance between the piston and cylinder and some taper in either the piston or cylinder, as illustrated in Fig. 5-18. The clearance permits leakage across the piston, and the taper causes a lateral force to be exerted on the piston.

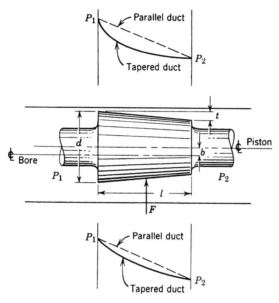

Figure 5-18   Tapered piston decentered in bore.

An examination of Fig. 5-18 provides insight concerning the generation of the lateral force. As the piston moves upward the top duct decreases in area, increasing the flow resistance. The area is smallest at the lip, causing a large initial negative pressure gradient. The pressure gradient is more uniform throughout the remaining piston length because the flow resistance is uniformly decreased. The initial negative pressure gradient is smaller in the lower duct because the entrance area is larger. Because the pressure is greater in the lower duct, there is a net lateral force on the piston, tending to push the piston to the top wall. The expression for this lateral force can be shown to be [4]

$$F = \frac{\pi l d t (P_1 - P_2)}{4b} \left[\frac{2C + t}{\sqrt{(2C + t)^2 - 4b^2}} - 1 \right] \qquad (5\text{-}95)$$

where C is the radial clearance at the large end with the piston centered. This force acts to decenter the piston and drive it to a wall. Substituting $b = C$ in (5-95) and simplifying, the force holding the piston against the sleeve is

$$\frac{F}{ld(P_1 - P_2)} = \frac{\pi}{4}\left[\frac{t}{C}\right]\left[\frac{2 + (t/C)}{\sqrt{4(t/C) + (t/C)^2}} - 1\right] \tag{5-96}$$

and is plotted in Fig. 5-19. From the curve we note that the right side of (5-96) has a maximum value of 0.27 occurring when $t/C = 0.9$. Thus the

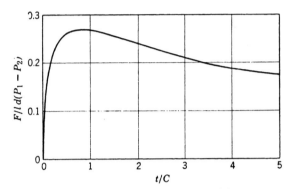

Figure 5-19   Normalized lateral force.

hydraulic locking force becomes

$$F \le 0.27 ld(P_1 - P_2) \tag{5-97}$$

and is extremely large for reasonable parameter values (if $\Delta P = 1000$ psi and $l = d = 1$ in., then $F \le 270$ lb). Therefore $t/C \ll 1$ to prevent large lateral forces. Because this is difficult to achieve it is reasonably safe to conclude that dimensional control of the taper is an impractical method of reducing and controlling lateral forces.

If the higher pressure is at the small end of the piston (Fig. 5-18) the lateral force acts to center the piston. If there is a distinct high pressure side on the piston, an intentional taper to that side could be used to obtain a centering force to prevent hydraulic lock. However, this is not practical, especially on spool valves, because the taper direction would be different on the various valve lands, making manufacture most difficult.

These lateral forces are commonly compensated by cutting peripheral grooves on the lands of valve spools and on pistons. These grooves allow flow around the spool periphery from high to low pressure areas, thereby tending to equalize these pressures (Fig. 5-20) and center the piston. These *balancing, equalizing,* or *centering grooves,* as they are often called, in effect short circuit any localized pressure buildup.

Sweeney [5, 6] reports that one groove at the piston center reduces the lateral force to 40% of that for an ungrooved piston. Three grooves equally spaced reduce the force to 6% and seven grooves, to 2.7%. Both the depth and width of the grooves should be at least ten times the clearance, and the sides of the grooves should be perpendicular to the bore to prevent wedging of dirt particles. The grooves should be placed at the

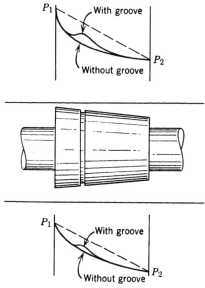

Figure 5-20   Effect of pressure equalizing grooves.

high pressure side of the land or piston and a minimum of three grooves used. If there is no distinct high pressure side, the grooves should be equally spaced across the valve land or piston.

Unless the land or piston is purposely tapered in the proper direction, these grooves should always be used on close fitting pistons or valves because their cost is negligible at the time of manufacture and, in addition to reducing lateral forces, they provide a reservoir for any dirt particles which might cause sticking or friction between the valve or piston and sleeve or cylinder. Centering grooves are recommended for use in any fluid such as air, fuel, or oil because the lateral forces discussed are independent of the lubricating properties of the fluid.

As discussed in Section 3-3, the laminar leakage flow past a piston touching the cylinder wall is 2.5 times that of a centered piston. Because equalizing grooves tend to center the piston, the leakage flow is also reduced and is another reason for the use of such grooves.

5-8 SPOOL VALVE DESIGN

Referring to Fig. 5-1a, there are several critical dimensions that must be maintained in the manufacture of spool valves. The widths of the porting lands must match the corresponding widths in the sleeve, and the distance between the lands must match the corresponding dimensions in the sleeve. In addition, close tolerances must be held on the radial dimension between spool and sleeve and on the squareness of the land edges. These five dimensions are held to tolerances of ±0.0001 in. or better in high performance valves, but in some cases tolerances on the order of 0.0003 in. may be acceptable. These tolerances are important because they have a pronounced effect on the flow gain and pressure sensitivity near null.

One of the first choices to be made is whether a three- or four-way valve is required. Three-way valves have advantages of only three critical dimensions and only one actuator line. However, such valves are restricted to use with pistons and have about half the pressure sensitivity and half the motor-load resonant frequency of a four-way combination, as discussed in Section 5-5. The particular application usually dictates the choice, and four-way valves are used in the majority of cases.

Another factor to be considered is that of the number of lands to be placed on the spool. Referring to Figs. 5-1a, b, and c, it is apparent that the number of critical dimensions are the same in each case. The two-land valve has the advantage of being shorter in length and somewhat simpler in construction due to the absence of additional lands. However, it has two disadvantages, the first being that the valve is statically unbalanced. Referring to Fig. 5-1a, imagine that an orifice is inserted in the top return line to illustrate that the resistance to flow of the two return lines are not identical. When such a valve is displaced from null, the pressures acting on the outside faces of the two lands will be different because of the fluid flowing, and the resultant force will open the valve further. For such a valve, the center position actually represents an unstable condition. The second disadvantage is that the valve lands may lodge in the sleeve openings if the ports consume a major portion of the periphery of the spool. The two-land type of construction should be avoided unless its drawbacks are well understood. It is possible to correct both of these disadvantages by placing two additional lands on the spool, Fig. 5-1c, to provide sealing and centering functions; however, these lands lengthen the spool. The three-land valve of Fig. 5-1b is also statically balanced and, because it has an intermediate number of lands, is the most widely used.

The shape of the ports depend on system requirements. In some special cases it is desirable to shape the ports to obtain a particular flow gain curve, as illustrated in Fig. 5-21. Round ports, although simple to machine,

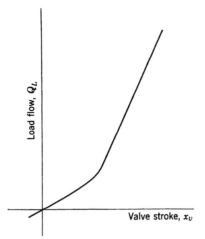

Figure 5-21   Nonlinear flow gain characteristic.

have the characteristic shown in Fig. 5-22 and are quite nonlinear. With such ports the flow gain near null is very low and increases away from null. However, nonlinear flow gain characteristics should be avoided because they offer little benefit when balanced against the many drawbacks. One should bear in mind that for a system to be stable in all modes of operation the system must be initially designed and adjusted to be stable in the region where the valve flow gain is largest. Because all servo performance parameters (error, bandwidth, stiffness, etc.) become worse with lower gain, it is apparent that servo performance is sacrificed in regions where the flow gain is less than maximum. This sacrifice in performance is unnecessary if the flow gain is reasonably linear. In most high performance systems performance cannot be sacrificed and use of rectangular ports to obtain a linear flow gain is essential. Therefore linear ports are optimum from the viewpoint of maintaining good servo performance over a broad operating range. Systems that use a nonlinear flow gain curve and are adjusted for stability in the low gain region near null usually become unstable when the valve is displaced to give larger flows. System instability at larger flows (i.e., at higher piston or motor velocities) is typical of systems using circular ports. The effect of nonlinear ports on system performance should be well understood before they are used in a system.

If system requirements dictate linear ports, which is usually the case, then the type of valve center (Fig. 5-2) must be established. Because a linear flow gain is desirable, the critical center valve is the usual choice. However, as discussed in Section 5-1, open center valves are useful in applications where the valve must be at the null position in a high temperature environment for extended periods of time and a continuous flow is

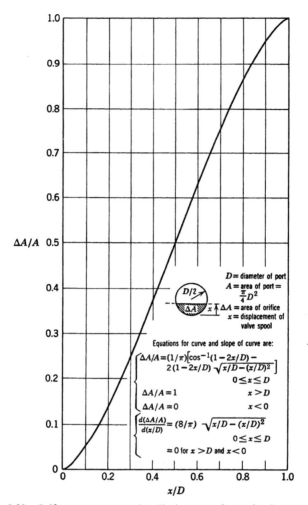

Figure 5-22   Orifice area versus valve displacement for a circular port in a flat plate.

required to maintain reasonable fluid temperatures. Open center valves are required in constant flow (as opposed to constant pressure) systems. The major disadvantage of open center valves is the increased flow gain near null. Because systems using such valves must be adjusted for stability at null, servo performance is reduced away from null because of the lowered flow gain. Another disadvantage is the power loss at null due to center flow.

Closed center valves are seldom desirable because of the deadband characteristic at null. Such a region of very low gain results in lack of control and a tendency for the output device to wander in position. If the output wanders to the edge of the deadband region, the loop will begin to have gain and initiate corrective action. In some cases this behavior may appear as a very low frequency of oscillation or periodic wandering and as such might be classified as unstable. However, this phenomenon should not be confused with the conventional definition of stability, that is, in the sense of Nyquist's or Routh's criterion, where the system is unstable usually because of excess (rather than lack of) gain. Once again we note the desirability of a linear flow gain.

A critical center four-way spool valve with rectangular ports is by far the most common servo valve. Such a valve is characterized by its area gradient w and maximum stroke x_{vm}. These valves are universally rated by the load flow obtainable at maximum stroke with a specified valve pressure drop (usually 1000 psi). This rated load flow can be computed from (5-33). Therefore

$$Q_L\big|_{\text{rated}} = C_d w x_{vm} \sqrt{\frac{1}{\rho} P_v\big|_{\text{rated}}} \qquad (5\text{-}98)$$

Using values of of $\rho = 0.78 \times 10^{-4}$ lb-sec^2/in.^{4} for petroleum base fluids, $C_d \approx 0.61$, and assuming the standard valve drop of 1000 psi, we find that the valve flow rating in gpm becomes

$$Q_L\big|_{\text{rated}} = 574\, w x_{vm} \quad \text{gal/min} \qquad (5\text{-}99)$$

where w = valve area gradient, in.^{2}/in.

x_{vm} = maximum valve stroke, in.

It should be understood that various values for load flow and valve pressure drop $(P_s - P_L)$ are possible as shown by the pressure-flow curves in Fig. 5-4. The rated load flow is simply an attempt to represent the valve capacity by a single number. A better way to rate such valves would be to give the area gradient and maximum stroke because all important performance parameters follow from these two quantities. The maximum valve area $w x_{vm}$ would also make a more meaningful valve rating because it does not require choosing a "rated" valve drop, as in the conventional method. When selecting a valve the pressure-flow curve for maximum stroke, that is, $x_v/x_{vm} = 1$ in Fig. 5-4, should encompass all load flow and pressures encountered in operation. The maximum load flow occurs when stroke is maximum and full supply pressure is dropped across the valve (i.e., $P_L = 0$).

Valve size is related to maximum orifice area, $w x_{vm}$. It should be clear that $w x_{vm}$ is the maximum area of each orifice and as such is an attribute of the valve as a whole. It is apparent that many combinations of area gradient and maximum stroke can result in a given maximum area. The question naturally arises as to what particular values of w and x_{rm} make the best combination. Therefore, let us discuss what values are desirable and the physical limitations on these quantities.

The valve area gradient is the principal parameter in the null flow gain (5-41) and, consequently, has a direct influence on system stability. The null flow gain can be computed with confidence and usually ranges from 10 to 5000 (or more) in.^{3}/sec/in. Fundamentally, the flow gain must be compatible with gains of other components in a system to yield the required loop gain. The null flow gain must be determined in this context. The required area gradient is then found from (5-41). Establishing the flow gain is very important if it is the most convenient method of varying system loop gain. In hydromechanical servos it is often the only method of gain control in the loop. In such systems a change in loop gain level requires a new valve with a different gradient and can become an expensive method of achieving system stability if trial and error adjustments are made. Supply pressure can be adjusted to make small changes in flow gain; however, large changes in supply pressure for loop gain adjustment purposes is not recommended because this pressure should be chosen from load considerations. In electrohydraulic servos a change in electronic amplifier gain is most convenient to alter loop gain level so that there is less emphasis on the valve gain. The electronic amplifier is usually designed to have a gain range compatible with a selected servo valve and other components to achieve loop gain requirements.

Once the area gradient w is selected, the minimum value of the spool diameter is determined because ports may be at most full periphery of the spool. Use of two or four rectangular ports symmetrically placed in the spool sleeve periphery is the most commonly used method of obtaining very low area gradients. The area gradient is then the total width of all slots at a particular orifice. Notched or beveled spools are less frequently used methods of obtaining low area gradients.

The maximum valve stroke can also be used to vary valve size. Electrohydraulic servovalves have strokes ranging from 0.005 to 0.010 in. for small valves up to 10 gpm capacity to 0.015 to 0.030 in. for 50 gpm valves and may be as long as 0.10 to 0.15 in. for large 100 to 200 gpm three-stage valves. Hydromechanical servovalves have much longer strokes in the same flow capacities because of the longer stroke capability of stroking sources and the requirement for low area gradients to get system stability. In general, the longer the stroke the better. Longer strokes give better resolution near null and improved performance with dirty fluids because the valve is open a larger percentage of the time. Increased operating time away from null position allows dirt particles to be flushed from the orifices and reduces silting problems. Silting is a condition where dirt particles in the fluid collect or dam up at the orifices and, especially at small openings, may eventually block them. The servo then loses control for lack of gain and the output wanders from position. The increase in error may stroke the valve enough to open the particle dam. The servo loop will then regain control, but the process may repeat as the valve silts up again. The time for a valve to silt depends on the particle size and quantity in relation to the orifice opening and will vary from a fraction of a minute to an hour or more if the valve is held at null.

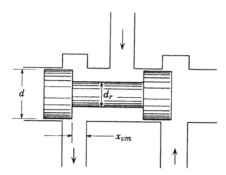

Figure 5-23   Typical passage in spool valves.

It is possible to obtain a design relation between w and x_{vm} from a consideration of flow saturation and valve strength. With reference to Fig. 5-23, the spool rod diameter should be at least $\frac{1}{2}d$ for adequate strength in the spool. The spool passage areas should be at least four times the maximum orifice areas to prevent flow saturation and to ensure that the orifices are the controlling restrictions; that is,

$$\frac{\pi}{4}(d^2 - d_r^2) > 4wx_{vm}$$

Letting $d_r = \frac{1}{2}d$ and simplifying, the criterion for maximum stroke becomes

$$x_{vm} < 0.147 \frac{d^2}{w} \tag{5-100}$$

The condition is hardest to meet when full periphery ports are used. The area gradient for such ports is $w = \pi d$ and the maximum valve stroke to prevent flow saturation becomes $0.047d$ or about 5% of the spool diameter.

In summary, the maximum valve area wx_{vm} should be large enough to give the required maximum flow and horsepower to the load. The pressure-flow curve for x_{vm} should include all values of Q_L and P_L in the system duty cycle. If the area gradient is too large (wide ports), the valve stroke is short and will encourage silting, sticking, and other problems related to a dirty fluid. On the other hand, large strokes have larger flow forces and are limited by the force and travel capability of driving units. This is especially so in electrohydraulic servovalves in which strokes are limited by the torque motor displacement capability to values usually less than 0.020 in. The area gradient should be given more attention than stroke in determining valve size for systems in which valve gain is used to vary system gain.

5-9 FLAPPER VALVE ANALYSIS AND DESIGN

Relatively loose tolerances with the corresponding reduction in cost and insensitivity to dirt makes the flapper valve attractive in low power applications in which the power loss due to leakage is permissible. Nearly all two-stage valves have a flapper-valve first stage which constitutes their major application. The pressure-flow curves of flapper valves are relatively linear, and the performance of these devices are quite predictable and dependable. The purpose of this section is to determine the pressure-flow curves, null coefficients, and flow forces for both single and double jet configurations and to develop relations useful in design.

Single-Jet Flapper Valve

Let us first consider the single-jet or three-way flapper valve (Fig. 5-24). A fixed upstream orifice A_0 is used to cause a control pressure P_c which can be modulated by moving the flapper. A capillary could also be used for this purpose, but the valve would be sensitive to temperature because of viscosity changes. An orifice is universally used for this reason. The curtain area rather than the nozzle area is the controlling restriction at the flapper. It is convenient to first establish the working range of the control pressure by obtaining the blocked-load characteristic. By continuity of flow

$$Q_1 = Q_2 + Q_L \tag{5-101}$$

where the flows are given by the orifice equation as

$$Q_1 = A_0 C_{d0} \left[\frac{2}{\rho} (P_s - P_c) \right]^{\frac{1}{2}} = \frac{\pi}{4} D_0{}^2 C_{d0} \left[\frac{2}{\rho} (P_s - P_c) \right]^{\frac{1}{2}} \tag{5-102}$$

$$Q_2 = A_f C_{df} \left[\frac{2}{\rho} P_c \right] = \pi D_N (x_{f0} - x_f) C_{df} \left(\frac{2}{\rho} P_c \right)^{\frac{1}{2}} \tag{5-103}$$

The quantity x_{f0} is the equilibrium flapper position about which the flapper is displaced an amount x_f. If there is no flow to the load, that is, the load

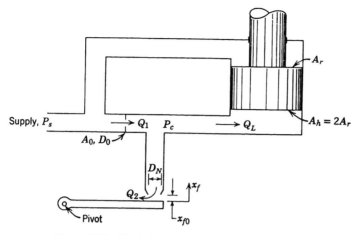

Figure 5-24   Single jet flapper valve with piston load.

is blocked, then $Q_1 = Q_2$ and the orifice flows can be equated and the result simplified to yield

$$\frac{P_c}{P_s} = \left[1 + \left(\frac{C_{df}A_f}{C_{d0}A_0} \right)^2 \right]^{-1} \qquad (5\text{-}104)$$

This equation for the blocked-load characteristic is plotted in Fig. 5-25, and it is quite apparent from the curve that adequate modulation of the control pressure requires an equilibrium value of about 0.5 to $0.6P_s$. Furthermore, it can be shown that maximum pressure sensitivity is achieved

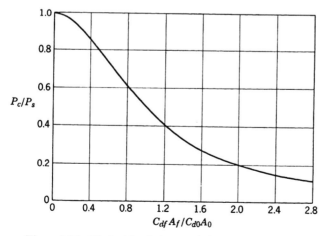

Figure 5-25   Blocked load characteristic of single jet flapper valve.

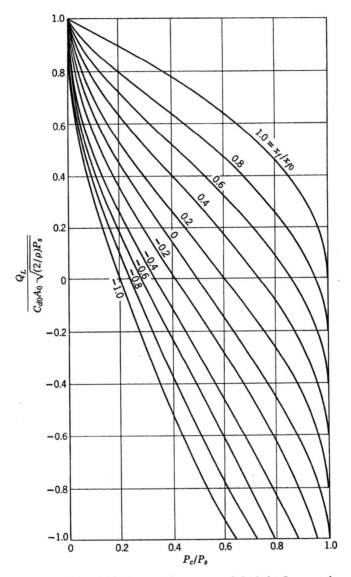

Figure 5-26 Pressure-flow curves of single jet flapper valve.

if $P_c = 0.5P_s$. Therefore an equilibrium control pressure of $0.5P_s$ is considered a design criterion. This criterion requires that the orifice ratio at the null point be

$$\frac{C_{df}A_f}{C_{do}A_0} = 1 = \frac{C_{df}\pi D_N x_{f0}}{C_{do}A_0} \tag{5-105}$$

and is a good design objective for both three- and four-way flapper valves. This criterion also requires that the pistons used with three-way valves have an area ratio of 2.

The equation for the pressure-flow curves is obtained by substituting (5-102) and (5-103) into (5-101). After algebraic simplification, the result is

$$\frac{Q_L}{C_{do}A_0\sqrt{(2/\rho)P_s}} = \left(1 - \frac{P_c}{P_s}\right)^{\frac{1}{2}} - \frac{C_{df}\pi D_N x_{f0}}{C_{do}A_0}\left(1 - \frac{x_f}{x_{f0}}\right)\left(\frac{P_c}{P_s}\right)^{\frac{1}{2}} \tag{5-106}$$

If the design criterion of (5-105) is followed, and it should be, then

$$\frac{Q_L}{C_{do}A_0\sqrt{(2/\rho)P_s}} = \left(1 - \frac{P_c}{P_s}\right)^{\frac{1}{2}} - \left(1 - \frac{x_f}{x_{f0}}\right)\left(\frac{P_c}{P_s}\right)^{\frac{1}{2}} \tag{5-107}$$

which is the final expression for the pressure-flow curves for a single-jet flapper valve and is plotted on a nondimensional basis in Fig. 5-26.

Equation (5-107) may be differentiated and the partials evaluated at $x_f = Q_L = 0$ and $P_c = \frac{1}{2}P_s$ to obtain the null coefficients. Therefore,

$$K_{q0} = \left.\frac{\partial Q_L}{\partial x_f}\right|_0 = C_{df}\pi D_N \left(\frac{P_s}{\rho}\right)^{\frac{1}{2}} \tag{5-108}$$

$$K_{p0} = \left.\frac{\partial P_c}{\partial x_f}\right|_0 = \frac{P_s}{2x_{f0}} \tag{5-109}$$

$$K_{c0} = -\left.\frac{\partial Q_L}{\partial P_c}\right|_0 = \frac{2C_{df}\pi D_N x_{f0}}{\sqrt{\rho P_s}} \tag{5-110}$$

The leakage or center flow at null is given by

$$Q_c = C_{df}\pi D_N x_{f0}\left(\frac{P_s}{\rho}\right)^{\frac{1}{2}} \tag{5-111}$$

and constitutes a power loss.

The fluid flow striking the flapper causes an unbalanced force on the flapper. The expression for this force is given by (5-130) with P_1 replaced by P_c. Therefore

$$F_1 = P_c A_N \left[1 + \frac{16 C_{df}^2}{D_N^2} (x_{f0} - x_f)^2 \right] \qquad (5\text{-}112)$$

where $A_N = (\pi/4) D_N^2$ is the nozzle area. The term $(x_{f0} - x_f)$ is simply the clearance between flapper face and nozzle. For small clearances the force on the flapper due to flow impingement is approximately the static pressure P_c acting on the projected nozzle area. As the flapper clearance increases, the velocity pressure increases giving a greater force. It can be shown [2] that the force on the flapper will approach twice the blocked nozzle force, that is, $2 P_c A_N$, when the flapper is moved a large distance from the nozzle. However, such large clearances are not encountered in flapper valves because good design requires that $x_{f0}/D_N < \frac{1}{16}$ and this criterion makes the second term in the bracket of (5-112) much less than unity.

Evaluating the derivative of (5-112) at the null point $x_f = 0$ and $P_c = \frac{1}{2} P_s$, we obtain

$$\left. \frac{dF_1}{dx_f} \right|_0 = -4 \pi C_{df}^2 P_s x_{f0} \qquad (5\text{-}113)$$

which is the spring gradient of the flow force. Thus, we see that flow impingement on the flapper acts as a *negative* spring. This has a destabilizing effect and requires that the source driving the flapper have a larger positive gradient to compensate. This negative gradient is quite small and usually causes no difficulty because three-way flapper valves are normally used in hydromechanical servos where the source actuating the flapper is quite stiff.

Because the flow force acts to statically unbalance the flapper, a small steady-state system error is required to provide a balancing force. It is usual practice to balance this force either with a spring or a small piston on the opposite side of the flapper with P_c acting on the piston area to prevent the error.

Double-Jet Flapper Valve

The double-jet or four-way flapper valve is shown in Figs. 5-27 and 5-30. The flow equations are

$$Q_L = Q_1 - Q_2 = C_{d0} A_0 \left[\frac{2}{\rho} (P_s - P_1) \right]^{\frac{1}{2}} - C_{df} \pi D_N (x_{f0} - x_f) \left(\frac{2}{\rho} P_1 \right)^{\frac{1}{2}}$$

$$(5\text{-}114)$$

$$Q_L = Q_4 - Q_3 = C_{df} \pi D_N (x_{f0} + x_f) \left(\frac{2}{\rho} P_2 \right)^{\frac{1}{2}} - C_{d0} A_0 \left[\frac{2}{\rho} (P_s - P_2) \right]^{\frac{1}{2}}$$

$$(5\text{-}115)$$

Using Equation 5-105, these equations reduce to

$$\frac{Q_L}{C_{d0}A_0\sqrt{P_s/\rho}} = \left[2\left(1 - \frac{P_1}{P_s}\right)\right]^{\frac{1}{2}} - \left(1 - \frac{x_f}{x_{f0}}\right)\left(\frac{2P_1}{P_s}\right)^{\frac{1}{2}} \qquad (5\text{-}116)$$

$$\frac{Q_L}{C_{d0}A_0\sqrt{P_s/\rho}} = \left(1 + \frac{x_f}{x_{f0}}\right)\left(\frac{2P_2}{P_s}\right)^{\frac{1}{2}} - \left[2\left(1 - \frac{P_2}{P_s}\right)\right]^{\frac{1}{2}} \qquad (5\text{-}117)$$

These two equations combined with the relation

$$P_L \equiv P_1 - P_2 \qquad (5\text{-}118)$$

completely define the pressure-flow curves for the two-jet flapper valve.

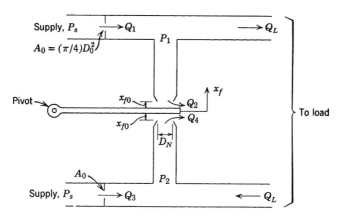

Figure 5-27 Double jet flapper valve.

Unfortunately, these equations cannot be combined into a single relation in a simple manner, but must be solved in the manner indicated in the footnote on page 81. The nondimensional plot in Fig. 5-28 will result, and as can be seen, the curves are quite linear. The blocked-load characteristic is given by (5-104) for each side of the valve by interpreting P_c as P_1 or P_2 and subtracting to get P_L. The resulting curve is shown in Fig. 5-29. As before, equilibrium values for P_1 and P_2 should be $0.5P_s$ with the design criterion given by (5-105).

The leakage or center flow at null is given by

$$Q_c = 2C_{df}\pi D_N x_{f0}\left(\frac{P_s}{\rho}\right)^{\frac{1}{2}} \qquad (5\text{-}119)$$

which is twice that of a single-jet flapper valve.

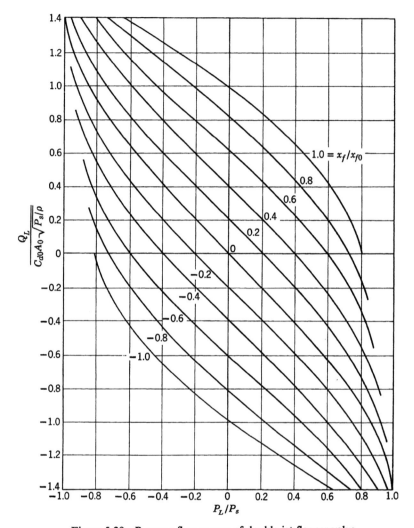

Figure 5-28   Pressure-flow curves of double jet flapper valve.

Let us now determine the valve coefficients at null, that is, at the point where $x_f = Q_L = P_L = 0$ and $P_1 = P_2 = P_s/2$. The linearized form of (5-116) is

$$\Delta Q_L = \frac{\partial Q_L}{\partial x_f} \Delta x_f + \frac{\partial Q_L}{\partial P_1} \Delta P_1 \tag{5-120}$$

Evaluating the derivatives of (5-116) at null and using (5-105), we find that the required partials are

$$\frac{\partial Q_L}{\partial x_f}\bigg|_0 = C_{df}\pi D_N \left(\frac{P_s}{\rho}\right)^{1/2} \tag{5-121}$$

$$\frac{\partial Q_L}{\partial P_1}\bigg|_0 = -\frac{2C_{df}\pi D_N x_{f0}}{\sqrt{\rho P_s}} \tag{5-122}$$

In a similar fashion (5-117) may be linearized about null to obtain

$$\Delta Q_L = C_{df}\pi D_N \left(\frac{P_s}{\rho}\right)^{1/2}\Delta x_f + \frac{2C_{df}\pi D_N x_{f0}}{\sqrt{\rho P_s}}\Delta P_2 \tag{5-123}$$

Addition of (5-120) and (5-123) and combining with $\Delta P_L = \Delta P_1 - \Delta P_2$ yields

$$\Delta Q_L = C_{df}\pi D_N \left(\frac{P_s}{\rho}\right)^{1/2}\Delta x_f - \frac{C_{df}\pi D_N x_{f0}}{\sqrt{\rho P_s}}\Delta P_L \tag{5-124}$$

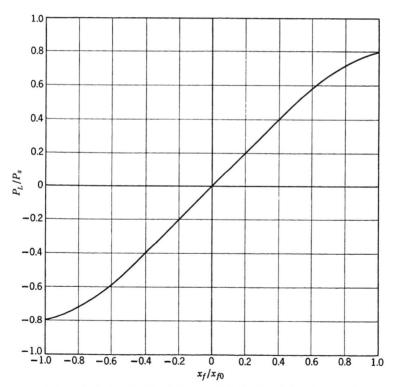

Figure 5-29   Blocked load characteristic for double jet flapper valve.

This is the linearized form of the pressure-flow equation of the two-jet flapper valve for operation at null. The null coefficients are obtained directly from this equation.

$$K_{q0} \equiv \frac{\Delta Q_L}{\Delta x_f}\bigg|_{\Delta P_L = 0} = C_{df}\pi D_N \left(\frac{P_s}{\rho}\right)^{\frac{1}{2}} \tag{5-125}$$

$$K_{p0} \equiv \frac{\Delta P_L}{\Delta x_f}\bigg|_{\Delta Q_L = 0} = \frac{P_s}{x_{f0}} \tag{5-126}$$

$$K_{c0} \equiv -\frac{\Delta Q_L}{\Delta P_L}\bigg|_{\Delta x_f = 0} = \frac{C_{df}\pi D_N x_{f0}}{\sqrt{\rho P_s}} \tag{5-127}$$

Comparing these relations with those of the single-jet flapper valve, we note that the flow gains are identical but the pressure sensitivity is doubled. It is apparent that increased pressure sensitivity is achieved at the expense of increased null leakage. The physical symmetry of the double-jet flapper valve results in reduced null shift due to temperature and supply pressure variations. Another advantage over the single-jet arrangement is the fact that the pressure force acting on the flapper is balanced at null. This force for the two-jet flapper valve is approximately given (neglecting the velocity pressure) by $P_L A_N$. This force does result in inherent feedback proportional to load pressure, but it is small without special attention and is usually neglected.

Flow Forces on Flapper Valves

Thus far it has been stated that the significant force on the flapper is that resulting from the static pressure acting on the nozzle area projected onto the flapper. The velocity or dynamic pressure also acts on the flapper, and it is of interest to include this effect. Consider the flapper-nozzle configuration in Fig. 5-30 in which the flapper is mounted on a torsion rod (torsion spring) with a torsion gradient of K_a in.-lb/rad. This is the case when the flapper is driven by a torque motor where the torque motor armature also serves as the flapper. If the flapper is driven from some other source, then K_a can be considered the equivalent torsion spring constant of the flapper and its driving source. Using Bernoulli's equation, we give the force F_1 by

$$F_1 = (P_1 + \tfrac{1}{2}\rho u_1^2)A_N \tag{5-128}$$

where u_1 is the fluid velocity at the plane of the nozzle diameter and is given by

$$u_1 = \frac{Q_2}{A_N} = \frac{C_{df}\pi D_N(x_{f0} - x_f)\sqrt{(2/\rho)p_1}}{\pi D_N^2/4} = \frac{4C_{df}(x_{f0} - x_f)\sqrt{(2/\rho)P_1}}{D_N} \tag{5-129}$$

Combining these two equations, we obtain

$$F_1 = P_1\left[1 + \frac{16C_{df}^2(x_{f0} - x_f)^2}{D_N^2}\right]A_N \qquad (5\text{-}130)$$

Similar reasoning leads to the equation for the force F_2.

$$F_2 = P_2\left[1 + \frac{16C_{df}^2(x_{f0} + x_f)^2}{D_N^2}\right]A_N \qquad (5\text{-}131)$$

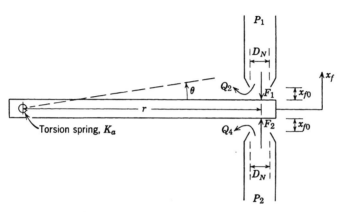

Figure 5-30   Enlarged view of nozzles.

Now the net force acting on the flapper is the difference of these forces.

$$F_1 - F_2 = (P_1 - P_2)A_N + 4\pi C_{df}^2[(x_{f0} - x_f)^2 P_1 - (x_{f0} + x_f)^2 P_2] \qquad (5\text{-}132)$$

Using (5-118) and approximating $P_1 \approx P_2 \approx P_s/2$, the steady-state value, we obtain the relation

$$F_1 - F_2 = P_L A_N + 4\pi C_{df}^2 x_{f0}^2 P_L + 4\pi C_{df}^2 x_f^2 P_L - 8\pi C_{df}^2 x_{f0} P_s x_f \qquad (5\text{-}133)$$

Good flapper valve design requires that $x_{f0}/D_N < \frac{1}{16}$, which makes the second term negligible compared with the first term. The third term can also be neglected because $P_L \approx x_f \approx 0$ near null where most operation occurs. Thus (5-133) can be approximated by

$$F_1 - F_2 = P_L A_N - (8\pi C_{df}^2 P_s x_{f0})x_f \qquad (5\text{-}134)$$

Now, the equation of motion of the flapper in Fig. 5-30 can be written

$$T_d = J_a\frac{d^2\theta}{dt^2} + K_a\theta + (F_1 - F_2)r \qquad (5\text{-}135)$$

where T_d is the driving torque on the flapper such as that which would be obtained from a torque motor. For small angles

$$\tan \theta = \frac{x_f}{r} \approx \theta \qquad (5\text{-}136)$$

A combination of the last three equations yields

$$T_d = \frac{J_a}{r}\frac{d^2x_f}{dt^2} + rP_LA_N + \left[\frac{K_a}{r^2} - (8\pi C_{df}^{\;2}P_sx_{f0})\right]rx_f \qquad (5\text{-}137)$$

Thus the flow force due to fluid impingement acts as a negative spring on the flapper. To get a feel for values, this negative spring rate is about 30 lb/in. if the supply pressure is 1000 psi and the null flapper clearance is 0.003 in.* The stiffness of the source driving the flapper, K_a/r^2, must be greater than this negative spring rate to prevent the possibility of flapper instability. However, this is not the only criterion for stability. The expression for P_L should also be included in (5-137), and Routh's stability criterion applied [7]. However, because P_L depends on the load on the flapper valve, over-all stability must be determined from an analysis of the system in which the valve is placed.

Flapper Valve Design

The design of flapper valves is straightforward. The flow gain is determined from system requirements, as discussed in Section 5-8. Once the null flow gain is established, the nozzle diameter is found from (5-125). Therefore

$$D_N = \frac{K_{q0}}{\pi C_{df}\sqrt{P_s/\rho}} \qquad (5\text{-}138)$$

The null flapper clearance x_{f0} should be as small as possible to achieve largest pressure sensitivity and smallest null leakage. However, it must be large enough to permit passage of the dirt particle sizes expected in the fluid. Values in the range of 0.001 to 0.005 in. are used. A maximum value for the null clearance can be established by requiring the curtain area to be less than one fourth the nozzle area. This criterion ensures that

---

* One source [8] reports the experimental negative spring rate to be three times the theoretical value. Rounding of the inside edge of the nozzle or reattachment of the radial jet onto the land of the nozzle face or a large land in relation to flapper clearance (see Fig. 5-31) could increase the contraction coefficient and give a discharge coefficient of $C_{df} = 0.85$ to 1 instead of the usual 0.61 or so. This could increase the negative spring rate substantially since it is proportional to $C_{df}^{\;2}$. Definitive investigations of flapper flow forces, and especially of the negative spring rate, are needed.

the curtain area is the controlling orifice. Therefore,

$$\pi D_N x_{f0} < \left(\frac{1}{4}\right)\frac{\pi D_N^2}{4}$$

Simplifying, we obtain

$$x_{f0} \leq \frac{D_N}{16} \tag{5-139}$$

Thus the maximum null clearance is $D_N/16$. Once D_N and x_{f0} are determined, the value for D_0 may be solved from (5-105) if the discharge coefficients are known. Therefore*

$$D_0 = 2\left(\frac{C_{df}}{C_{d0}} D_N x_{f0}\right)^{1/2} \tag{5-140}$$

The upstream orifice is usually of the "short tube" type because length to diameter ratios of 2 to 4 are common. Consequently, the discharge coefficient C_{d0} is greater than that of a sharp-edged orifice and usually is in the range 0.8 to 0.9. The discharge coefficient C_{df} of the annular curtain area between nozzle and flapper is more difficult to establish because the geometry is more complex. Experimental results of Lichtarowicz and Markland [9] (Fig. 5-31) indicate that the land-length to gap ratio should be less than 2, that is, $l < 2x_{f0}$, for this orifice to be considered sharp edged with a discharge coefficient of about 0.6. However, Feng [7] reports a discharge coefficient of about this value with much less favorable geometry ($D_N = 0.0208$ in., $l = 0.0251$ in., and $x_{f0} < 0.0032$ in.). Land lengths up to 0.010 in. are in common use despite warnings to seek a sharp-edged orifice. A sharp-edged orifice is preferred at the flapper because it is less influenced by temperature, has more predictable behavior, and is less susceptible to flow instabilities caused by flow reattaching to a longer land length. The bevel angle α is not critical if greater than 30° because it does not materially affect the discharge coefficient. It has been the author's experience to expect a discharge coefficient ratio C_{df}/C_{d0} of about 0.8, and this value is recommended for preliminary design purposes. When the discharge coefficients are measured, some minor adjustments in the orifice diameters may be necessary.

* This value of D_0 is based on selecting the equilibrium value of P_c (or P_1 and P_2 for the 2-jet valve) to be $P_{c0} = 0.5P_s$ (see page 119). This value gives maximum pressure sensitivity and has been used in deriving most relations in this section. Actually, satisfactory performance can be achieved for equilibrium values in the range $0.3P_s$ to $0.7P_s$, corresponding to values of D_0 in the range

$$1.62(C_{df}D_N x_{f0}/C_{d0})^{1/2} \qquad \text{to} \qquad 2.48(C_{df}D_N x_{f0}/C_{d0})^{1/2}.$$

The larger value of D_0 corresponding to $0.7P_s$ is often preferred to pass a more contaminated fluid but it also results in larger center flow and larger flapper flow forces.

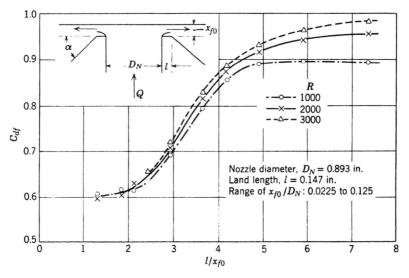

Figure 5-31   Variation of C_{df} with land/gap ratio. The Reynolds number is defined by $R = Q/\pi D_N \nu$ (from "Calculation of Potential Flow with Separation in a Right-angled Elbow with Unequal Branches," by A. Lichtarowicz and E. Markland, *Journal of Fluid Mechanics*, vol. 17, Pt. 4, Dec., 1963, Cambridge University Press, Cambridge, England).

It may be of interest to tabulate some compatible values of the quantities involved. If dirt-passing performance is favored, the maximum null clearance of

$$x_{f0} = \frac{D_N}{16} \tag{5-141}$$

is recommended. Substituting this value into (5-140) and assuming $C_{df}/C_{d0} = 0.8$, we have

$$D_0 = 0.44 D_N \tag{5-142}$$

Selecting three null clearance values, we tabulate the results in Table 5-1. The center flow was computed from (5-119) with $P_s = 1000$ psi and tabulated to demonstrate the large increase in this flow with increasing null

Table 5-1   Compatible Values of x_{f0}, D_N, and D_0

Null Clearance x_{f0} (in.)	Nozzle Diameter D_N (in.)	Upstream Orifice Diameter D_0 (in.)	Center Flow Q_c (gpm)
0.001	0.016	0.007	0.06
0.002	0.032	0.014	0.23
0.003	0.048	0.021	0.52

clearance. It should be emphasized that Table 5-1 contains recommended design values based on the use of (5-141). The general design relations are (5-139) and (5-140).

REFERENCES

[1] Feng, Tsun-Ying, "Characteristics of Unevenly Underlapped Four-Way Hydraulic Servo Valves," *ASME Paper* No. 56-*A*-140, November 1956.

[2] Blackburn, J. F., G. Reethof, and J. L. Shearer, *Fluid Power Control*. New York: Technology Press of M.I.T. and Wiley, 1960.

[3] Clark, R. N., "Compensation of Steady State Flow Forces in Spool Type Hydraulic Valves," *ASME Trans.*, November 1957, 1784–1788.

[4] Blackburn, J. F., "Contributions to Hydraulic Control 5 (Lateral Forces on Hydraulic Pistons)," *ASME Trans.*, 75, 1953, 1175–1180.

[5] Sweeney, D. C., "Preliminary Investigation of Hydraulic Lock," *Engineering*, 172, 1951, 513–516, 580–582.

[6] Manham, J., and D. C. Sweeney, "An Investigation of Hydraulic Lock," Institution of Mechanical Engineers, London, May 13, 1955.

[7] Feng, Tsun-Ying, "Static and Dynamic Control Characteristics of Flapper-Nozzle Valves," ASME Trans., *J. Basic Eng.*, September 1959, 275–284.

[8] Williams, L. J., "High Performance Single-stage Servovalve," Technical Bulletin 106, Moog Servocontrols, Inc., East Aurora, New York.

[9] Lichtarowicz, A., and E. Markland, "Calculation of Potential Flow with Separation in a Right-Angled Elbow with Unequal Branches," *J. Fluid Mech.*, Cambridge University Press, 17, Pt. 4, December 1963, 596–606.

6

Hydraulic Power Elements

Hydraulic actuation devices may be linear or rotary and are usually referred to as pistons or motors, respectively. These two actuation devices may be controlled by a pump or a valve giving four basic hydraulic power elements and two basic over-all systems: pump controlled and valve controlled. Such hydraulic power elements, simply a combination of the components discussed in the last two chapters, are the principal power device in all hydraulic systems.

The pump controlled system consists of a variable delivery pump supplying fluid to an actuation device. The fluid flow is controlled by the stroke of the pump to vary output speed, and the pressure generated matches the load. It is usually difficult to close couple the pump to the actuator and this causes large contained volumes and slow response. The valve controlled system consists of a servovalve controlling the flow from a hydraulic power supply to an actuation device. The hydraulic power supply is usually a constant pressure type (as opposed to constant flow) and there are two basic configurations. One consists of a constant delivery pump with a relief valve to regulate pressure, whereas the other is much more efficient because it uses a variable delivery pump with a stroke control to regulate pressure.

The relative advantages of the two types of system may be summarized as shown on page 133.

The features of each system tend to complement the other so that application requirements would dictate the choice to be made. Generally there is not a cost advantage to either because the need for a replenishing arrangement and a stroke servo for the pump controlled system offsets the costly servo valve and heat exchangers required for the valve controlled system. However, the faster response capability of valve controlled systems—both to valve and load inputs—makes this arrangement preferred in the majority of applications in spite of its lower theoretical maximum operating efficiency of 67%. In low power applications where the inefficiency is comparatively less important, use of valve controlled systems

Pump Controlled	*Valve Controlled*
1. Slow response because pressures must be built up, contained volumes are large, and the stroke servo has comparatively slow response.	1. Fast response to valve and load inputs because contained volumes are small and supply pressure is constant.
2. Much more efficient since both pressure and flow are closely matched to load requirements.	2. Less efficient because supply pressure is constant regardless of load, and leakages are greater.
3. Bulky power element size makes application difficult if pump is close coupled to actuator.	3. Small and light power element but a bulky hydraulic power supply is required.
4. Auxiliary pump and valving required to provide oil for replenishing and cooling.	4. Oil temperature builds up because of inefficiency which necessitates heat exchangers.
5. An electrohydraulic servo is generally required to stroke the pump which increases system cost and complexity.	5. Several valve-controlled systems can be fed from a single hydraulic power supply.

is nearly universal. Applications which require large horsepowers for control purposes usually do not require fast response so that a pump controlled system is preferred because of its superior theoretical maximum operating efficiency of 100%.

The purpose of this chapter is to define parameters and determine the transfer functions of several basic hydraulic power element combinations with various loads. These analyses yield a description of the dynamic performance, knowledge of which is absolutely essential in the rational design of hydraulic control systems.

6-1 VALVE CONTROLLED MOTOR

The hydraulic power element composed of a servovalve controlled rotary motor is probably the most widely used combination. A thorough analysis of this combination to obtain dynamic performance will give results fundamental to design and to other power element combinations. Because many of the equations are nonlinear, especially that describing the pressure-flow curves of the valve, a linearized analysis must be used. Machine computation could be used to solve the nonlinear equations, but the result would apply only to a given system because numerical values must be assumed for coefficients. Even if nonlinear solutions were easily obtained, all would not be well because design procedures, performance specifications, and the interpretation of test data are based on linear theory. Therefore, linearized analysis is essential, but care must be taken to investigate all operating points. If allowance is made for the range of values

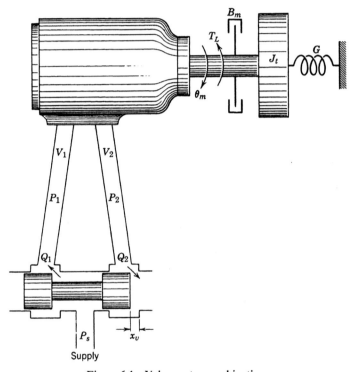

Figure 6-1   Valve-motor combination.

which a parameter can assume due to different operating points, then a linearized analysis is not unduly restrictive.

Consider the valve-motor combination shown schematically in Fig. 6-1 and let us first determine the servovalve equations. Assuming the servovalve orifices are matched and symmetrical, the pressures in the lines will rise above and below $P_s/2$ by equal amounts so that the pressure drops across the two valve orifices are identical. Hence the valve coefficients for both forward and return flows are the same. Therefore, assuming constant supply pressure, the linearized servovalve flow equations are

$$Q_1 = K_q x_v - 2K_c P_1 \qquad (6\text{-}1)$$

$$Q_2 = K_q x_v + 2K_c P_2 \qquad (6\text{-}2)$$

where Q_1, Q_2 = forward and return flows, in.^{3}/sec

$\quad\quad$ P_1, P_2 = forward and return pressures, psi

$\quad\quad\quad$ x_v = valve displacement from neutral, in.

$\quad\quad\quad$ K_q = valve flow gain, in.^{3}/sec/in.

$\quad\quad\quad$ K_c = valve flow-pressure coefficient, in.^{3}/sec/psi

The valve coefficients were discussed in Chapter 5 for various valve types. The flow-pressure coefficient for each port in Fig. 6-1 is twice that for the valve as a whole because K_c was defined with respect to P_L and a change in P_L is twice that which occurs across a port. Adding the servovalve flow equations gives

$$Q_L = K_q x_v - K_c P_L \tag{6-3}$$

where $\quad\quad$ $Q_L = \dfrac{Q_1 + Q_2}{2} = \text{load flow, in.}^3/\text{sec}$ $\quad\quad\quad$ (6-4)

$$P_L = P_1 - P_2 = \text{load pressure difference, psi}$$

This is the usual form of the linearized flow equation of the servovalve. The load flow represents the average of the flows in the lines and cannot be interpreted as being equal to the flow in each line unless external leakage is zero. As discussed in Section 4-3, the concept of load flow is useful since it reduces the number of flow variables required.

Let us now turn to the motor chambers and assume that the pressure in each chamber is everywhere the same and does not saturate or cavitate, fluid velocities in the chambers are small so that minor losses are negligible, line phenomena are absent, and temperature and density are constant. Application of the continuity equation (3-53) to each motor chamber yields

$$Q_1 - C_{im}(P_1 - P_2) - C_{em}P_1 = \frac{dV_1}{dt} + \frac{V_1}{\beta_e}\frac{dP_1}{dt} \tag{6-5}$$

$$C_{im}(P_1 - P_2) - C_{em}P_2 - Q_2 = \frac{dV_2}{dt} + \frac{V_2}{\beta_e}\frac{dP_2}{dt} \tag{6-6}$$

where C_{im} = internal or cross-port leakage (i.e., leakage from one motor line to the other) coefficient of motor, in.^{3}/sec/psi

$\quad\quad$ C_{em} = external leakage (i.e., leakage from each motor line to case drain) coefficient of motor, in.^{3}/sec/psi

$\quad\quad$ β_e = effective bulk modulus of system (includes oil, entrapped air, and mechanical compliance of chambers), psi

$\quad\quad$ V_1 = volume of forward chamber (includes servovalve, connecting line or manifold, motor passages, and volume swept out by pistons or vanes), in.^{3}

$\quad\quad$ V_2 = volume of return chamber (includes servovalve, connecting line or manifold, motor passages, and volume swept out by pistons or vanes), in.^{3}

$\quad\quad$ t = time, sec

The volume in each motor chamber is not constant but varies in a discontinuous sawtoothed fashion with shaft rotation, as illustrated in Fig. 6-2. This is a characteristic of all types of motors and may be deduced from an examination of a chamber. For example, consider the high pressure chamber of a piston type motor. The chamber volume changes at the porting instants because the piston completing its power stroke has accumulated a large head volume, and it is replaced by a piston starting a power stroke with a small head volume. Hence there is an instantaneous

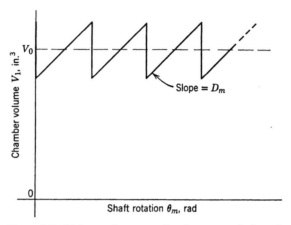

Figure 6-2   Volume of a motor chamber versus shaft angle.

change in volume equal to the area of one piston times its stroke. These discontinuities occur each time a piston undergoes transition from high to the low pressure chamber so that the frequency is equal to the number of pistons times the motor speed. An identical but opposite change in volume occurs simultaneously in the other motor chamber. The two chamber volumes may be expressed by

$$V_1 = V_0 + f_v(\theta_m) \tag{6-7}$$
$$V_2 = V_0 - f_v(\theta_m) \tag{6-8}$$

where V_0 = average contained volume of each motor chamber (includes servovalve, connecting line or manifold, and motor passages), in.^{3}

$f_v(\theta_m)$ = sawtoothed variation in each chamber volume, in.^{3}

θ_m = angular position of motor shaft, rad

D_m = volumetric displacement of motor, in.^{3}/rad

It should be apparent that the volume of a chamber must depend on shaft rotation. Because there is no direct connection between the chambers, a continuous flow through the motor is achieved only if the volume of one chamber steadily increases while the other chamber volume decreases with shaft position. This is physically possible because a piston cylinder fills with fluid, and the trapped fluid is then transported to the other chamber where it empties. A succession of such pistons forms a sort of bucket brigade which permits continuous flow. The rate at which the volume increases (or decreases) with shaft position is therefore the motor displacement. Hence the time derivatives of (6-7) and (6-8) are given by

$$\frac{dV_1}{dt} = \frac{df_v(\theta_m)}{dt} = D_m \frac{d\theta_m}{dt} = -\frac{dV_2}{dt} \tag{6-9}$$

and represent the theoretical flow to and from the motor, respectively. Equation 6-9 overlooks the discontinuities in Fig. 6-2 and assumes constant motor displacement. Although the latter is not true for many motors, the variation in displacement is usually quite small and may be neglected.

Adding (6-7) and (6-8), we obtain

$$V_t = V_1 + V_2 = 2V_0 \tag{6-10}$$

where V_t = total contained volume of both chambers, in.[3]

V_t is the total volume of oil under pressure P_1 and P_2 and, therefore, is often called the *total compressed volume*. The name *total contained volume* is also used and emphasizes the fact that it is the volume confined within the motor as opposed to the volume $f_v(\theta_m)$ which is used to produce shaft rotation. V_t is the total of the volumes of: (a) both motor lines of the servovalve with spool centered, (b) both connecting lines or in the manifold between servovalve and motor, and (c) both motor chambers and associated passageways.

It is desirable and possible to express the continuity equations in a more useful form. Substituting (6-7) through (6-9) into (6-5) and (6-6) and then adding them, we have

$$Q_L = D_m \frac{d\theta_m}{dt} + \left(C_{im} + \frac{C_{em}}{2}\right)(P_1 - P_2) + \frac{V_0}{2\beta_e}\frac{d(P_1 - P_2)}{dt}$$

$$+ \frac{f_v(\theta_m)}{2\beta_e}\left(\frac{dP_1}{dt} + \frac{dP_2}{dt}\right) \tag{6-11}$$

Let us now examine the last term on the right in (6-11). This term must be zero for a linearized analysis. It may be neglected by assuming that $|f_v(\theta_m)| \ll V_0$ or by differentiating (5-18) and (5-19), which are assumed to be applicable, to show that $dP_1/dt + dP_2/dt = 0$. Therefore, (6-11) can now be Laplace transformed to give

$$Q_L = D_m s\theta_m + C_{tm}P_L + \frac{V_t}{4\beta_e}sP_L \tag{6-12}$$

where $C_{tm} = C_{im} + \dfrac{C_{em}}{2}$ = total leakage coefficient of motor, in.^{3}/sec/psi

s = Laplace operator, sec^{-1}

Equation 6-12 is the basic form of the continuity equation for all hydraulic actuators. In fact, it is customary to write this equation directly, and (6-3) as well, without detailed considerations of the flows in each chamber. The flow proportional to pressure derivative is known as the *compressibility flow*. Thus the load flow Q_L is consumed by leakage, flow to displace the actuator, and flow stored due to compressibility.

The final basic relation is the torque balance equation for the motor which, Laplace transformed, is*

$$T_g = (P_1 - P_2)D_m = J_t s^2 \theta_m + B_m s \theta_m + G \theta_m + T_L \qquad (6\text{-}13)$$

where T_g = torque generated or developed by motor, in.-lb

J_t = total inertia of motor and load (referred to motor shaft), in.-lb-sec^2

B_m = viscous damping coefficient (would include motor internal damping) of load, in.-lb-sec

G = torsional spring gradient of load, in.-lb/rad

T_L = arbitrary load torque on motor, in.-lb

Static and coulomb friction loads would also be present to some degree but must be neglected in a linearized analysis [1].

Equations 6-3, 6-12, and 6-13 define the valve-motor combination. Let us now combine them to obtain physically interpretable results and an over-all transfer function. Combining and simplifying yields

$$\theta_m = \frac{\dfrac{K_q}{D_m} x_v - \dfrac{K_{ce}}{D_m{}^2}\left(1 + \dfrac{V_t}{4\beta_e K_{ce}}s\right)T_L}{\dfrac{V_t J_t}{4\beta_e D_m{}^2}s^3 + \left(\dfrac{K_{ce}J_t}{D_m{}^2} + \dfrac{B_m V_t}{4\beta_e D_m{}^2}\right)s^2 + \left(1 + \dfrac{B_m K_{ce}}{D_m{}^2} + \dfrac{G V_t}{4\beta_e D_m{}^2}\right)s + \dfrac{K_{ce}G}{D_m{}^2}}$$

$$(6\text{-}14)$$

where $K_{ce} = K_c + C_{tm} = K_c + C_{im} + C_{em}/2$ = total flow-pressure coefficient, in.^{3}/sec/psi.

---

* For many rotary motors there is an internal friction torque proportional to the sum of the line pressures and sign dependent on velocity (see Section 4-3). Although the pressure sum is nearly constant, that is, $P_1 + P_2 = P_s$, with servovalve control so that this friction torque is constant, it must be omitted from a linear analysis because of the nonlinear dependence on velocity. Qualitatively, this friction torque would act to increase motor damping at small inputs. Inclusion of this term would require a nonlinear analysis [1].

Equation 6-14 gives the motor response to both valve position and load torque inputs. The system characteristic equation is a cubic. It should be emphasized that J_t, B_m, and G are lumped coefficients representing the total acceleration, velocity, position dependent torques, respectively, on the motor so that (6-14) is quite general. Because usual loads are often simpler, some special cases of (6-14) are of interest. The quantity $D_m{}^2/K_{ce}$ is a damping coefficient due primarily to the valve and is usually very much greater than B_m. Hence the term $B_m K_{ce}/D_m{}^2$ can be neglected compared with unity in (6-14). Furthermore, let spring loads be absent, that is, $G = 0$, for the present, so that (6-14) reduces to

$$\theta_m = \frac{\dfrac{K_q}{D_m} x_v - \dfrac{K_{ce}}{D_m{}^2}\left(1 + \dfrac{V_t}{4\beta_e K_{ce}} s\right) T_L}{s\left(\dfrac{s^2}{\omega_h{}^2} + \dfrac{2\delta_h}{\omega_h} s + 1\right)} \tag{6-15}$$

where $\omega_h = \sqrt{\dfrac{2\beta_e D_m{}^2}{V_0 J_t}} = \sqrt{\dfrac{4\beta_e D_m{}^2}{V_t J_t}}$ $\qquad$ (6-16)

$$\delta_h = \frac{K_{ce}}{D_m}\sqrt{\frac{\beta_e J_t}{V_t}} + \frac{B_m}{4D_m}\sqrt{\frac{V_t}{\beta_e J_t}} \tag{6-17}$$

ω_h = hydraulic undamped natural frequency, rad/sec
δ_h = hydraulic damping ratio, dimensionless

Equation 6-15 gives the dynamic response of the motor with a predominantly inertia load. The first term in the numerator can be identified as the no load speed and the second term gives the speed drop due to load. The two transfer functions resulting from the two inputs are

$$\frac{\Delta\theta_m}{\Delta x_v} = \frac{\dfrac{K_q}{D_m}}{s\left(\dfrac{s^2}{\omega_h{}^2} + \dfrac{2\delta_h}{\omega_h} s + 1\right)} \tag{6-18}$$

due to valve position and

$$\frac{\Delta\theta_m}{\Delta T_L} = \frac{-\dfrac{K_{ce}}{D_m{}^2}\left(1 + \dfrac{V_t}{4\beta_e K_{ce}} s\right)}{s\left(\dfrac{s^2}{\omega_h{}^2} + \dfrac{2\delta_h}{\omega_h} s + 1\right)} \tag{6-19}$$

due to load torque.

Equation 6-18 is the usual form of the transfer function for a valve-motor combination. Because it is commonly used in the analysis and design of hydraulic servos, a detailed knowledge of its parameters is essential. The transfer function contains an integration which simply means that motor speed is proportional to valve position for slowly varying inputs. We also note that the gain constant K_q/D_m is the valve flow gain divided by motor displacement. Variations in the gain constant are due almost solely to changes in flow gain. The largest flow gain occurs at null and is reduced when load is applied. Using the usual design rule of $P_L = \frac{2}{3}P_s$ for maximum load conditions, the valve gain would be reduced to $\sqrt{P_s - P_L}/\sqrt{P_s} = \sqrt{\frac{1}{3}}$, or 57.7% of the no-load value. There is an additional gain reduction of 50% if the valve is underlapped and stroked away from null. Because a drop in gain does not adversely affect stability in usual servos, it is customary to use the null flow gain in (6-18) for design purposes. However, it is worth emphasizing that flow gain reductions of 50% or more may be encountered. If the servo loop is or tends to behave like a conditionally stable system, then such a reduction could cause stability or slow response problems.

The parameter ω_h is the natural frequency due to interaction of the inertia with the trapped oil springs and is very important because it establishes the over-all speed of response of the valve-motor combination. It should be apparent that the trapped volume of oil in each motor chamber causes a torsional spring with a rate of $\beta_e D_m^2/V_0$ in.-lb/rad. Because these two springs add, the total hydraulic spring rate is*

$$K_h = \frac{2\beta_e D_m^2}{V_0} = \frac{4\beta_e D_m^2}{V_t} \tag{6-20}$$

It is this hydraulic spring coupled with the total inertia which causes ω_h. We shall see in later chapters that ω_h sets an upper bound on the speed of response of hydraulic servos and systems. If fast responses are desired, then ω_h must be large.

Typically, the computed values of ω_h are some 40% or so higher than measured values. To account for this discrepancy, which at present has not been fully resolved, some have conveniently misinterpreted V_0 to mean V_t. With twice the correct volume, the frequency is reduced by $1/\sqrt{2}$

---

* The hydraulic spring rate K_h is not a spring in the usual sense of giving a *static* shaft deflection when a *static* load torque is applied. Such an interpretation would be correct only if the chamber volumes were perfectly sealed. However, this is never possible because a motor with no leakage paths *and* a perfect (i.e., with $K_c = 0$) valve would be required. K_h is simply a useful concept in computing hydraulic natural frequencies and in interpreting dynamic response (for an example, see Section 11-2). As explained in the discussion of Fig. 6-4, there is a region of frequencies where the motor response to applied loads acts as a spring with rate K_h. Hence, in some sense K_h can be thought of as a "dynamic" spring.

and agreement, although not justified, is achieved. The quantities D_m, V_0 or V_t, and J_t in (6-16) are well defined and do not vary significantly in value. However, as discussed in Section 2-5, it is difficult to determine values for the effective bulk modulus and it is reasonable to assume that this quantity is the source of the discrepancy. The author has found that $\beta_e = 100,000$ psi yields good results, but test data, of course, is preferred.

The remaining parameter in (6-18) is the damping ratio δ_h. Usually, the load damping coefficient B_m contributes very little so that (6-17) can be approximated

$$\delta_h = \frac{K_{ce}}{D_m}\left(\frac{\beta_e J_t}{V_t}\right)^{\frac{1}{2}} \tag{6-21}$$

Furthermore, the leakage coefficients of the motor are usually much smaller than K_c of the valve so that $K_{ce} \approx K_c$ is a good approximation for many cases. Intentional cross-port leakage paths, usually a needle valve or a coiled capillary tube inserted across motor lines, are often used to increase the damping ratio and/or make it more constant. The leakage coefficients for such paths add directly to K_{ce}. The drawbacks of such paths are the increased power loss and the decreased pressure sensitivity which will result in greater errors when constant and friction loads are applied to hydraulic servos.

All the quantities which determine δ_h are fairly constant except for K_c. This quantity varies widely with valve stroke, as may be deduced from an examination of (5-45) and (5-36) for the critical center valve. The flow-pressure coefficient is minimum at null, giving the lowest damping ratio. As the valve stroke is increased, giving increased motor speeds, the damping ratio increases rapidly to a value in excess of unity, and the roots of the quadratic factor become real with one increasing and the other decreasing from the value ω_h. This behavior is graphically seen in Fig. 6-3, which is a continuous plot of the amplitude response of a valve-motor combination with the valve oscillated about neutral at different amplitudes. Small input signals must be used to avoid pressure saturation when making sinusoidal tests. As described in Section 6-4, this phenomenon will cause the amplitude plot to be attenuated below the predicted linear response.

The null value for K_c is usually used in (6-18) because system stability is most critical when damping ratios are lowest. Measured damping ratios at null are always larger than computed values and this is primarily due to internal motor friction. This friction adds considerable damping at null because the motor velocity reversals about zero velocity cause reversals in the friction torque; however, this effect decreases at higher velocities because reversals no longer occur. Measured damping ratios for null operation are at least 0.1 or 0.2 and are often much higher.

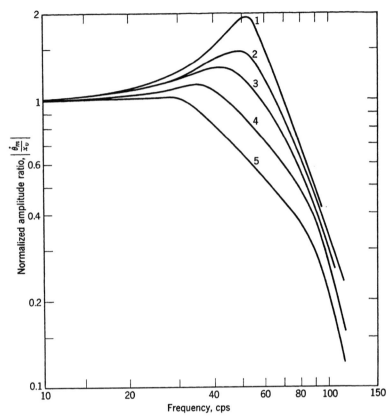

Figure 6-3   Measured frequency response of a valve-motor combination. System constants were $V_t = 17$ in.^{3}, $J_t = 0.043$ in lb-sec^2, $D_m = 0.486$ in.^{3}/rad, and $K_{ce} \approx 0.004$ in.^{3}/sec/psi. Assuming $\beta_e = 100{,}000$ psi, the computed natural frequency and damping ratio are 57.3 cps and 0.131. Measured values are 53 cps and 0.25. Valve input amplitudes (zero to peak) for curves 1 through 5 are 0.001 in., 0.0015 in., 0.002 in., 0.003 in., and 0.005 in., respectively.

It is perhaps worth repeating that small variations (perhaps 2 or 3 to 1) in the gain constant and natural frequency and very large variations (perhaps 20 or 30 to 1) in the damping ratio of the valve-motor dynamics can and do occur. The large damping ratios are obtained at higher motor speeds and loads. These parameter variations in the valve-motor transfer function cause the frequency response to float around, so to speak, as the operating point is changed. These variations must always be considered in the design and in the analysis and interpretation of test results.

The other transfer function (6-19) gives the dynamic response of the motor to load torques. A computation of the transient response due to load is usually omitted in system design because it is generally not critical or of particular interest, it cannot be altered appreciably by design, it does not affect system stability, and it is difficult to specify. However, occasions may arise which require that the dynamic compliance $\Delta\theta_m/\Delta T_L$ (or dynamic stiffness, $\Delta T_L/\Delta\theta_m$) of the valve-motor combination be computed. These two quantities are negative because a load torque causes a drop in motor speed. Also note from the definitions of δ_h and ω_h that the numerator break frequency in (6-19) can be written

$$\frac{4\beta_e K_{ce}}{V_t} = 2\delta_h\omega_h \qquad (6\text{-}22)$$

and will vary widely because of variations in δ_h. This break frequency is usually lower than ω_h at neutral because δ_h is small.

A sketch of the amplitude portion of the valve-motor dynamic stiffness, the inverse of (6-19), for a sinusoidal load torque is shown in Fig. 6-4. At low frequencies the combination acts like a viscous damper with damping coefficient $D_m{}^2/K_{ce}$ because the flows which result from the pressure difference due to load are impeded by the small leakage paths. In the frequency range between $2\delta_h\omega_h$ and ω_h there is not sufficient time to permit leakage flows so that the stiffness becomes that of the trapped oil spring K_h. At frequencies above ω_h the inertia load impedes angular motion, and stiffness increases with the second power of frequency. Variations in the quantities K_{ce}, δ_h, and ω_h, as previously discussed, will cause considerable shifting of the plot in Fig. 6-4 so that only a loose interpretation is justified.

The static stiffness of the valve-motor is often of interest. It should be apparent that the static position stiffness, that is, the stiffness with $\omega = 0$ in (6-19), is zero since a static load torque will cause a leakage flow and continuous shaft rotation. However, a static velocity compliance can be defined as the ratio of the drop in speed to applied load torque and is a useful quantity. This quantity is $K_{ce}/D_m{}^2$ rad/sec/in.-lb for the valve-motor and should be as low as possible for the combination to possess high stiffness. The fact that this quantity is low is one reason for the wide use of hydraulic control.

Let us now consider the case in which the motor drives a spring type load. Torsion spring loads are quite unusual on motors because motors are selected for their unlimited travel capacity and such a load, since the torque required increases with shaft position, would eventually stall the motor. Spring loads are much more common with piston type actuators because of their limited travel. However, it is conceivable that such a load could exist if the spring gradient G is low and/or the motor rotation is limited.

Returning to (6-14), we see that if the coefficients are numerically specified, then the cubic can be factored. However, certain approximations can be made which will yield literal factors of this cubic and will lead to a better understanding of spring loads. The load viscous damping coefficient

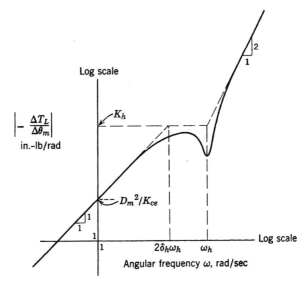

Figure 6-4   Dynamic stiffness of valve-motor versus frequency.

B_m is generally negligible so that (6-14) may be written

$$\theta_m = \frac{\dfrac{K_q}{D_m} x_v - \dfrac{K_{ce}}{D_m^{\,2}}\left(1 + \dfrac{s}{2\delta_h\omega_h}\right)T_L}{\dfrac{s^3}{\omega_h^{\,2}} + \dfrac{2\delta_h}{\omega_h} s^2 + \left(1 + \dfrac{G}{K_h}\right)s + \dfrac{K_{ce}G}{D_m^{\,2}}} \qquad (6\text{-}23)$$

The cubic characteristic equation may be approximately factored to obtain

$$\theta_m = \frac{\dfrac{K_q}{D_m} x_v - \dfrac{K_{ce}}{D_m^{\,2}}\left(1 + \dfrac{s}{2\delta_h\omega_h}\right)T_L}{\left(s + \dfrac{K_{ce}G}{D_m^{\,2}}\right)\left(\dfrac{s^2}{\omega_h^{\,2}} + \dfrac{2\delta_h}{\omega_h} s + 1\right)} \qquad (6\text{-}24)$$

provided that the following conditions, obtained by expanding the approximate factors and equating coefficients with the given cubic, are met:

$$\frac{G}{K_h} \ll 1 \qquad (6\text{-}25)$$

$$\left(\frac{K_{ce}\sqrt{GJ_t}}{D_m^{\,2}}\right)^2 \ll 1 \qquad (6\text{-}26)$$

These two conditions require that the hydraulic spring rate be much greater than the load spring rate and that the damping coefficient of the valve-motor $D_m^{\,2}/K_{ce}$ be much greater than the critical damping value of the load spring and inertia combination. These conditions are nearly always met but should be checked in each case.

Comparing (6-15) and (6-24), we note that the principal dynamic effect of a spring load is to replace the motor integration with a low frequency lag at $K_{ce}G/D_m^{\,2}$ rad/sec. If the load spring constant is decreased, this break frequency becomes lower and the lag factor becomes a better approximation of an integration. Full integrating performance is achieved when $G = 0$. An integration in a servo loop is very desirable because of the low steady-state errors and better control of stability.

It is often desirable to represent the three basic equations (6-3), (6-12), and (6-13) in block diagram fashion. If a nonlinear representation is desired, (6-3) is replaced by the appropriate equation for the pressure-flow curves of the servovalve. Two possible block diagrams can be made depending on whether motor position is computed from flow or torque considerations (Fig. 6-5). Both diagrams are perfectly valid, and the one best suited to the situation should be used. The diagram based on load flow solution (Fig. 6-5a) is usually preferred and is best suited to situations where dynamics play a small role because J_t can be made zero; however, a differentiator is required in an analog simulation. The diagram based on load pressure solution (Fig. 6-5b) is best suited to analog computer simulation where the dynamics are dominant, that is, large inertia J_t and leakage coefficient K_{ce}.

6-2 VALVE CONTROLLED PISTON

The combination of servovalve and piston is also a common hydraulic power element. Because the analysis of this combination closely parallels that given in Section 6-1, we shall merely outline the major relations. Therefore, a reading of Section 6-1 is a recommended prerequisite to the present discussion.

The valve-piston combination is shown schematically in Fig. 6-6. The servovalve orifices are assumed matched and symmetrical so that the valve

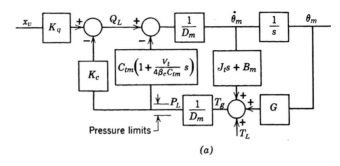

(a)

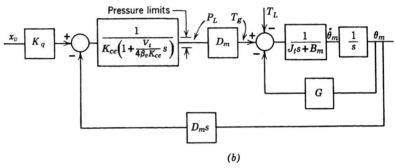

(b)

Figure 6-5   Block diagrams of a valve controlled motor: (a) diagram based on obtaining motor position from flow, (b) diagram based on obtaining motor position from torques.

flows are described by (6-3) and (6-4). Applying the continuity equation to each of the piston chambers yields

$$Q_1 - C_{ip}(P_1 - P_2) - C_{ep}P_1 = \frac{dV_1}{dt} + \frac{V_1}{\beta_e}\frac{dP_1}{dt} \tag{6-27}$$

$$C_{ip}(P_1 - P_2) - C_{ep}P_2 - Q_2 = \frac{dV_2}{dt} + \frac{V_2}{\beta_e}\frac{dP_2}{dt} \tag{6-28}$$

where V_1 = volume of forward chamber (includes valve, connecting line, and piston volume), in.^{3}

V_2 = volume of return chamber (includes valve, connecting line, and piston volume), in.^{3}

C_{ip} = internal or cross-port leakage coefficient of piston, in.^{3}/sec/psi

C_{ep} = external leakage coefficient of piston, in.^{3}/sec/psi

The other symbols were defined in Section 6-1.

The volumes of the piston chambers may be written

$$V_1 = V_{01} + A_p x_p \tag{6-29}$$

$$V_2 = V_{02} - A_p x_p \tag{6-30}$$

where A_p = area of piston, in.^{2}
x_p = displacement of piston, in.
V_{01} = initial volume of forward chamber, in.^{3}
V_{02} = initial volume of return chamber, in.^{3}
In contrast with the rotary motor, the initial chamber volumes are not

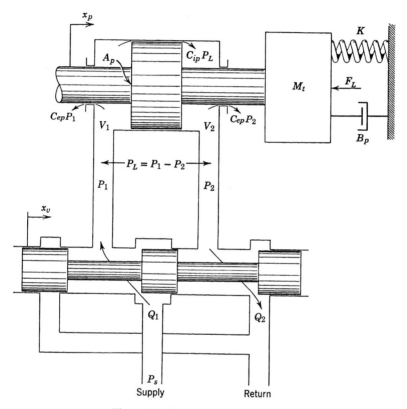

Figure 6-6   Valve-piston combination.

necessarily equal for the piston. However, it will be assumed that the piston is centered such that these volumes are equal, that is,

$$V_{01} = V_{02} = V_0 \qquad (6\text{-}31)$$

If this assumption is not made the analysis becomes much more involved, and the results are quite difficult to interpret. Experience has shown that stability problems are more acute when the piston is centered so that the assumption of equal chamber volumes should be conservative. It is apparent from Fig. 6-6 that the sum of the two volumes is constant and independent of piston position. Therefore,

$$V_t = V_1 + V_2 = V_{01} + V_{02} = 2V_0 \qquad (6\text{-}32)$$

where V_t = total volume of fluid under compression in both chambers, in.^{3}

In a manner similar to the development of (6-12), the volume and continuity expressions can be combined to yield

$$Q_L = A_p s x_p + C_{tp} P_L + \frac{V_t}{4\beta_e} s P_L \qquad (6\text{-}33)$$

where $C_{tp} = C_{ip} + (C_{ep}/2)$ = total leakage coefficient of piston, in.^{3}/sec/psi,

which is the usual form of the continuity equation.

The final equation arises by applying Newton's second law to the forces on the piston. The resulting force equation, Laplace transformed, is

$$F_g = A_p P_L = M_t s^2 x_p + B_p s x_p + K x_p + F_L \qquad (6\text{-}34)$$

where F_g = force generated or developed by piston, lb
M_t = total mass of piston and load referred to piston, lb-sec^2/in.
B_p = viscous damping coefficient of piston and load, in.-lb-sec
K = load spring gradient, lb/in.
F_L = arbitrary load force on piston, lb

Equations 6-3, 6-33, and 6-34 are the three basic equations and may be solved simultaneously to obtain

$$x_p = \frac{\dfrac{K_q}{A_p} x_v - \dfrac{K_{ce}}{A_p^2}\left(1 + \dfrac{V_t}{4\beta_e K_{ce}} s\right) F_L}{\dfrac{V_t M_t}{4\beta_e A_p^2} s^3 + \left(\dfrac{K_{ce} M_t}{A_p^2} + \dfrac{B_p V_t}{4\beta_e A_p^2}\right) s^2 + \left(1 + \dfrac{B_p K_{ce}}{A_p^2} + \dfrac{K V_t}{4\beta_e A_p^2}\right) s + \dfrac{K_{ce} K}{A_p^2}}$$

$$(6\text{-}35)$$

where $K_{ce} = K_c + C_{ip} + C_{ep}/2$ = total flow-pressure coefficient, in.^{3}/sec/psi

Equation 6-35 gives the piston response to both valve position and load force inputs and is analogous, term by term, to that of (6-14) for the valve-motor combination. This equation is quite general and is applicable to any four-way valve and piston combination whether it be a power output device or a two-stage servovalve. For a two-stage servovalve, such as a flapper valve driving a spool valve, the spool valve acts as a piston being controlled by the flapper valve.

If the valve-piston combination is a power output device, very often spring loads are absent. Also, $B_p K_{ce}/A_p^2$ is usually much smaller than unity. Under these conditions (6-35) reduces to

$$x_p = \frac{\dfrac{K_q}{A_p} x_v - \dfrac{K_{ce}}{A_p^2}\left(1 + \dfrac{V_t}{4\beta_e K_{ce}} s\right) F_L}{s\left(\dfrac{s^2}{\omega_h^2} + \dfrac{2\delta_h}{\omega_h} s + 1\right)} \qquad (6\text{-}36)$$

where $\omega_h = \sqrt{\dfrac{4\beta_e A_p^2}{V_t M_t}}$ = hydraulic natural frequency, rad/sec   (6-37)

$$\delta_h = \frac{K_{ce}}{A_p}\sqrt{\frac{\beta_e M_t}{V_t}} + \frac{B_p}{4A_p}\sqrt{\frac{V_t}{\beta_e M_t}} = \text{damping ratio, dimensionless}$$

$$(6\text{-}38)$$

The two trapped oil springs, each with a spring gradient of $\beta_e A_p^2/V_0$ lb/in., add and interact with the mass M_t to give the hydraulic natural frequency. Although it is difficult to show mathematically, experience indicates that these springs may be added even when the initial chamber volumes are not the same. In this event the hydraulic natural frequency would be given by

$$\omega_h = \left[\frac{\beta_e A_p^2}{M_t}\left(\frac{1}{V_{01}} + \frac{1}{V_{02}}\right)\right]^{1/2} \qquad (6\text{-}39)$$

This equation indicates that the lowest natural frequency occurs when $V_{01} = V_{02}$, that is, when the piston is centered. As the piston moves to either end of its stroke, the spring rate of the smaller volume dominates and the natural frequency is increased.

Usually force loads are omitted in system designs so that only the transfer function from input is of interest. This transfer function is

$$\frac{x_p}{x_v} = \frac{\dfrac{K_q}{A_p}}{s\left(\dfrac{s^2}{\omega_h^2} + \dfrac{2\delta_h}{\omega_h} s + 1\right)} \qquad (6\text{-}40)$$

As discussed in Section 6-1, variations in the gain constant K_q/A_p, the hydraulic natural frequency ω_h, and especially in the damping ratio δ_h occur and cause considerable shifting in the frequency response with different operating points. Hence the frequency response is quite elastic and this must always be kept in mind during design or in viewing test data.

If $B_p K_{ce}/A_p{}^2 \ll 1$ and K is sufficiently small, that is, conditions similar to (6-25) and (6-26) are satisfied, the cubic characteristic will have approximate factors so that (6-35) becomes

$$x_p = \frac{\dfrac{K_q}{A_p} x_v - \dfrac{K_{ce}}{A_p{}^2}\left(1 + \dfrac{V_t}{4\beta_e K_{ce}} s\right) F_L}{\left(s + \dfrac{K_{ce}K}{A_p{}^2}\right)\left(\dfrac{s^2}{\omega_h{}^2} + \dfrac{2\delta_h}{\omega_h} s + 1\right)} \qquad (6\text{-}41)$$

This equation is useful if spring loads, often the case for pistons, are present, and it shows that such loads cause a low frequency lag at $K_{ce}K/A_p{}^2$ rad/sec which becomes a true integration only when $K = 0$.

The discussion given in Section 6-1 on the response due to load inputs applies as well to the valve-piston combination.

6-3 THREE-WAY VALVE CONTROLLED PISTON

Three-way valve-piston combinations find their greatest applications as power elements in hydromechanical position servos such as those used in aircraft controls and tracer controlled machine tools. This combination is often used in aircraft servos which perform sensing and/or computation functions.

In this section we shall determine the transfer function for this element, shown schematically in Fig. 5-10. Assuming constant supply pressure, the valve flow equation was derived in Section 5-5 as

$$Q_L = K_q x_v - K_c P_c \qquad (6\text{-}42)$$

where P_c = control pressure on head area of piston, psi.

Application of the continuity equation to the control volume in Fig. 5-10 yields

$$Q_L + C_i(P_s - P_c) = \frac{dV_h}{dt} + \frac{V_h}{\beta_e} \frac{dP_c}{dt} \qquad (6\text{-}43)$$

where C_i = cross-port leakage coefficient, in.^{3}/sec/psi
V_h = head chamber volume, in.^{3}
The other symbols have been defined in preceding sections.

The head chamber volume is given by

$$V_h = V_0 + A_h x_p \qquad (6\text{-}44)$$

where V_0 = initial head chamber volume, in.^{3}
A_h = head side area, in.^{2}
A_r = rod side area, in.^{2}

If we assume small piston motion such that $|A_h x_p| \ll V_0$, (6-43) and (6-44) can be combined and Laplace transformed to give

$$Q_L + C_i P_s = A_h s x_p + C_i P_c + \frac{V_0}{\beta_e} s P_c \qquad (6\text{-}45)$$

Assuming mass and spring load, the Laplace transformed force equation for the piston is

$$P_c A_h - P_s A_r = M_t s^2 x_p + K x_p \qquad (6\text{-}46)$$

where M_t = total mass of piston and load, lb-sec^2/in.
      K = load spring gradient, lb/in.
Other load terms can be included, as illustrated in preceding sections.

Equations 6-42, 6-45, and 6-46 describe the system. If (6-45) and (6-46) are linearized for small changes in the variables, then P_s can be made zero in these equations because it is constant. Therefore these three equations can be combined to obtain

$$\frac{\Delta x_p}{\Delta x_v} = \frac{\dfrac{K_q}{A_h}}{\dfrac{s^3}{\omega_h{}^2} + \dfrac{2\delta_h}{\omega_h} s^2 + \left(1 + \dfrac{K}{K_h}\right)s + \dfrac{(K_c + C_i)K}{A_h{}^2}} \qquad (6\text{-}47)$$

where $\omega_h = \sqrt{\dfrac{K_h}{M_t}} = \sqrt{\dfrac{\beta_e A_h{}^2}{V_0 M_t}}$ = hydraulic undamped natural frequency, rad/sec

$$\delta_h = \frac{(K_c + C_i)}{2A_h}\sqrt{\frac{\beta_e M_t}{V_0}} = \text{damping ratio, dimensionless}$$

$$K_h = \frac{\beta_e A_h{}^2}{V_0} = \text{hydraulic spring rate, lb/in.}$$

If the load spring rate is sufficiently small so that conditions similar to (6-25) and (6-26) are met, (6-47) can be factored to yield

$$\frac{\Delta x_p}{\Delta x_v} = \frac{\dfrac{K_q}{A_h}}{\left[s + \dfrac{(K_c + C_i)K}{A_h{}^2}\right]\left(\dfrac{s^2}{\omega_h{}^2} + \dfrac{2\delta_h}{\omega_h} s + 1\right)} \qquad (6\text{-}48)$$

Thus, a load spring causes a lag at $(K_c + C_i)K/A_h{}^2$ rad/sec as discussed in

Section 6-1. If the load spring rate is zero, (6-48) reduces to

$$\frac{\Delta x_p}{\Delta x_v} = \frac{\dfrac{K_q}{A_h}}{s\left(\dfrac{s^2}{\omega_h^2} + \dfrac{2\delta_h}{\omega_h}s + 1\right)} \qquad (6\text{-}49)$$

Comparing this result to that obtained for the four-way valve controlled piston (6-40), we note that the undamped natural frequency and damping ratio are both lowered by a factor of $1/\sqrt{2}$. This is due to the fact that only one volume contributes an oil spring because only one line is controlled. If both lines are controlled, as with a four-way valve, two trapped oil springs would be obtained. Therefore, if all other quantities are equal, the dynamic response of a four-way valve controlled piston is superior.

6-4 PUMP CONTROLLED MOTOR

Pump controlled motors are the preferred power element in applications which require considerable horsepower for control purposes because of their high maximum operating efficiency which can approach 90% in practice. However, the comparatively slow response of these elements limit their use in high performance systems.

The basic pump controlled motor, often called a *hydrostatic* or *hydraulic transmission*, is shown schematically in Fig. 6-7. A variable displacement pump, driven by a constant speed power source and capable of reversing the direction of flow, is directly connected to a fixed displacement hydraulic motor. Hence, the motor speed and direction of rotation may be controlled by varying the pump stroke.

A *replenishing supply* is required to replace leakage losses from each line and to establish a minimum pressure in each line. This auxiliary supply is a constant pressure type with low capacity because only leakage flows are supplied. The replenishing pressure P_r is set low to keep power losses at a minimum and to keep line pressure low when loads are applied to the motor. The replenishing supply prevents cavitation and air entrainment because it pressurizes each line and helps dissipate heat by providing cooler fluid to replace the leakage.

Safety relief valves are used in the lines to protect the system from damage due to pressure peaks. These valves establish an upper limit to the line pressures and are set to operate above normal operating pressures. These valves must respond rapidly and have a large capacity because they must pass the maximum pump flow in an extreme overload. These valves should be connected across the lines so that overload flow is dumped to the other line to help prevent cavitation.

The stroke control on the pump may have linear or angular motion, usually the latter, and requires a small electrohydraulic servo for positioning. The stroke control servo consists of an electrohydraulic servovalve, a piston for actuation, appropriate electrical feedback devices, and electronic error amplifiers. The capacity of the replenishing supply may be increased to supply the stroke control servo.

During normal operation the pressure in one line will be at replenishing pressure, and the other line pressure will modulate to match the load. The

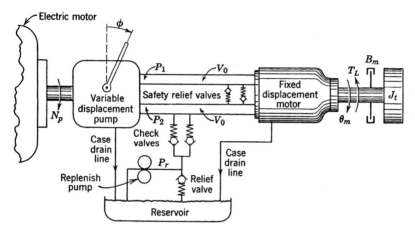

Figure 6-7   Schematic of a pump controlled motor.

two lines will switch functions if the load dictates a pressure reversal. It is possible for both line pressures to vary simultaneously if transients are rapid and load reversals occur. However, we shall assume that only one line pressure varies at a time because this case is more common.*

Referring to Fig. 6-7, consider P_1 to be the high pressure line, that is, $P_1 > P_r$ and $P_2 = P_r$, as a sign convention to write equations. Since P_2 will not occur in the analysis to follow, a load reversal would require a negative value for P_1 from a mathematical viewpoint. It should be apparent, however, that this condition is physically achieved by the lines reversing their functions. Let us assume all pressures are uniform, no line losses or dynamics, negligible minor losses, no pressure saturation, constant fluid density and temperature, constant chamber volumes in pump and motor, constant replenishing pressure, constant pump speed, zero

---

* If both line pressures vary, a similar analysis can be made and will show that the hydraulic natural frequency is larger by a factor of $\sqrt{2}$ because two trapped oil springs are then present [2].

case drain pressure, and laminar leakage flows. Although the assumptions are many, none of them is unduly restrictive. Under these conditions, the continuity equation for the forward chamber can be written as

$$D_p N_p - C_{ip}(P_1 - P_r) - C_{ep}P_1 - C_{im}(P_1 - P_r)$$
$$- C_{em}P_1 - D_m \frac{d\theta_m}{dt} = \frac{V_0}{\beta_e} \frac{dP_1}{dt} \quad (6\text{-}50)$$

where D_p = volumetric displacement of pump, in.³/rad
   D_m = volumetric displacement of motor, in.³/rad
   N_p = pump speed (assumed constant), rad/sec
   θ_m = angular position of motor shaft, rad
   V_0 = average volume of forward chamber (includes one side of pump and motor, connecting line, and incidental volumes associated with the check and safety relief valves), in.³
   P_1 = forward chamber pressure, psi
   P_2 = return chamber pressure, psi
   P_r = replenishing pressure, psi
   β_e = effective bulk modulus of system (includes oil, entrapped air, and mechanical compliance of chambers), psi
   C_{ip} = internal or cross-port leakage coefficient of pump, in.³/sec/psi
   C_{im} = internal or cross-port leakage coefficient of motor, in.³/sec/psi
   C_{ep} = external leakage (i.e., leakage from a line to case drain) coefficient of pump, in.³/sec/psi
   C_{em} = external leakage (i.e., leakage from a line to case drain) coefficient of motor, in.³/sec/psi
The displacement of the pump is described by

$$D_p = k_p \phi \quad (6\text{-}51)$$

where ϕ = pump stroke angle, degrees
   k_p = displacement gradient of pump control, in.³/rad/deg
These two equations can be combined and Laplace transformed to yield

$$k_p N_p \phi + C_{it}P_r = D_m s\theta_m + C_t P_1 + \frac{V_0}{\beta_e} s P_1 \quad (6\text{-}52)$$

where $C_{it} = C_{ip} + C_{im}$ = total internal leakage coefficient, in.³/sec/psi
   $C_{et} = C_{ep} + C_{em}$ = total external leakage coefficient, in.³/sec/psi
   $C_t = C_{it} + C_{et}$ = total leakage coefficient, in.³/sec/psi
   s = Laplace operator, sec⁻¹

Assuming lumped constants to describe the load, Newton's second law is used to obtain the torque balance equation. Therefore,

$$T_g = (P_1 - P_r)D_m = J_t \frac{d^2\theta_m}{dt^2} + B_m \frac{d\theta_m}{dt} + \frac{\dot{\theta}_m}{|\dot{\theta}_m|}(P_1 + P_r)C_f D_m + T_L$$

$$(6\text{-}53)$$

where T_g = torque generated or developed by motor, in.-lb

J_t = total inertia of motor and load (referred to motor shaft), in.-lb-sec^2

B_m = total viscous damping coefficient (includes internal motor damping), in.-lb-sec

C_f = internal motor friction coefficient, dimensionless

T_L = arbitrary load torque on motor, in.-lb

This equation includes the internal friction torque (see Section 4-3) due to friction between motor elements in relative motion and proportional to the sum of the two line pressures. However, this term is a nonlinearity and must be made zero to proceed with a linear analysis (see Section 10-8 and Reference 1). This assumption is hard to justify, except for mathematical expediency, even though C_f is usually small (less than 0.1). Qualitatively, the friction term would act to increase the motor damping at high pressures and/or small amplitude inputs. Letting $C_f = 0$ and Laplace transforming gives

$$P_1 D_m = J_t s^2 \theta_m + B_m s \theta_m + P_r D_m + T_L \qquad (6\text{-}54)$$

Equations 6-52 and 6-54 describe the system. Because P_r is assumed constant, terms containing this quantity may be omitted from these two equations for a linearized analysis. Therefore these equations may then be combined to yield

$$\Delta\theta_m = \frac{\dfrac{k_p N_p}{D_m}\Delta\phi - \dfrac{C_t}{D_m{}^2}\left(1 + \dfrac{V_0}{\beta_e C_t}s\right)\Delta T_L}{\dfrac{V_0 J_t}{\beta_e D_m{}^2}s^3 + \left(\dfrac{C_t J_t}{D_m{}^2} + \dfrac{B_m V_0}{\beta_e D_m{}^2}\right)s^2 + \left(1 + \dfrac{B_m C_t}{D_m{}^2}\right)s} \qquad (6\text{-}55)$$

where the prefix Δ denotes a small change in the variable. For usual cases $B_m C_t / D_m{}^2 \ll 1$ and (6-55) can be simplified to give

$$\Delta\theta_m = \frac{\dfrac{k_p N_p}{D_m}\Delta\phi - \dfrac{C_t}{D_m{}^2}\left(1 + \dfrac{V_0}{\beta_e C_t}s\right)\Delta T_L}{s\left(\dfrac{s^2}{\omega_h{}^2} + \dfrac{2\delta_h}{\omega_h}s + 1\right)} \qquad (6\text{-}56)$$

where $\omega_h = \sqrt{\dfrac{\beta_e D_m{}^2}{V_0 J_t}}$ = hydraulic undamped natural frequency, rad/sec

$$(6\text{-}57)$$

$$\delta_h = \frac{C_t}{2D_m}\sqrt{\frac{\beta_e J_t}{V_0}} + \frac{B_m}{2D_m}\sqrt{\frac{V_0}{\beta_e J_t}} = \text{damping ratio, dimensionless} \quad (6\text{-}58)$$

Equations 6-15 and 6-56 should be examined to compare the dynamic performance of valve versus pump controlled motors. Because these two equations are identical in form, there is no basic difference in the response of these two combinations. However, there is considerable difference in the value and variation of corresponding parameters. The undamped natural frequency of the pump controlled motor is lower for two basic reasons. Because only one line is controlled the trapped oil spring rate is one half that of a valve controlled motor and reduces the natural frequency by a factor of $1/\sqrt{2}$. Also, the chamber volume V_0 is larger with a pump because it is usually difficult to close couple pump to motor and because the pump volume is larger than that of a comparable valve. The damping ratio is much more constant for the pump controlled motor because leakage coefficients are quite constant compared with valve flow-pressure coefficients. The damping ratios are also lower so that pump controlled systems are almost always underdamped. Intentional cross-port leakage paths, at the sacrifice of power, and/or internal pressure feedback loops are often necessary to achieve satisfactory damping ratios. The gain constants $k_p N_p/D_m$ and $C_t/D_m{}^2$ are also much more constant than corresponding quantities for the valve controlled motor. In summary, the dynamic response of a pump controlled motor is more predictable and less subject to variation with operating point compared with a valve controlled motor. However, the over-all response is low because of the lower natural frequency and the slow response of the stroke control servo.

Two basic transfer functions can be obtained from (6-56). The transfer function from stroke input is

$$\frac{\Delta\theta_m}{\Delta\phi} = \frac{\dfrac{k_p N_p}{D_m}}{s\left(\dfrac{s^2}{\omega_h{}^2} + \dfrac{2\delta_h}{\omega_h}s + 1\right)} \tag{6-59}$$

and the system dynamic compliance is given by

$$\frac{\Delta\theta_m}{\Delta T_L} = \frac{-\dfrac{C_t}{D_m{}^2}\left(1 + \dfrac{V_0}{\beta_e C_t}s\right)}{s\left(\dfrac{s^2}{\omega_h{}^2} + \dfrac{2\delta}{\omega_h}s + 1\right)} \tag{6-60}$$

Equation 6-59 is the transfer function commonly used for system design. Therefore, any variation in its parameters is of interest. Both the gain constant and damping ratio are quite constant. Variation in the natural frequency similar to that of the valve controlled motor can be expected. Computed and measured performance should agree fairly well for a pump controlled motor because it is a reasonably linear element. However, in making tests, care must be taken to use an input small enough so that pressure saturation does not occur. For sinusoidal inputs the pressure required for acceleration increases with the square of frequency and may reach the values set by the safety relief valves. Pressure saturation causes the amplitude portion of the frequency response to be highly attenuated below the expected linear response. The appearance of such plots often suggests that additional dynamics are present. However, this is the effect of pressure saturation and the expected response will probably be obtained with a smaller input amplitude.

A discussion of the load response, given by (6-60), similar to that given in Section 6-1 can be made. The pump controlled motor is not as dynamically stiff as the valve system because of the lower natural frequency and damping ratio and larger chamber volume.

The analysis presented in this section is also applicable to a pump controlled piston. For such a case the symbols θ_m and D_m are replaced by x_p and A_p which are the piston displacement and area, respectively.

6-5 VALVE CONTROLLED MOTOR WITH LOAD HAVING MANY DEGREES OF FREEDOM

Thus far we have assumed the load on the power element is such that it can be represented by single lumped parameters. However, in many cases the drive system has many degrees of freedom and cannot be so simply described. Many drive systems have structural resonances which are lower than the hydraulic natural frequency. Such structural resonances then dominate the frequency response of the valve-motor combination, and this fact must be considered in system design because servo bandwidths are limited by the lowest natural frequency that occurs in the loop. Drive systems associated with large antennas, gimbals, shake tables, and machine tools are typical examples in which high system bandwidths are required and structural resonances often dominate and limit over-all system performance.

In this section we outline the case illustrated in Fig. 6-8, in which the valve-motor power element is coupled to a drive system and a load which has many degrees of freedom. The transfer function for the power element will be formulated and some examples will be given to illustrate the significance of the drive dynamics.

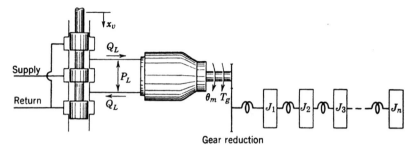

Figure 6-8 Valve controlled motor coupled to a load having n degrees of freedom.

The linearized pressure-flow curves of the servovalve are described by (6-3). Continuity of flows in the motor chambers results in (6-12). These two equations are applicable here and can be combined to give

$$K_q x_v = D_m s\theta_m + K_{ce}P_L + \frac{V_t}{4\beta_e} sP_L \qquad (6\text{-}61)$$

where all symbols are defined in Section 6-1. The torque generated by the motor is

$$T_g = D_m^{'}P_L \qquad (6\text{-}62)$$

The final equation is that which describes the load dynamics as seen by the generated torque of the motor. Let us imagine that the motor acts as a torsional exciter and applies a sinusoidal torque to the load and that the angular position of the shaft is measured. The ratio of shaft position to applied torque is then the transfer function which describes the load dynamics. In general, such a transfer function will have the following form:

$$\frac{\theta_m}{T_g} = \frac{\left(\dfrac{s^2}{\omega_a^2} + \dfrac{2\delta_a}{\omega_a}s + 1\right)\left(\dfrac{s^2}{\omega_b^2} + \dfrac{2\delta_b}{\omega_b}s + 1\right)\cdots}{J_t s^2 \left(\dfrac{s^2}{\omega_1^2} + \dfrac{2\delta_1}{\omega_1}s + 1\right)\left(\dfrac{s^2}{\omega_2^2} + \dfrac{2\delta_2}{\omega_2}s + 1\right)\left(\dfrac{s^2}{\omega_3^2} + \dfrac{2\delta_3}{\omega_3}s + 1\right)\cdots} \qquad (6\text{-}63)$$

where θ_m = angular position of motor shaft, rad
T_g = torque generated by motor, in.-lb
J_t = total inertia of motor and load (referred to motor), in.-lb-sec^2
$\omega_a, \omega_b, \ldots$ = undamped natural frequencies of numerator quadratics, rad/sec
$\omega_1, \omega_2, \ldots$ = undamped natural frequencies of denominator quadratics, rad/sec
$\delta_a, \delta_b, \ldots$ = damping ratios of numerator quadratics, dimensionless
$\delta_1, \delta_2, \ldots$ = damping ratios of denominator quadratics, dimensionless

This equation assumes that the dynamic compliance of structural elements within the motor can be neglected and that the motor housing is rigidly attached to a solid base so that that the dynamic compliance of the motor mounting can be neglected.

Because mechanical structures are continuous systems, they are described by partial differential equations of formidable complexity, and it is not presently possible to predict all quantities involved in (6-63).* Hence, the dynamics must be determined by measurement and this fact detracts from the usefulness of analysis. However, structural characteristics are often similar for a given type of application so that the drive dynamics need not be measured for each particular application.

A number of observations concerning (6-63) can be made which will aid in understanding the dynamics involved.

1. ω_1, ω_2, ... are the undamped natural frequencies of modes of vibration; ω_1 is the frequency of the fundamental mode; ω_2 is the frequency of the second mode, etc.; ω_a, ω_b, ... are the undamped natural frequencies of nodes of vibration. Excitation at forcing frequencies corresponding to denominator and numerator natural frequencies results in points of maximal and minimal motion at the place where the response is measured.

2. The free s^2 term in the denominator indicates that at low frequencies, where all the quadratic factors approximate unity, the applied torque is consumed in accelerating the total inertia J_t of the system reflected to the motor.

3. Because a structure is a passive system it must be stable. Therefore, all poles of the drive dynamics appear in the left half of the s-plane; that is, all roots have negative real parts.

4. The number of denominator quadratic factors corresponds to the number of degrees of freedom. Physical realizability requires that the degree of the numerator be less than the denominator. ω_a is generally the lowest of all the frequencies.

5. If deflection is measured at the same point and in the same direction as the applied torque, the frequencies are interspersed so that

$$\omega_a < \omega_1 < \omega_b < \omega_2 < \omega_c < \omega_3 \cdots$$

In this event the phase lag contributed by the quadratic factors never exceeds 180°. For this case the transfer is called the *driving point transfer function*.

---

* In recent years rather sophisticated digital computer programs have been developed which compute mode shapes and natural frequencies of complex structures [3, 4]. Continued work in this direction may ultimately lead to the prediction of complete dynamic performance.

6. If deflection is measured at a point other than where the torque is applied, the transfer function can become nonminimum phase. Thus, any amount of phase lag is possible from nondriving point transfers.

It is convenient symbolically to combine (6-62) and (6-63). Therefore

$$\theta_m = \frac{G_L(s)}{s^2} \frac{D_m}{J_t} P_L \qquad (6\text{-}64)$$

where $G_L(s)$ is the ratio of the quadratic factors in (6-63).

Equations 6-61 and 6-64 describe the system and can be combined by eliminating P_L to yield

$$\frac{\theta_m}{x_v} = \frac{\dfrac{K_q}{D_m} G_L(s)}{s\left[\dfrac{s^2}{\omega_h{}^2} + \dfrac{2\delta_h}{\omega_h} s + G_L(s)\right]} \qquad (6\text{-}65)$$

where ω_h and δ_h are defined by (6-16) and (6-21) and can be identified as the hydraulic natural frequency and damping ratio, respectively, for the power element with a simple lumped parameter load.

Several comments can be made from an examination of (6-65):

1. If modal and nodal frequencies are very high or absent, $G_L(s) = 1$ and (6-65) reduces to

$$\frac{s\theta_m}{x_v} = \frac{\dfrac{K_q}{D_m}}{\dfrac{s^2}{\omega_h{}^2} + \dfrac{2\delta_h}{\omega_h} s + 1}$$

which is also the result if a simple lumped inertia load is assumed.

2. If the structural resonances are greater than ω_h, then $G_L(s) \approx 1$ at the frequency $\omega = \omega_h$ so that the hydraulic resonance is the lowest and therefore the dominant natural frequency. Equation 6-65 then approximates to

$$\frac{s\theta_m}{x_v} \approx \frac{\dfrac{K_q}{D_m} G_L(s)}{\dfrac{s^2}{\omega_h{}^2} + \dfrac{2\delta_h}{\omega_h} s + 1}$$

in which the quadratic factors of $G_L(s)$ appear also as factors of $s\theta_m/x_v$.

3. If the hydraulic resonance is much greater than structural resonances, $G_L(s)$ becomes dominant in the denominator of (6-65) and this equation

approaches

$$\frac{s\theta_m}{x_v} \approx \frac{K_q}{D_m}$$

Therefore structural resonances are minimized by large values of ω_h.

4. If ω_h is in between structural natural frequencies, then the structural modes lower in frequency than ω_h tend to be suppressed, removed, or minimized in the over-all response $s\theta_m/x_v$. Thus a higher ω_h can "pull up" lower structural frequencies.

5. Zeros of $G_L(s)$ are also zeros of $s\theta_m/x_v$.

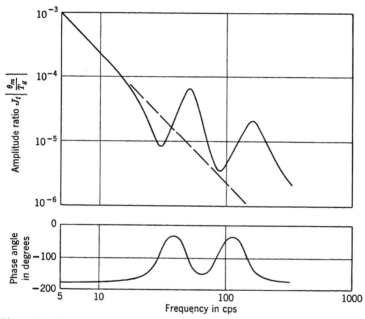

Figure 6-9   Frequency response of a drive system having many degrees of freedom.

Therefore we can conclude that the hydraulic resonance is roughly the dominant resonance regardless of whether it is lower than the structural resonances, at least as far as the transfer to motor shaft position is concerned. This conclusion is not valid if motion is sensed at some other position in the drive system.

As a numerical example, let us assume that $G_L(s)$ consists of two numerator quadratic factors at frequencies $f_a = 30$ cps and $f_b = 90$ cps and two denominator quadratics at frequencies $f_1 = 50$ cps and $f_2 = 150$ cps. Let the damping ratios be 0.1 for these quadratics. With these numerical values the dynamic compliance of the drive system becomes that shown in Fig. 6-9. Now let us choose hydraulic natural frequencies of 20, 50, and 120 cps which bracket and coincide with the lowest structural mode at 50 cps. Choosing $\delta_h = 0.2$ we can evaluate (6-65) and plot the three resulting frequency response functions shown in Fig. 6-10. It is clear from these plots that the structural resonances become less important as ω_h is increased. When ω_h is lowest it dominates the response but appears at a still lower frequency (about 15 cps in Fig. 6-10) in the over-all response.

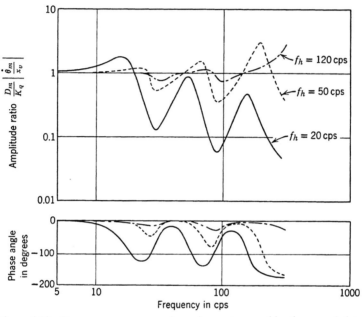

Figure 6-10   Frequency responses of a valve-motor combination coupled to a drive system having many degrees of freedom.

It is perhaps worth repeating that these conclusions cannot be drawn if displacement is measured at a point in the drive system other than shaft rotation. In such a case the structural response from the applied torque to the point of measurement would have to be obtained and used in the analysis rather than θ_m/T_g. The area of complex load dynamics deserves more attention and study than it has been given in the past. Perhaps the formulation given in this section will suggest more work in this area.

6-6   PRESSURE TRANSIENTS IN POWER ELEMENTS

Pressure transients, commonly referred to as pressure surges, frequently occur in hydraulic systems. These pressure peaks may become substantially higher than steady-state values and generate noise and/or cause damage to system components if safe stress levels are exceeded. Therefore identification of physical situations that can cause pressure surges, computing the magnitude of such surges, and methods of limiting surges are essential considerations in successful system design.

In general, pressure peaks are difficult to predict even in simple circuits. However, two fairly common physical situations give predictable results, and familiarity with these situations is considered fundamental in dealing with more complex circuits. The first situation, usually called *waterhammer* and discussed in Section 3-7, arises when a column of fluid flowing in

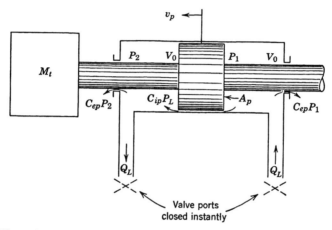

Figure 6-11   A piston being decelerated due to a sudden valve closure.

a line is suddenly stopped due to a valve closure. The pressure peak which is generated is controlled by keeping fluid velocities low. The second situation, the subject of the present section, arises when a hydraulic actuator is suddenly stopped. Of course, it is possible for both situations to occur simultaneously, generating a complex transient. However, such cases are extremely difficult to treat in a simple manner.

Let us examine Fig. 6-11 to gain a clear understanding of the physics involved when a piston or motor is suddenly stopped. Consider the piston to be moving at a velocity and let both lines be instantly closed, thereby trapping fluid in each chamber. Because of inertia the piston continues to move and compresses the fluid in the return chamber, which causes the pressure to increase rapidly above the steady-state level (Fig. 6-12). Simultaneously, the pressure in the forward chamber is decreased below the steady-state level and may cavitate, that is, approach zero absolute pressure. The piston will come to a stop and reverse direction when the kinetic energy of the moving mass is stored in the two fluid springs as potential energy. This oscillatory behavior will continue until leakage losses dissipate the energy involved. Because the forward chamber pressure becomes zero for a large transient, the kinetic energy is stored as potential energy, mainly in the fluid spring of the return chamber, and we can obtain an approximate expression for the pressure peak by equating these energies.

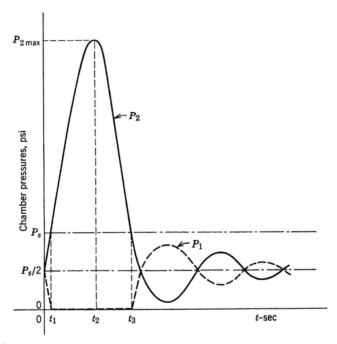

Figure 6-12   A large pressure transient.

Therefore

$$\tfrac{1}{2}M_t v_{p0}^2 = \tfrac{1}{2}K_h x_{p0}^2 \qquad (6\text{-}66)$$

where $K_h = \beta_e A_p^2/V_0$ = hydraulic fluid spring gradient of return chamber, lb/in.

x_{p0} = overtravel of piston, in.

v_{p0} = initial velocity of piston, in./sec

M_t = total mass of piston and load, lb-sec²/in.

V_0 = volume of return chamber (normally one half of the total contained volume of the system), in.³

β_e = effective bulk modulus of the system, lb/in.²

A_p = piston area, in.²

$P_{2\,max}$ = peak value of return pressure, psi

The displacement of the fluid spring is given by

$$x_{p0} = \frac{P_{2\max}A_p}{K_h} \tag{6-67}$$

Combining and solving for $P_{2\max}$ gives

$$P_{2\max} = v_{p0}\left(\frac{\beta_e M_t}{V_0}\right)^{\frac{1}{2}} \tag{6-68}$$

as an approximate value because zero damping is assumed.

Let us now develop more exact expressions for $P_{2\max}$ assuming that the load consists only of inertia so that

$$(P_1 - P_2)A_p = P_L A_p = M_t\frac{dv_p}{dt} \tag{6-69}$$

where v_p = instantaneous piston velocity, in./sec

A typical pressure transient for a valve controlled actuator is shown in Fig. 6-12. In the interval $0 \leq t \leq t_1$, both line pressures vary equal but opposite amounts from $P_s/2$ until the forward chamber pressure becomes zero and the return pressure becomes P_s. In the interval $t_2 \leq t \leq t_3$, the forward pressure is zero. The system is piecewise linear in the two intervals and may be solved accordingly. Analysis of the first interval is required to establish the initial conditions for the second interval.

During the first interval, assuming zero leakage flows and equal chamber volumes, (6-33) and (6-69) can be combined and solved for a step change in flow from $v_{p0}A_p$ to zero at $t = 0$ and subject to the initial conditions of $P_2 = P_s$ and $P_2 = 0$ at $t = 0$. The result is

$$P_2(t) = P_s\left(\frac{1}{2} + \frac{P_R}{\sqrt{2}}\sin\omega_h t\right) \tag{6-70}$$

where $P_R = \frac{v_{p0}}{P_s}\sqrt{\frac{\beta_e M_t}{V_0}}$ = ratio of approximate peak pressure to supply pressure, dimensionless $\tag{6-71}$

$\omega_h = \sqrt{\frac{2\beta_e A_p^2}{V_0 M_t}}$ = conventional hydraulic natural frequency, rad/sec $\tag{6-72}$

Substituting $P_2 = P_s$ in (6-70) we obtain the time t_1 as

$$\omega_h t_1 = \sin^{-1}(\sqrt{2}P_R)^{-1} \tag{6-73}$$

Combining with (6-70), we can establish the end conditions at time t_1 as

$$P_2(t_1) = P_s \tag{6-74}$$

$$\dot{P}_2(t_1) = \frac{1}{\sqrt{2}}\omega_h P_s\sqrt{P_R^2 - \tfrac{1}{2}} \tag{6-75}$$

Let us now turn to the second time interval. Because $P_1 = 0$ and $Q_L = 0$, the continuity equation for the return chamber is

$$A_p v_p - (C_{ip} + C_{ep} + K_c)P_2 = \frac{V_0}{\beta_e} \frac{dP_2}{dt} \qquad (6\text{-}76)$$

where C_{ip} = internal (cross-port) leakage coefficient, in.³/sec/psi
C_{ep} = external leakage coefficient, in.³/sec/psi
K_c = flow-pressure coefficient of servovalve, in.³/sec/psi
Combining (6-69) and (6-76), we have

$$\frac{1}{\omega_2^{\,2}} \frac{d^2 P_2}{dt^2} + \frac{2\delta_2}{\omega_2} \frac{dP_2}{dt} + P_2 = 0 \qquad (6\text{-}77)$$

where $\omega_2 = \dfrac{\omega_h}{\sqrt{2}}$ = natural frequency with one fluid spring, rad/sec

$$(6\text{-}78)$$

$$\cdot\delta_2 = \frac{(C_{ip} + C_{ep} + K_c)}{2A_p} \sqrt{\frac{\beta_e M_t}{V_0}} = \text{a damping ratio, dimensionless}$$

$$(6\text{-}79)$$

Even though δ_2 differs somewhat from the conventional value defined in Section 6-2, the conventional damping ratio can be used without great error.

We must now solve (6-77) by using the initial conditions given by (6-74) and (6-75) for $\delta_2 < 1$ and $\delta_2 > 1$. For $\delta_2 < 1$, the time response can be found to be

$$\frac{P_2(t)}{P_s} = \frac{e^{-\delta_2 \omega_2 (t - t_1)}}{\sqrt{1 - \delta_2^{\,2}}} \left[\left(\sqrt{P_R^{\,2} - \tfrac{1}{2}} + \delta_2\right) \sin \sqrt{1 - \delta_2^{\,2}}\,\omega_2(t - t_1) \right.$$

$$\left. + \sqrt{1 - \delta_2^{\,2}} \cos \sqrt{1 - \delta_2^{\,2}}\,\omega_2(t - t_1) \right] \quad (6\text{-}80)$$

Differentiating and equating to zero, we obtain the time to peak t_2 as

$$\omega_2(t_2 - t_1) = \frac{1}{\sqrt{1 - \delta_2^{\,2}}} \tan^{-1} \left[\frac{\sqrt{1 - \delta_2^{\,2}}\sqrt{P_R^{\,2} - \tfrac{1}{2}}}{1 + \delta_2 \sqrt{P_R^{\,2} - \tfrac{1}{2}}} \right] \qquad (6\text{-}81)$$

and substitution into (6-80) yields $P_{2\,\text{max}}$ for $\delta_2 < 1$ as

$$\frac{P_{2\,\text{max}}}{P_s} = e^{-\delta_2 \omega_2 (t_2 - t_1)} \sqrt{\tfrac{1}{2} + P_R^{\,2} + 2\delta_2 \sqrt{P_R^{\,2} - \tfrac{1}{2}}} \qquad (6\text{-}82)$$

For $\delta_2 > 1$ a similar exercise yields

$$\frac{P_{2\,max}}{P_s} =$$

$$\left[\frac{1 + (\delta_2 - \sqrt{\delta_2{}^2 - 1})\sqrt{P_R{}^2 - \tfrac{1}{2}}}{1 + (\delta_2 + \sqrt{\delta_2{}^2 - 1})\sqrt{P_R{}^2 - \tfrac{1}{2}}}\right]^{\frac{\delta_2 - \sqrt{\delta_2{}^2 - 1}}{2\sqrt{\delta_2{}^2 - 1}}} \times \left(1 + \frac{\sqrt{P_R{}^2 - \tfrac{1}{2}}}{\delta_2 + \sqrt{\delta_2{}^2 - 1}}\right)$$

$$(6\text{-}83)$$

for the maximum return pressure.

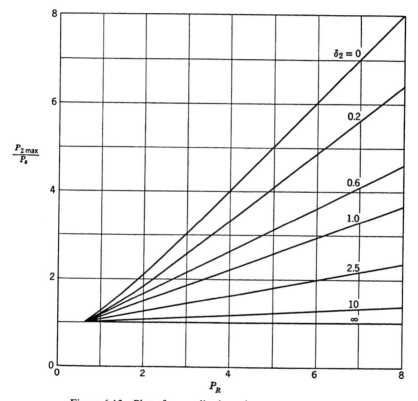

Figure 6-13   Plot of normalized maximum return pressure.

Equations 6-82 and 6-83 are plotted in Fig. 6-13. Thus the pressure peak is determined by computing P_R and δ_2 and referring to Fig. 6-13. We note that large pressure peaks are associated with large values of P_R. For large P_R and small δ_2 (usually the case) (6-81) approximates to

$$\omega_2(t_2 - t_1) \approx \frac{\pi}{2} \qquad (6\text{-}84)$$

and from (6-73) we obtain $t_1 \approx 0$. Thus the peak pressure occurs approximately $\pi/2\omega_2$ sec (one fourth the undamped natural period) after the valve is closed. For large P_R and small δ_2 (6-82) can be approximated by

$$P_{2\,max} \approx v_{p_0}\left(\frac{\beta_e M_t}{V_0}\right)^{1/2} e^{-(\pi/2)\delta_2} \qquad (6\text{-}85)$$

which agrees with the result obtained by equating energies (6-68) if $\delta_2 = 0$.

Let us now investigate methods of reducing the peak pressure $P_{2\,max}$. Of course, the most obvious solution would be to control the deceleration of the motor so that large peaks are not generated in the first place. However, this would require a fairly elaborate and costly control. Because $P_{2\,max}$ depends only on P_R and δ_2, either of these quantities can be controlled to limit peak pressures. Examining (6-71), we note that the quantities which make up P_R are fixed by other considerations. V_0 is desired small and β_e large to keep the hydraulic natural frequency high. Accumulators could be used to alter the value of β_c/V_0 when the accumulator charge pressure is exceeded. However, accumulators are bulky and it is preferred that the kinetic energy be dissipated rather than stored. Therefore, we must conclude that no parameter of P_R can be effectively controlled to limit pressure peaks. However, δ_2 can be increased by inserting a relief valve across the lines (a similar relief valve is required to protect the other chamber) with a static characteristic, idealized as a straight line, illustrated in Fig. 6-14. This safety relief valve dissipates the kinetic energy and helps prevent cavitation by dumping fluid to the forward chamber.

A piecewise linear analysis is required to establish the peak pressure with the safety relief. However, if we assume the relief valve is set to open at P_s, that is, $P_r = P_s$, then the relief valve will open at t_1 and the previous analysis is applicable. Under these conditions the quantity δ_2 given by (6-79) now becomes

$$\delta_2 = \frac{(K_r + K_c + C_{ip} + C_{ep})}{2A_p}\left(\frac{\beta_e M_t}{V_0}\right)^{1/2} \qquad (6\text{-}86)$$

where K_r = slope of relief valve characteristic, in.^{3}/sec/psi

P_r = relief valve pressure setting, psi

Because K_r is quite large, there is a massive increase in the damping ratio when the relief valve is opened. For a given P_R we can select a δ_2-value from Fig. 6-13 that will yield a satisfactory pressure peak. The required K_r-value would then be computed from (6-86).

Let us now determine the flow capacity and speed of response required of safety relief valves. From Fig. 6-14 we can write

$$Q_{r\,max} = K_r(P_{2\,max} - P_r) \qquad (6\text{-}87)$$

Because $K_r \gg K_c + C_{ip} + C_{ep}$ and $P_s = P_r$, (6-71), (6-86), and (6-87) can be combined to yield

$$\frac{Q_{r\,max}}{v_{p0}A_p} = \frac{2\delta_2}{P_R}\left(\frac{P_{2\,max}}{P_s} - 1\right)$$

(6-88)

Now, for large P_R (where pressure peaks are most severe) and for large

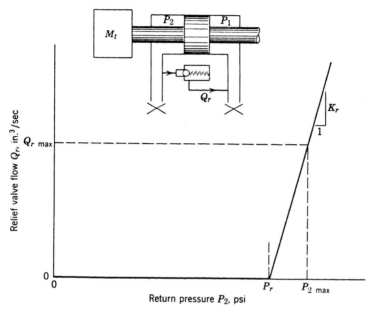

Figure 6-14   Static characteristic of a safety relief valve.

δ_2 (due to relief valve being used) (6-83) can be approximated by

$$\frac{P_{2\,max}}{P_s} \approx 1 + \frac{P_R}{2\delta_2}$$

(6-89)

A combination of (6-87) and (6-89) yields

$$\frac{Q_{r\,max}}{v_{p0}A_p} \approx 1$$

(6-90)

Hence the relief valve capacity must be large enough to accommodate full motor flow $v_{p0}A_p$ if P_R is large. This is because the piston acts as a pump with output of $v_{p0}A_p$ in.^{3}/sec during the initial portion of the pressure transient. To maintain P_2 at some constant level the relief valve must bypass the generated flow. In fact one can determine $P_{2\,max}$ from the relief valve characteristic because $Q_{r\,max} \approx v_{p0}A_p$. If P_R is small the relief valve capacity can be less because the piston has slowed appreciably by the time t_1.

From Fig. 6-12 it is apparent that the relief valve must be fast enough to open by t_1. Hence the relief valve bandwidth should be many times (perhaps 10) larger than ω_h if P_R is large. If P_R is small (near unity), the relief valve bandwidth is less critical because t_1 is larger.

The discussion presented thus far is equally applicable to rotary motors. The parameters x_{p0}, v_{p0}, A_p, and M_t are replaced by θ_{m0}, $\dot{\theta}_{m0}$, D_m and J_t, respectively,

where θ_{m0} = overtravel of motor after valve closure, rad

$\dot{\theta}_{m0}$ = initial velocity of motor, rad/sec

D_m = volumetric displacement of motor, in.^{3}/rad

J_t = total inertia of motor and load referred to motor, in.-lb-sec^2

For the case of a pump controlled motor, discussed in Section 6-4, only one line pressure varies at a time, and a transient analysis can be made to obtain the peak pressure generated due to a sudden valve closure. The result is

$$P_{2\,max} = \dot{\theta}_{m0}\left(\frac{\beta_e J_t}{V_0}\right)^{1/2} e^{-(\pi/2)\delta_h} \tag{6-91}$$

where δ_h is given by (6-58). Cross-port safety relief valves, illustrated in Fig. 6-7, are recommended to limit pressure peaks.

Thus far it has been assumed that the valve closure occurs instantaneously. If the valve is slowly closed the pressure peak, of course, is reduced. In such cases one might assume that the valve is closed in such a manner that motor deceleration is constant. The maximum pressure would then be

$$P_{2\,max} = \frac{J_t \dot{\theta}_{m0}}{D_m t_c} \tag{6-92}$$

where t_c = valve closure time, sec

However, the largest pressure peak always occurs with an instant valve closure. Therefore, if a system design is satisfactory for instant valve closures it follows that it is satisfactory for other closures.

6-7 NONLINEAR ANALYSIS OF VALVE CONTROLLED ACTUATORS

So far in the analyses of valve-actuator combinations we have assumed small excursions in the system variables about some particular operating point where the valve coefficients K_q and K_c are evaluated. This has been done for mathematical expediency because otherwise the differential equations are highly nonlinear and no simple and useful result can be obtained. However, it should be clear that such an assumption, that is, of a linearized analysis, must conform with physical facts to a substantial degree. Otherwise the worth of such analyses is highly questionable and certainly would not form a sound basis for design procedures.

Linearized analyses are usually made at the null operating point (i.e., in the vicinity of $x_v = 0$, $P_L = 0$, and $Q_L = 0$) because damping is lowest and stability is most adversely affected. Excursions in the variables x_v, P_L, and Q_L are kept small say, for example, less than 5% of their full range values. It is obvious, therefore, that an analysis technique is desirable which permits larger excursions in these variables. This would produce more exact solutions to compare with the results of linearized analyses and, if correlation were good, would improve our confidence in linearized analysis. However, the highly nonlinear nature of the servovalve pressure-flow curves prevent hand solutions. Of course, exact solutions are possible for full range of the variables by using digital and/or analog computer simulations [5, 6, 7, 8]. However, the following treatment should improve our confidence in linearized analyses without resorting to such machine solutions.

Suppose the power element is the valve-piston combination shown in Fig. 6-6. It will become apparent that an analogous treatment can be made for a valve-motor combination. Let us assume a critical center servovalve for which the pressure-flow curves are given by (5-33). Therefore,

$$Q_L = C_d \left(\frac{P_s}{\rho}\right)^{1/2} w x_v \left(1 - \frac{x_v}{|x_v|}\frac{P_L}{P_s}\right)^{1/2} \tag{6-93}$$

Assuming the piston is centered such that $V_1 = V_2 = V_t/2$, the continuity equation is given by (6-33), which can be written

$$Q_L = A_p v_p + C_{tp} P_L + \frac{V_t}{4\beta_e}\frac{dP_L}{dt} \tag{6-94}$$

where $v_p = dx_p/dt$ = piston velocity, in./sec

Assuming a simple lumped mass load, typical of many systems, the equation of piston motion becomes

$$P_L A_p = M_t \frac{dv_p}{dt} \tag{6-95}$$

Equating (6-93) and (6-94) and substituting in P_L from (6-95) yields

$$\frac{C_d}{A_p}\left(\frac{P_s}{\rho}\right)^{1/2} w x_v \left(1 - \frac{x_v}{|x_v|}\frac{M_t}{P_s A_p}\frac{dv_p}{dt}\right)^{1/2} = \frac{V_t M_t}{4\beta_e A_p^2}\frac{d^2 v_p}{dt^2} + \frac{C_{tp} M_t}{A_p^2}\frac{dv_p}{dt} + v_p \tag{6-96}$$

This nonlinear differential equation describes the valve-piston combination. Machine computation is necessary to obtain solutions. However, some insight into its nature can be gained by assuming $P_L/P_s \ll 1$ so that the approximation

$$\left(1 - \frac{x_v}{|x_v|}\frac{P_L}{P_s}\right)^{1/2} \approx 1 - (\tfrac{1}{2})\frac{x_v}{|x_v|}\frac{P_L}{P_s} \tag{6-97}$$

can be used. This approximation is within 10% for values of P_L/P_s as high as 0.6. This assumption is not unrealistic because many systems are

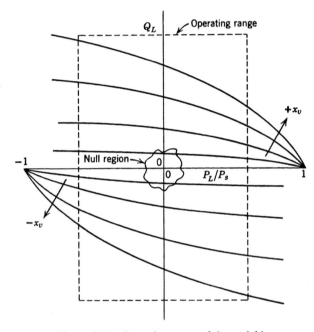

Figure 6-15   Operating range of the variables.

designed such that $P_L/P_s < \tfrac{2}{3}$ to prevent reduction in valve flow gain and the associated deterioration in servo performance (see Section 8-1). Using this approximation for the radical in (6-96) it reduces to

$$\frac{C_d w}{A_p}\left(\frac{P_s}{\rho}\right)^{1/2} x_v = \frac{V_t M_t}{4\beta_e A_p^2}\frac{d^2 v_p}{dt^2} + \left[\frac{C_d w \, |x_v|}{2P_s}\left(\frac{P_s}{\rho}\right)^{1/2} + C_{tp}\right]\frac{M_t}{A_p^2}\frac{dv_p}{dt} + v_p \tag{6-98}$$

We can write this equation in the form

$$\frac{C_d w}{A_p}\left(\frac{P_s}{\rho}\right)^{1/2} x_v = \frac{1}{\omega_h^2}\frac{d^2 v_p}{dt^2} + \frac{2\delta_h}{\omega_h}\frac{dv_p}{dt} + v_p \tag{6-99}$$

where $\omega_h = \sqrt{\dfrac{4\beta_e A_p^2}{V_t M_t}}$ = hydraulic natural frequency, rad/sec

$$\delta_h = \left[\frac{C_d w\,|x_v|}{2P_s}\sqrt{\frac{P_s}{\rho}} + C_{tp}\right]\frac{M_t \omega_h}{2A_p^2} = \text{hydraulic damping ratio,}$$

$$\text{dimensionless}$$

It is important to note that we have made no restrictions on the ranges of x_v and Q_L. We have restricted only P_L/P_s, but it can make rather large excursions (say to 0.6). Thus, the operating range is as illustrated in Fig. 6-15, which can hardly be considered restrictive. However, from (6-99) we see that the hydraulic natural frequency ω_h is constant and does not vary with the operating point. Thus the hydraulic natural frequency is a concept which is useful over a very wide range of operation. However, the damping ratio varies with $|x_v|$, which strongly depends on the operating point. If the input valve motion is sinusoidal about null, that is, $x_v = X_v \sin \omega t$, the average value $(0.636X_v)$ or rms value $(0.707X_v)$ could be used for $|x_v|$.

From this analysis we can conclude that a second order (i.e., a quadratic) representation of a hydraulic power element is valid over a fairly wide range. The hydraulic natural frequency is constant but the damping ratio must be thought of as varying with valve position. This analysis helps to relieve our anxiety about the usefulness of linearized analyses and increases our confidence in designs based on such analyses.

REFERENCES

[1] Merritt, H. E., and J. T. Gavin, "Friction Load on Hydraulic Servos," *Proc. Natl. Conf. Indl. Hydraulics*, **16**, 1962, 174.

[2] Blackburn, J. F., G. Reethof, and J. L. Shearer, *Fluid Power Control*. New York: Technology Press of M.I.T. and Wiley, 1960, Chapter 15.

[3] Sevcik, J. K., "System Vibration and Static Analysis," *ASME Paper No. 63-AHGT-57.*

[4] Long, G. W., and J. R. Lemon, "Structural Dynamics in Machine-Tool Chatter," *ASME Paper No. 64-WA/PROD-12.*

[5] Royle, J. K., "Inherent Non-Linear Effects on Hydraulic Control Systems with Inertia Loading," *Proc. Inst. Mech. Eng.* (London), **173**, No. 9, 1959.

[6] Butler, R., "A Theoretical Analysis of the Response of a Loaded Hydraulic Relay," *Proc. Inst. Mech. Eng.* (London), **173**, No. 16, 1959.

[7] McCloy, D., and H. R. Martin, "Some Effects of Cavitation and Flow Forces in the Electro-Hydraulic Servomechanism," *Proc. Inst. Mech. Eng.* (London), **178**, Pt. 1, No. 21, 1963–1964.

[8] Rausch, R. G., "The Analysis of Valve-Controlled Hydraulic Servomechanisms," *Bell System Tech. J.*, **38**, No. 6, 1959, 1513–1549.

7

Electrohydraulic Servovalves

The versatility of electrical devices makes them ideal for feedback measuring and for signal amplification and manipulation. On the other hand, power output and compactness of hydraulic actuators makes them ideally suited as power devices. In any combination of electric and hydraulic devices there must of necessity be an element which bridges the gap. This interface connection in control systems is achieved by an electrohydraulic servovalve. Such a servovalve converts low power electrical signals into motion of a valve which in turn controls the flow and/or pressure to a hydraulic actuator.

The electrohydraulic servovalve, probably the youngest of the standard hydraulic components, made its appearance in the latter 1940s to satisfy aerospace needs for a fast response servo control system. Electrohydraulic servo systems then in use were not significantly faster than electrical servos because they were lacking an element which could rapidly translate electrical signals into hydraulic flows. These early servovalves were actuated by small electric servomotors and the associated large time constants were such that the servovalve was often the slowest element in the control loop and limited system performance. In the early 1950s permanent magnet torque motors having fast response gained favor as a method of stroking valves and the electrohydraulic servovalve took its present form.

Because the electrohydraulic servovalve connects the electronic and hydromechanical portions of a system, it is often the focal point of discussion when system troubles occur. Also, some knowledge of electronics and magnetics, as well as mechanics and hydraulics, is required for an understanding of such servovalves. Electric and hydraulic performance specifications must be met by these valves. Therefore it is essential that system analysts and designers be thoroughly familiar with the various types and performance characteristics of electrohydraulic servovalves.

7-1 TYPES OF ELECTROHYDRAULIC SERVOVALVES

These servovalves may be broadly classified as either *single-stage* or *two-stage*. Single-stage servovalves consist of a torque motor which is directly attached to and positions a spool valve. Because torque motors have limited power capability, this in turn limits the flow capacity of single-stage servovalves and may also lead to stability problems in some applications.

Two-stage servovalves have a hydraulic preamplifier (first stage) which substantially multiplies the force output of the torque motor to a level sufficient to overcome considerable flow forces, stiction forces, and forces resulting from acceleration or vibration. Two-stage servovalves overcome both disadvantages of single-stage servovalves, and this is the main reason for their existance. Flapper, jet pipe, and spool valves find use as a first stage valve while the second stage valve is almost universally spool type.

Two-stage servovalves may be classified by the type of feedback used which are *spool position, load pressure*, and *load flow feedback*. Each type of feedback gives a distinct shape to the steady state pressure-flow curves, as illustrated in Fig. 7-1. Position feedback two-stage servovalves are by far the most common, and three basic types of construction can be identified, depending on how the spool position is sensed; these are *direct* feedback, *force* feedback, and *spring centered* spool. With direct position feedback the main spool follows the first stage valve in a one-to-one relation; this type construction is sometimes referred to as a hydraulic follower. In force feedback servovalves the main spool position is converted to a force by a (feedback) spring and this force is balanced at the torque motor armature against the torque due to input current. Both constructions are widely used and will be discussed in Sections 7-4 and 7-5. The third basic type uses stiff springs at the spool ends to center the spool against the pressure differential of the pilot stage. This type construction is less frequently used.

Two-stage servovalves with load pressure feedback attempt to control load pressure difference. Internal passages direct the load pressure back to the first stage where the resulting force is balanced against that due to the input current. A special form of these servovalves, called the *dynamic pressure feedback servovalve*, has characteristics of a position feedback servovalve at low frequencies and characteristics of a pressure feedback servovalve at higher frequencies. These servovalves are useful in some applications to increase the damping of valve-actuator combinations. Two-stage servovalves using flow feedback have some device which senses flow and converts it to a force which is then balanced at the torque motor armature with the force due to input current. Combinations of flow and

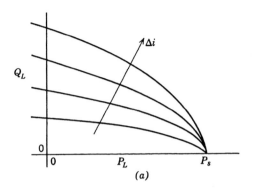

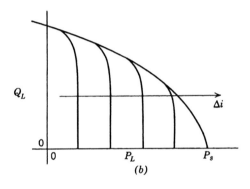

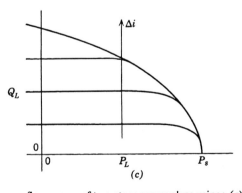

Figure 7-1   Pressure-flow curves of two-stage servovalves using: (a) position feedback; (b) load pressure feedback; (c) load flow feedback.

pressure feedback can be used to produce a valve called *pressure-flow servovalve* which has characteristics in between the extremes of either pressure or flow feedback. The pressure-flow curves of these servovalves are quite linear compared with the usual square-root type curve. These pressure and flow feedback servovalves, and their combinations, are used only in special applications which require the particular pressure-flow curves of these valves. The vast majority of applications use the more conventional direct and force feedback servovalves.

Another type of two-stage servovalve which has been used to some degree is the *acceleration switching servovalve*. A more descriptive title is *pulse length modulated* (PLM) servovalve. This servovalve utilizes a high frequency pulse length modulated wave generated in an electronic driver amplifier as the input signal to the torque motor. At zero signal to the driver amplifier the output wave to the servovalve is a symmetrical square wave so that no net spool motion results. As the signal is increased there is a time unbalance of the wave such that the positive pulse lengths are longer than the negative pulse lengths. Thus the valve spool dwells more time in one direction than in the other and produces a net flow to the load. However, the load flow has small variations at the switching frequency which must be filtered by the power element to achieve smooth output motion. Construction of the PLM servovalve is identical to that of conventional two-stage servovalves except that there is no feedback. Therefore, at least theoretically, the *rate* of output flow is proportional to the input current signal so that the PLM servovalve is an integrating device. However, flow forces on the second stage spool cause a large lag to result rather than a true integration. This large lag, which behaves effectively as an integration, must be compensated by a lead in servo loops to achieve stability.

Relaxed tolerances, higher reliability in adverse environments, insensitivity to contamination and vibration, and low cost are advantages claimed for PLM servovalves. However, the disadvantages of filtering the switching frequency, more complex electronic driver amplifier, undesirable noise and mechanical vibration induced by the switching frequency, and higher null leakage have limited the use of PLM servovalves to well below their anticipated potential.

7-2 PERMANENT MAGNET TORQUE MOTORS

Electric servomotors, magnetostrictive devices, piezoelectric crystals, proportional solenoids [1], a-c torque motors, and moving coil devices (similar to loudspeaker motors) have been or are being used to stroke servovalves from an electrical signal. However, the permanent magnet torque motor illustrated in Fig. 7-2 is by far the most widely used electro-mechanical device for this purpose. These devices may produce rotary motion (torque motor) or translational motion (force motor), and the torque or force produced is proportional to input current. Torque (i.e., rotary) motors are more common, and the particular bridge type magnetic circuit shown in Fig. 7-2 has been almost universally accepted.

Torque (or force) motors may be broadly classified as being wet or dry, depending on whether they are intended for use immersed in the fluid.

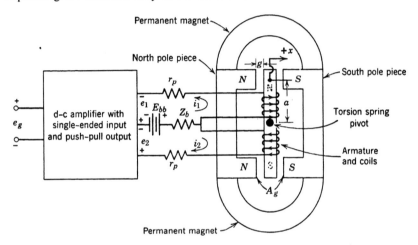

Figure 7-2   Schematic of permanent magnet torque motor being driven from an amplifier.

Dry units seal the output member to prevent fluid from entering the volumes around the armature, coils, and magnets. Wet units have the disadvantage that ferrous contaminants in the fluid are attracted by the permanent magnet and may accumulate in the air gaps and ultimately prevent operation; however, these units are cooled by the fluid.

The torque motor (Fig. 7-2) consists of an armature mounted on a torsion pin pivot and suspended in the air gap of a magnetic field. Two pole pieces, one permanently polarized north and the other south by the permanent magnet, form the framework around the armature and provide paths for the magnetic flux. When current is made to flow through the armature coils, the armature ends become polarized and each end is attracted to one pole piece and repelled by the other. A torque is thus produced on the armature and it would continue to rotate were it not for the restraining torque of the torsion spring on which the armature is mounted. Thus the torsion spring causes the output motion to be proportional to current. The servovalve to be positioned may be mechanically

attached to either armature end. Because the static and dynamic characteristics of torque motors play a major role in servovalve design and performance, an investigation of these characteristics is essential.

Torque Motor Analysis

Referring to Fig. 7-2, the two coils on a torque motor are generally
supplied from a push-pull source.* A voltage E_{bb} in the amplifier driving
the torque motor establishes a quiescent current I_0 in each coil but there
is no net torque on the armature because the currents oppose each other.
An increase in the input to the amplifier causes the current in one coil to
increase as the current in the other coil decreases simultaneously by the
same amount. Hence the current in the two coils may be written

$$i_1 = I_0 + i \tag{7-1}$$

$$i_2 = I_0 - i \tag{7-2}$$

where i_1, i_2 = current in each coil, respectively,

I_0 = constant quiescent current in each coil, amp

i = signal current in each coil, amp

Δi = current difference in the two coils, amp

The flux in the armature, and consequently the developed torque, is
proportional to the difference in current in the two coils. Subtracting
these equations yields

$$i_1 - i_2 = 2i \equiv \Delta i \tag{7-3}$$

---

* Another popular connection is to drive the coils in parallel from a single-ended
amplifier (usually transistorized). This offers increased reliability because operation is
still possible if one coil opens. Analysis is similar to the push-pull case except for the
voltage equation. If i is the total current (i.e., $i/2$ in each coil) and E_0 is the open circuit
output voltage and R_0 is the output resistance of the amplifier, then the voltage applied
to each coil is $E_0 - iR_0$. Hence, the voltage equation for each coil is

$$E_0 - iR_0 = R_c\left(\frac{i}{2}\right) + K_b\frac{d\theta}{dt} + L_c\frac{d(i/2)}{dt} + M\frac{d(i/2)}{dt}$$

Because the two coils are close coupled, the self and mutual inductance are approximately equal (i.e., $M \approx L_c$). Therefore Laplace transforming yields

$$E_0 = \left(R_0 + \frac{R_c}{2}\right)i + K_b s\theta + L_c si$$

Comparing with (7-21), we note that the push-pull analysis can be used if μe_g is interpreted as E_0, Δi is interpreted as i, and $r_p/2$ is interpreted as R_0 in all the equations.

Although most treatments use signal current as the input, we shall use differential current for a parameter because this quantity is used in performance curves and measurements. Care should be taken to distinguish clearly between signal and differential current; they are simply related by (7-3). The quiescent current I_0 is usually about one half the maximum differential current. Thus for maximum signal input to the amplifier, the current in one torque motor coil will be about zero so that the maximum differential current will occur in one coil. With such a quiescent current level, which is required because of amplifier characteristics, the efficiency is only 50%. This is a minor point, however, because the electrical power involved is trivial compared with hydraulic losses. The torque motor coils can also be operated in series or in parallel from a single-ended source. Parallel operation has the advantage of increased reliability in the event one coil fails.

The signal voltages for push-pull operation are

$$e_1 = e_2 = \mu e_g \qquad (7\text{-}4)$$

where e_1, e_2 = signal voltages from amplifier, volts
μ = amplifier gain for each side, dimensionless
e_g = signal voltage input to amplifier, volts
The voltage equations for each coil circuit are

$$E_{bb} + e_1 = i_1(Z_b + R_c + r_p) + i_2 Z_b + \frac{N_c}{10^8} \frac{d\phi_a}{dt} \qquad (7\text{-}5)$$

$$E_{bb} - e_2 = i_2(Z_b + R_c + r_p) + i_1 Z_b - \frac{N_c}{10^8} \frac{d\phi_a}{dt} \qquad (7\text{-}6)$$

where R_c = resistance of each coil, ohms
N_c = number of turns in each coil
ϕ_a = total magnetic flux through the armature, lines (maxwells)
r_p = internal resistance (plate resistance) of amplifier in each coil circuit, ohms
E_{bb} = constant voltage required for quiescent current, volts
Z_b = impedance in common line of coils, ohms
Subtracting (7-6) from (7-5) and then combining with (7-3) and (7-4) yields

$$2\mu e_g = (R_c + r_p) \Delta i + \frac{2N_c}{10^8} \frac{d\phi_a}{dt} \qquad (7\text{-}7)$$

which is the fundamental voltage equation for the torque motor.

Now let us analyze the magnetic flux circuits to establish the flux in the air gaps and in the armature.* As illustrated in Fig. 7-3a, the flux paths are quite complex. An approximate but adequate analysis can be made by assuming that the four air gaps constitute the dominant reluctances in the circuit, that is, the reluctances of magnetic materials are negligible in

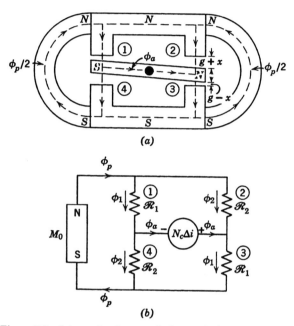

Figure 7-3   Schematic of magnetic flux paths in torque motor.

* It might prove helpful to review the basic magnetic circuit equation to avoid confusion concerning units. The fundamental relation for a magnetic circuit, analogous to Ohm's law for an electric circuit, is $M = \phi \mathscr{R}$,

where $M = Ni$ = magnetomotive force (mmf), amp-turns

ϕ = magnetic flux, lines (maxwells)

$\mathscr{R} = l/\mu A$ = reluctance of flux path, amp-turns/line

l = length of flux path, in.

A = area of path normal to flux, in.²

$\mu = \mu_r \mu_0$ = permeability

μ_r = relative permeability (i.e., compared with free space); for air $\mu_r = 1$

$\mu_0 = 0.4\pi(2.54) = 3.19$ = permeability of free space (value depends on system of units used and definition of magnetomotive force; the value given is for English units)

Circuit equations which result from principles analogous to Kirchhoff's first and second laws for electric circuits may also be written for magnetic circuits; that is, the sum of the flux is zero at any node and the sum of the magnetomotive forces around any loop must be zero.

comparison. Because of symmetry, we can assume that the reluctances of diagonally opposite air gaps are equal and therefore are given by

$$\mathfrak{R}_1 = \frac{g - x}{\mu_0 A_g} \tag{7-8}$$

$$\mathfrak{R}_2 = \frac{g + x}{\mu_0 A_g} \tag{7-9}$$

where $\mathfrak{R}_1$ = reluctance of gaps 1 and 3, amp-turn/line
   $\mathfrak{R}_2$ = reluctance of gaps 2 and 4, amp-turn/line
   g = length of each air gap at neutral, in.
   x = displacement of the armature tip (at the center of the pole face) from neutral position, in.
   A_g = pole face area at the air gaps, in.^{2}
   $\mu_0 = 3.19$ = permeability of free space (air) used with English units

With these assumptions the magnetic circuit can be represented by Fig. 7-3b. Because this circuit is a symmetrical bridge, it is apparent that the flux through opposite arms (i.e., through diagonally opposite air gaps) are equal. Because the mmf's around each loop must be zero, let us choose the loops containing opposite bridge arms and both mmf sources. Two such loops exist and we can write directly that

$$\phi_1 = \frac{M_0 + N_c \Delta i}{2\mathfrak{R}_1} = \frac{M_0 + N_c \Delta i}{2\mathfrak{R}_g (1 - x/g)} \tag{7-10}$$

$$\phi_2 = \frac{M_0 - N_c \Delta i}{2\mathfrak{R}_2} = \frac{M_0 - N_c \Delta i}{2\mathfrak{R}_g (1 + x/g)} \tag{7-11}$$

where ϕ_1 = magnetic flux through gaps 1 and 3, lines (maxwells)
   ϕ_2 = magnetic flux through gaps 2 and 4, lines (maxwells)
   ϕ_v = total flux through permanent magnets, lines (maxwells)
   M_0 = total mmf of all permanent magnets, amp-turns
   $\mathfrak{R}_g = g/\mu_0 A_g$ = reluctance of each air gap at neutral, amp-turns/ line
$N_c(i_1 - i_2) = N_c \Delta i$ = net mmf due to control currents, amp-turns

   It is more convenient to express M_0 in terms of the air-gap flux when the armature is at neutral. At neutral position (i.e., where $\Delta i = x = 0$), these two equations reduce to

$$\phi_{10} = \frac{M_0}{2\mathfrak{R}_g} = \phi_{20} \equiv \phi_g \tag{7-12}$$

where ϕ_g is the flux in each of the four air gaps when the armature is at neutral. Therefore the air gap flux relations become

$$\phi_1 = \frac{\phi_g + \phi_c}{1 - x/g} \tag{7-13}$$

$$\phi_2 = \frac{\phi_g - \phi_c}{1 + x/g} \tag{7-14}$$

where the quantity ϕ_c is the flux due to control currents and is defined by

$$\phi_c \equiv \frac{N_c \Delta i}{2 \Re_g} \tag{7-15}$$

Thus we see that two air gaps carry the sum of the permanent magnet and signal flux while the other two air gaps carry the difference.

Summing the magnetic flux at two nodes we obtain

$$\phi_p = \phi_1 + \phi_2 \tag{7-16}$$

$$\phi_a = \phi_1 - \phi_2 \tag{7-17}$$

The armature flux relation is of particular interest and can be combined with (7-13) and (7-14) to yield

$$\phi_a = \frac{2\phi_g(x/g) + 2\phi_c}{1 - x^2/g^2} \tag{7-18}$$

Because torque motors are designed so that $x/g < 1$ (usually less than one third so that $x^2/g^2 < 0.1$), (7-18) can be simplified to obtain

$$\phi_a = 2\phi_g \frac{x}{g} + \frac{N_c}{\Re_g} \Delta i \tag{7-19}$$

From geometrical considerations it is clear that

$$\tan \theta = \frac{x}{a} \approx \theta \tag{7-20}$$

where θ = angular deflection of armature, rad
a = radius of armature from pivot to center of pole face, in.

The approximation indicated is usually valid because deflections are small. Now (7-19) can be differentiated and substituted into (7-7) and the result can be Laplace transformed to give

$$2\mu e_g = (R_c + r_p) \Delta i + 2K_b s\theta + 2L_c s \Delta i \tag{7-21}$$

where $K_b = (2 \cdot 10^{-8})\left(\dfrac{a}{g}\right)N_c\phi_g$ = back (counter) emf constant for

$$\text{each coil, volts/rad/sec} \qquad (7\text{-}22)$$

$$L_c = (10^{-8})\frac{N_c^{\,2}}{\mathfrak{R}_g} = \text{self-inductance of each coil, henrys} \qquad (7\text{-}23)$$

This iş the final form of the basic voltage equation of the torque motor.*

Let us now determine the torque which acts on the armature because of the interaction of the permanent magnet and control flux in the air gaps. The fundamental force equation (see most any electrical engineering text) is

$$F = 1.38 \cdot 10^{-8}B^2A = 4.42 \cdot 10^{-8}\frac{\phi^2}{\mu_0 A} \qquad (7\text{-}24)$$

where F = attractive force between magnetized parallel surfaces separated
    by an air gap, lb
  $\phi = BA$ = magnetic flux in the gap, lines (maxwells)
  A = area normal to flux path, in.^{2}
  B = flux density in the air gap, lines/in.^{2}

Because the torques developed in the two air gaps at each end of the armature are in opposition, the net torque developed will be proportional to the difference of the squares of the fluxes. Using the foregoing relation, we obtain the total net torque developed on the armature.

$$T_d = 2a(\phi_1^{\,2} - \phi_2^{\,2})\frac{4.42 \cdot 10^{-8}}{\mu_0 A_g} \qquad (7\text{-}25)$$

The 2 multiplier is due to the fact that the same torque is developed by the pair of air gaps at the other end of the armature. Substituting in

---

* We note in examining each term of (7-21) that the power consumed by the resistive term is lost as heat and the power of the inductive term is stored magnetically. The only voltage term that contributes to the power output of the torque motor is the counter emf. Thus, the useful electric power input (for both coils) is

$$\text{hp}|_{\text{in}} = \frac{1}{746}\left(iK_b\frac{d\theta}{dt} + iK_b\frac{d\theta}{dt}\right)$$

Now, the mechanical output power is

$$\text{hp}|_{\text{out}} = \frac{T_d\,(d\theta/dt)}{12 \times 550} = \frac{K_t\,\Delta i\,(d\theta/dt)}{12 \times 550}$$

Equating these powers, and because $\Delta i = 2i$, we get $K_t = 8.847K_b$. Thus, the torque and counter emf constants are directly related. This relation can also be obtained from the basic definitions of K_t and K_b. Tests indicate that the proportionality factor is somewhat higher than 8.847.

(7-13) and (7-14), we obtain, after much algebraic manipulation,

$$T_d = \frac{(1 + x^2/g^2)K_t\,\Delta i + (1 + \phi_c^2/\phi_g^2)K_m\theta}{(1 - x^2/g^2)^2} \qquad (7\text{-}26)$$

where T_d = total torque developed on the armature due to electrical current input, in.-lb

$$K_t = (4.42 \cdot 10^{-8})4\left(\frac{a}{g}\right)N_c\phi_g = \text{torque constant of the torque motor (i.e., for each coil,) in.-lb/amp} \qquad (7\text{-}27)$$

$$K_m = (4.42 \cdot 10^{-8})8\left(\frac{a}{g}\right)^2\mathfrak{R}_g\phi_g^2 = \text{magnetic spring constant of torque motor, in.-lb/rad} \qquad (7\text{-}28)$$

Torque motors are designed so that $(x/g)^2 \ll 1$ and $(\phi_c/\phi_g)^2 \ll 1$ to improve linearity and static stability and prevent knockdown of the permanent magnets. Neglecting these quantities or for operation near null (7-26) becomes

$$T_d = K_t\,\Delta i + K_m\theta \qquad (7\text{-}29)$$

Applying Newton's second law to the armature we can write

$$T_d = J_a\frac{d^2\theta}{dt^2} + B_a\frac{d\theta}{dt} + K_a\theta + T_L \qquad (7\text{-}30)$$

where J_a = inertia of armature and any attached load, in.-lb-sec^2
B_a = viscous damping coefficient of mechanical armature mounting (usually negligible) and load, in.-lb-sec
K_a = mechanical torsion spring constant of armature pivot, in.-lb/rad
T_L = any arbitrary load torque on the armature, in.-lb
Combining with (7-29) and Laplace transforming yields

$$K_t\,\Delta i = J_a s^2\theta + B_a s\theta + (K_a - K_m)\theta + T_L \qquad (7\text{-}31)$$

Thus the magnetic spring constant appears as a negative spring, and it is apparent that it must be less than the mechanical spring rate for stability.

Static Performance Characteristics

After some algebraic juggling the developed torque given by (7-26) can be written in the following manner:

$$\frac{T_d}{(g/a)K_m} = \frac{(\alpha + \beta)(1 + \alpha\beta)}{(1 - \beta^2)^2} \qquad (7\text{-}32)$$

where $\alpha \equiv \dfrac{\phi_c}{\phi_g} = \dfrac{N_c}{2\phi_g \mathfrak{R}_g}\,\Delta i$ $\qquad\qquad\qquad\qquad$ (7-33)

$\beta \equiv \dfrac{x}{g} = \dfrac{a}{g}\,\theta$ $\qquad\qquad\qquad\qquad\qquad\quad$ (7-34)

This expression can be plotted in a dimensionless manner and is useful in the design of torque motors for force feedback servovalves. For static

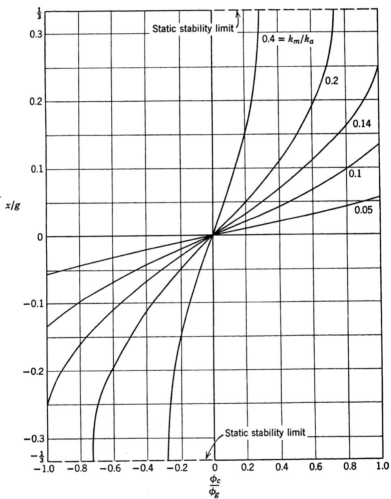

Figure 7-4 Normalized plot of the no load deflection versus current characteristic curve for a torque motor.

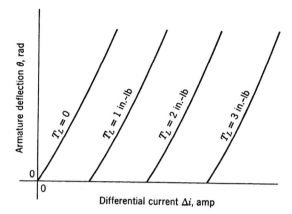

Figure 7-5   Typical static characteristic curves for a torque motor.

operation $T_d = K_a\theta + T_L$ which can be combined with (7-32) to yield

$$\frac{K_m}{K_a}(\alpha + \beta)(1 + \alpha\beta) = (1 - \beta^2)^2\beta + \frac{(1 - \beta^2)^2}{(g/a)K_a}T_L \qquad (7\text{-}35)$$

This equation describes the static characteristic curves of the torque motor with load. The no load curve is most useful. By letting $T_L = 0$ and solving for α the equation of this particular curve is

$$\alpha = \frac{1 + \beta^2}{2\beta}\left(\left\{1 + \frac{K_a}{K_m}\left[\frac{2\beta(1 - \beta^2)}{1 + \beta^2}\right]^2 - \left(\frac{2\beta}{1 + \beta^2}\right)^2\right\}^{1/4} - 1\right) \qquad (7\text{-}36)$$

and is plotted in Fig. 7-4 for various K_m/K_a ratios. Note that the armature ceases to be statically stable when $|x/g| > \frac{1}{3}$ regardless of the value for K_m/K_a. This limit is increased if the reluctance of the magnetic materials is appreciable [2]. Although the limit is independent of spring rates, the current required to cause static instability increases with K_m/K_a. It is also apparent that the slope or gain of this curve increases with the ratio K_m/K_a. However, linearity imposes constraints on the ranges of both ϕ_c/ϕ_g and K_m/K_a. Because ϕ_c/ϕ_g is kept well below unity to avoid knockdown of the permanent magnets, linearity is controlled mainly by the ratio of K_m/K_a. The static performance curves with load are also of interest and have the general shape illustrated in Fig. 7-5.

Dynamic Performance Characteristics

Equations 7-21 and 7-31, together with the torque motor parameters defined by (7-22), (7-23), (7-27), and (7-28), completely describe the dynamic characteristics of the torque motor. The parameters are computed from the design constants $\mathcal{R}_g$, N_c, g, a, and ϕ_g. The quantity ϕ_g is often expressed in terms of the flux density B_g in lines/in.^{2} at the gaps ($\phi_g = B_g A_g$). It is clear from the definitions of the parameters that they are related because

$$K_t = 8.84 K_b \tag{7-37}$$

$$K_m L_c = K_b K_t \tag{7-38}$$

Neglecting the mechanical damping B_a, we can combine (7-21) and (7-31) to yield

$$\theta = \frac{K_0 e_g - \left[\dfrac{1}{K_a\left(1 - \dfrac{K_m}{K_a}\right)}\right]\left(1 + \dfrac{s}{\omega_a}\right)T_L}{\dfrac{s^3}{\omega_a \omega_m{}^2\left(1 - \dfrac{K_m}{K_a}\right)} + \dfrac{s^2}{\omega_m{}^2\left(1 - \dfrac{K_m}{K_a}\right)} + \dfrac{s}{\omega_a\left(1 - \dfrac{K_m}{K_a}\right)} + 1} \tag{7-39}$$

where $K_0 = \dfrac{2K_t\mu}{(R_c + r_p)K_a(1 - K_m/K_a)}$ = static gain constant, rad/volt

$\omega_a = \dfrac{(R_c + r_p)}{2L_c}$ = armature circuit break frequency for each coil*, rad/sec

$\omega_m = \sqrt{\dfrac{K_a}{J_a}}$ = natural frequency of armature, rad/sec

The cubic characteristic equation can be easily factored if the coefficients are numerical. However, factors in literal form are desirable for design purposes. In this respect we are fortunate because the coefficients are so related that this cubic can be factored, with some labor, in a general manner and the roots can be represented graphically. Therefore, the final equation which gives the dynamic response of the torque motor to both voltage and load torque inputs is

$$\theta = \frac{K_0 e_g - \dfrac{1}{K_a\left(1 - \dfrac{K_m}{K_a}\right)}\left(1 + \dfrac{s}{\omega_a}\right)T_L}{\left(\dfrac{s}{\omega_r} + 1\right)\left(\dfrac{s^2}{\omega_0{}^2} + \dfrac{2\delta_0}{\omega_0}s + 1\right)} \tag{7-40}$$

* The total inductance in each coil circuit must include the mutual inductance of the other coil as well as the coil self inductance L_c. Because the two coils are in series for signal currents and are close coupled, the mutual inductance equals the self inductance. Thus the total inductance in each coil circuit is $L_c + M = 2L_c$, which explains the denominator of ω_a.

where ω_r, ω_0, δ_0 are determined from Figs. 7-6, 7-7, and 7-8, respectively. From these charts we note that the cubic always has a linear and an underdamped quadratic as factors. These factors are due to the electrical time constant $(1/\omega_a)$ of the armature circuit and to the mechanical resonance of the armature, respectively. However, break frequencies of the cubic are lower, that is, $\omega_r < \omega_a$ and $\omega_0 < \omega_m$. Because $K_m/K_a < \frac{1}{2}$ in

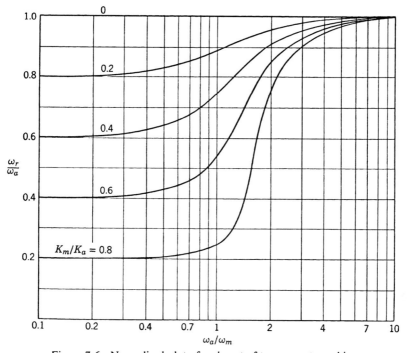

Figure 7-6   Normalized plot of real root of torque motor cubic.

practical torque motors, the following approximations are useful in preliminary design calculations.

$$\omega_r \approx \omega_a \quad \omega_0 \approx \omega_m \quad \delta_0 \approx \frac{1}{2}\left(\frac{K_m}{K_a}\right) \qquad (7\text{-}41)$$

Comparing with exact values from the charts, these approximations are satisfactory for small K_m/K_a.

Several conclusions can be drawn from (7-40) and the cubic solution charts:

1. Because the basic break frequency of each coil $R_c/2L_c$ is usually low, a high impedance driver (such as pentode vacuum tubes or the collector

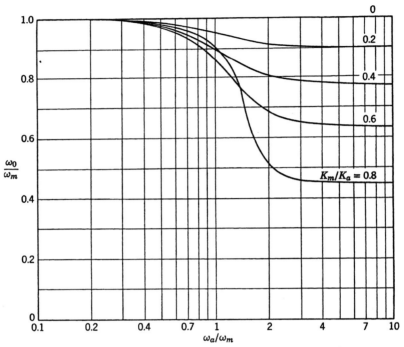

Figure 7-7   Normalized plot of natural frequency of quadratic factor of torque motor cubic.

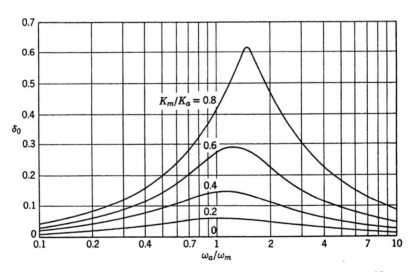

Figure 7-8   Plot of damping ratio of quadratic factor of torque motor cubic.

190

circuit of the transistors) which gives a large r_p or a current feedback amplifier must be used to obtain a large ω_a. The value of ω_a is then decreased to ω_r in the over-all response. Thus the amplifier design must be such to compensate the low electrical break frequency $R_e/2L_e$ of the armature circuit.

2. The natural frequency of the quadratic factor ω_0 is substantially independent of the magnetic spring constant at low ω_a/ω_m and is only slightly affected at higher ω_a/ω_m.

3. The damping ratio δ_0 is maximum when ω_a/ω_m is near unity. This is another reason for amplifiers which increase ω_a. The damping ratio increases directly with K_m/K_a. However, static linearity and hysteresis limit the value for K_m/K_a. Thus the value for K_m/K_a must be a compromise between requirements for linearity and damping. It is important to note that adequate damping requires both a compensated amplifier and a satisfactory value for K_m/K_a.

4. An increase in the air gap flux density B_g causes a direct increase in the damping ratio δ_0 and a slight lowering of the break frequencies ω_r and ω_0. The most noticeable affect on the frequency response is the increased δ_0. Increasing B_g also increases the electrical hysteresis.

Omitting the load torque input, the transfer function of the torque motor from voltage input is

$$\frac{\theta}{e_g} = \frac{K_0}{\left(\dfrac{s}{\omega_r} + 1\right)\left(\dfrac{s^2}{\omega_0{}^2} + \dfrac{2\delta_0}{\omega_0} s + 1\right)} \tag{7-42}$$

Torque motors and servovalves are usually specified with differential current as the input. The transfer function from current can be obtained from (7-31) as

$$\frac{\theta}{\Delta i} = \frac{\dfrac{K_t}{K_a\left(1 - \dfrac{K_m}{K_a}\right)}}{\dfrac{s^2}{\omega_m{}^2\left(1 - \dfrac{K_m}{K_a}\right)} + \dfrac{B_a}{K_a\left(1 - \dfrac{K_m}{K_a}\right)} s + 1} \tag{7-43}$$

Thus the response is that of a quadratic with a natural frequency slightly less than ω_m. The damping ratio is quite nebulous and often very low because it depends on structural damping in the torsion spring and on that obtained from the load attached to the torque motor. Current is used as an input by servovalve manufacturers to avoid specifying the amplifier characteristics. However, care must always be used in interpreting frequency response data because voltage is often used as an input but the results plotted in terms of Δi. This incorrectly assumes that voltage and current are directly related with no dynamics. Dividing (7-42) by (7-43) we obtain

$$\frac{\Delta i}{e_g} = \frac{\dfrac{2\mu}{R_c + r_p}\left[\dfrac{s^2}{\omega_m{}^2\left(1 - \dfrac{K_m}{K_a}\right)} + \dfrac{B_a}{K_a\left(1 - \dfrac{K_m}{K_a}\right)}s + 1\right]}{\left(\dfrac{s}{\omega_r} + 1\right)\left(\dfrac{s^2}{\omega_0{}^2} + \dfrac{2\delta_0}{\omega_0}s + 1\right)} \qquad (7\text{-}44)$$

as the relation between voltage and current. The numerator natural frequency is always somewhat lower than ω_0. Typical frequency response characteristics for a torque motor are illustrated in Figs. 7-9 and 7-10.

Torque motors offer a fast and effective method of stroking a valve when supplied from a properly designed amplifier. However, there are

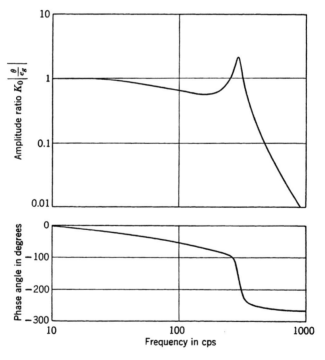

Figure 7-9   Frequency response of a torque motor from voltage input to armature rotation.

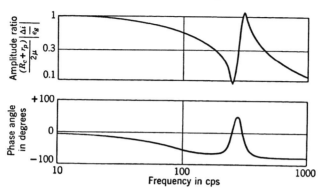

Figure 7-10   Frequency response of torque motor from input voltage to differential current.

some drawbacks. As with many electrical devices, magnetic hysteresis is unavoidable and can cause limit cycle oscillation in Type 1 servos which use lag compensation. Torque motors are also quite load sensitive. This makes their characteristics highly dependent on the type of load. From (7-40) we see that the steady-state stiffness of the torque motor to loads is

$$\frac{\Delta T_L}{\Delta \theta}\bigg|_{\omega=0} = K_a\left(1 - \frac{K_m}{K_a}\right) = K_a - K_m \qquad (7\text{-}45)$$

which is less than the mechanical stiffness K_a. Torque motor loading is especially severe in single-stage servovalves, as we shall see in the next section.

7-3   SINGLE-STAGE ELECTROHYDRAULIC SERVOVALVES

Single-stage electrohydraulic servovalves consist of a torque motor directly attached to a four-way spool valve. The spool valve is positioned by the torque motor and ports flow to a hydraulic actuator, as illustrated in Fig. 7-11. Although flapper and jet pipe valves can also be used to form single-stage valves, they would not be suitable for direct control of a load because of leakage characteristics.

Single-stage servovalves are relatively simple and inexpensive but have two major faults. The flow capacity is limited because steady state flow forces on the spool tend to stall the torque motor and limit the valve stroke. The other disadvantage is the fact that stability depends to a large extent on the load dynamics. Although this can be minimized by proper servovalve design, each case should be investigated to assure stability.

Although it is usually easier to investigate static first and then dynamic performance, this procedure is reversed here because a study of dynamic performance reveals the nature of the stability problem and suggests design criteria to minimize it. The dynamics of all the elements in this servovalve have been discussed, so that we need only summarize and combine the appropriate relations. Let us assume that the spool valve is a

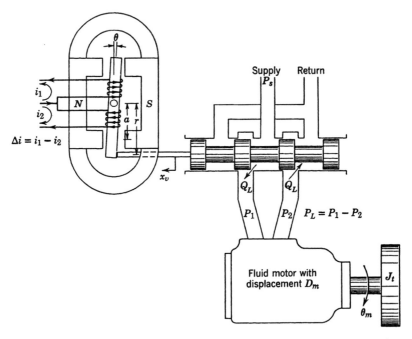

Figure 7-11   Schematic of a single stage electrohydraulic servovalve connected to a motor with inertia load.

four-way critical center type with rectangular ports. Therefore the following relations fron Section 5-3 are applicable:

$$Q_L = C_d w x_v \left[\frac{1}{\rho}(P_s - P_L)\right]^{\frac{1}{2}} \tag{7-46}$$

$$K_p = \frac{K_q}{K_c} = \frac{2(P_s - P_{L0})}{x_{v0}} \tag{7-47}$$

where x_{v0} and P_{L0} are evaluated at the particular operating point. Assuming the spool has no net damping length and flow forces are not compensated, the stroking force consists only of spool mass and steady state flow force. Therefore,

$$F_i = M_v \frac{d^2 x_v}{dt^2} + 0.43w(P_s - P_L)x_v \qquad (7\text{-}48)$$

Because P_L and x_v are both variables, the last term must be linearized. The equation can then be Laplace transformed to give

$$\Delta F_i = M_v s^2 \Delta x_v + 0.43w(P_s - P_{L0}) \Delta x_v - 0.43w x_{v0} \Delta P_L \qquad (7\text{-}49)$$

From the geometry in Fig. 7-11 we note for small deflections that

$$x_v \approx r\theta \qquad (7\text{-}50)$$

where r = radius arm of torque motor, in. Let us now consider the torque motor dynamics. It is clear that the mass and one of the flow force terms in (7-49) can be added to the mass and spring constants of the torque motor armature. The other flow force term can be treated as a load on the torque motor; that is,

$$\Delta T_L = -0.43r w x_{v0} \Delta P_L \qquad (7\text{-}51)$$

Therefore, using (7-40), we give the torque motor dynamics by

$$\theta = \frac{K_0 e_g - \dfrac{1}{K_{at}\left(1 - \dfrac{K_m}{K_{at}}\right)}\left(1 + \dfrac{s}{\omega_a}\right) T_L}{\left(\dfrac{s}{\omega_r} + 1\right)\left(\dfrac{s^2}{\omega_0{}^2} + \dfrac{2\delta_0}{\omega_0}s + 1\right)} \qquad (7\text{-}52)$$

where $K_{at} = K_a + 0.43r^2 w(P_s - P_{L0})$ = total spring constant, in.-lb/rad

$$K_0 = \frac{2K_t \mu}{(R_c + r_p)K_{at}(1 - K_m/K_{at})} = \text{gain constant, rad/volt}$$

$$\omega_m = \sqrt{\frac{K_{at}}{(J_a + r^2 M_v)}} = \text{torque motor natural frequency, rad/sec}$$

M_v = mass of spool, lb-sec²/in.

Other symbols are defined in Section 7-2. The quantities ω_r, ω_0, and δ_0 are obtained from Figs. 7-6, 7-7, and 7-8 by interpreting K_m/K_a as K_m/K_{at}.

Because stability depends on the load dynamics, there is the question of what particular load condition to assume. The most common load is inertia, so that

$$P_L D_m = J_t s^2 \theta_m \qquad (7\text{-}53)$$

The valve-motor transfer function for this condition is

$$\frac{\Delta \theta_m}{\Delta x_v} = \frac{\dfrac{K_q}{D_m}}{s\left(\dfrac{s^2}{\omega_h{}^2} + \dfrac{2\delta_h}{\omega_h} s + 1\right)} \tag{7-54}$$

where $\dfrac{2\delta_h}{\omega_h} = \dfrac{K_c J_t}{D_m{}^2}$ $\tag{7-55}$

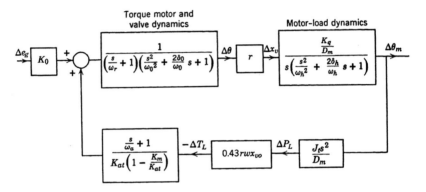

Figure 7-12 Block diagram of a single-stage electrohydraulic servovalve connected to a rotary motor with inertia load.

These relations were derived in Section 6-1 where symbols are defined. Although this load is relatively simple, it does closely resemble the majority of cases and results in a model simple enough to yield design criteria.

The equations given define the servovalve and load system and can be used to form the block diagram of Fig. 7-12. From this figure we see that a portion of the steady state flow force, namely, the third term on the right of (7-49), results in a positive feedback loop which is inherent in the combination of servovalve and motor with load. It is well known that positive feedback loops often present stability problems. The loop gain function is given by

$$GH = \frac{-K_1 s\left(1 + \dfrac{s}{\omega_a}\right)}{\left(\dfrac{s^2}{\omega_h{}^2} + \dfrac{2\delta_h}{\omega_h} s + 1\right)\left(\dfrac{s^2}{\omega_0{}^2} + \dfrac{2\delta_0}{\omega_0} s + 1\right)\left(\dfrac{s}{\omega_r} + 1\right)} \tag{7-56}$$

where K_1 is the open loop gain constant and has units of seconds. Using

(7-47), (7-55), and the definition of K_{at}, K_1 can be written

$$K_1 = \frac{0.43r^2 w K_q x_{v0} J_t}{(K_{at} - K_m)D_m^2} = \frac{2K_R(1 - P_{L0}/P_s)}{1 + K_R(1 - P_{L0}/P_s)} \left(\frac{2\delta_h}{\omega_h}\right) \qquad (7\text{-}57)$$

where $K_R = \dfrac{0.43r^2 w P_s}{(K_a - K_m)}$ = ratio of no load flow force spring rate to net torque motor spring rate, dimensionless

$$(7\text{-}58)$$

We must now determine the stability requirements for the loop in question.* If numerical values were known, we could apply Routh's criterion or make a Nyquist plot to determine stability. However, literal relations are desirable for design purposes. Unfortunately, such relations cannot be formulated in a simple manner for this loop without making approximations. In the majority of cases, ω_h is the lowest break frequency. If we assume that ω_h is dominant, (7-56) can be approximated by

$$GH \approx \frac{-K_1 s}{\dfrac{s^2}{\omega_h^2} + \dfrac{2\delta_h}{\omega_h} s + 1} \qquad (7\text{-}59)$$

Applying Routh's criterion to the characteristic equation $1 + GH = 0$, we obtain the following condition for stability

$$K_1 < \frac{2\delta_h}{\omega_h} \qquad (7\text{-}60)$$

Now, combining with (7-57) gives the final result for stability:

$$K_R\left(1 - \frac{P_{L0}}{P_s}\right) = \frac{0.43r^2 w(P_s - P_{L0})}{K_a - K_m} = \frac{r^2 K_f}{K_a - K_m} < 1 \qquad (7\text{-}61)$$

Thus the ratio of the flow force spring rate to the net spring rate of the torque motor must be less than unity for the combination of single-stage servovalve and motor to be stable. This potential stability problem is caused by the pressure effect of the flow force feeding back on the spool and forming a positive feedback loop.

A viscous friction load on the motor helps to stabilize the system and it is of interest to include this effect in the analysis. This requires adding a term $B_m s\theta_m$ to the right side of (7-53). Also, as shown in Section 6-1,

---

* It should be clear that we are discussing stability of the combination servovalve and motor and not stability of the electrohydraulic servo loop in which these elements would ultimately be placed. It is, of course, desirable that elements which form such a loop be stable in themselves.

addition of B_m will decrease slightly the gain constant and hydraulic natural frequency and increase the damping ratio of the motor-load transfer function. Let us assume the principal effect of B_m is to increase the damping ratio. Hence, we can simply replace $2\delta_h/\omega_h$ by $2\delta_h/\omega_h + V_0 B_m/ 2\beta_e D_m^2$ in (7-54), (7-56), and (7-57) (see Section 6-1 for symbol definition). Hence, the criterion for stability will become

$$K_1 < \frac{2\delta_h}{\omega_h} + \frac{V_0 B_m}{2\beta_e D_m^2} = \frac{2\delta_h}{\omega_h}\left(1 + \frac{B_m}{2\delta_h \omega_h J_t}\right) \qquad (7\text{-}62)$$

which can be combined with (7-57) to yield

$$K_R\left(1 - \frac{P_{L0}}{P_s}\right) < \frac{1 + B_m/2\delta_h \omega_h J_t}{1 - B_m/2\delta_h \omega_h J_t} \qquad (7\text{-}63)$$

Although $B_m/2\delta_h \omega_h J_t$ is usually small (less than unity), the right side of (7-63) is always greater than unity. Hence stability is improved with load friction. Several conclusions can be drawn from this analysis:

1. Because $P_{L0} \approx 0$ for many steady-state load conditions, we conclude from (7-61) and (7-63) that single-stage servovalves should be designed so that K_R is always less than unity. On the other hand, it is desirable that K_R be large to minimize the torque motor size. Because load viscous friction and leakage contribute to damping, it is possible for stability to result even if $K_R > 1$. Weighing all these factors, a value of $K_R \approx 1$ probably represents a satisfactory compromise and a recommended design objective. The reader is referred to the literature [3, 4, 5, 6] for further discussion of this complex problem.*

2. Referring to (7-56), we find that if $\delta_0 \gg 1$ the dynamics of the valve itself becomes dominant and controls stability. Commercial single-stage servovalves often have dashpots attached to the spool to provide the needed damping.

3. Spool valves with underlap have twice the flow force spring rate near null so that K_R should be less than $\frac{1}{2}$ for stability.

4. Because methods of flow force compensation are not effective at small valve openings, the criterion given for K_R should also be used with compensated spools.

5. Even though a simple load condition was assumed, many approximations had to be made to arrive at a useful result. With more complex

---

* McCloy and Martin [6] have made a nonlinear analysis of the single-stage servovalve, assuming no torque motor dynamics but including load viscous friction. When this friction is zero, they show (see Fig. 8 of their paper) that small valve strokes require $K_R < 1$ and large strokes require $K_R < 0.4$ for stability. A linear analysis, not valid for large signals, yields only that $K_R < 1$ for stability.

load dynamics, it would be extremely difficult to predict stability. Therefore a test with each load condition would be necessary to establish stability beyond any doubt. For this reason single-stage servovalves should not be used in systems with changing load dynamics. The advantages of low cost and simplicity of these servovalves are best utilized in servos which are produced in quantity. If applications are few in number, a two-stage servovalve should be used because of its more dependable stability.

Experimentally, the single-stage valve and motor instability occurs at the hydraulic natural frequency ω_h. A simple test to determine the pressure range of stable performance consists of connecting the servovalve to a motor with specified inertia and raising the supply pressure P_s until instability occurs at certain speeds. Instability will first occur in the middle of the motor speed range where B_m is a minimum. (B_m is large at low speeds due to coulomb friction and is large at high speeds due to viscous friction; it has a minimum value in between these speed extremes.) Referring to (7-58) and (7-63), we see that raising P_s increases K_R until instability occurs; K_R can be easily obtained experimentally by plotting curves of valve stroke x_v versus input current i with and without the oil pressure turned on. Substitution of the slopes of these curves into

$$K_R = \frac{\Delta x_v / \Delta i \big|_{\text{oil off}}}{\Delta x_v / \Delta i \big|_{\text{oil on}}} - 1$$

yields the value of K_R for the particular supply pressure selected when the oil is on [4].

The maximum value of GH in (7-59) can be shown to occur at $s = j\omega_h$ and is less than unity if the criterion in (7-60) is satisfied. Because $|GH| < 1$, the feedback portion of the loop in Fig. 7-12 has negligible effect on dynamic response and the over-all transfer function of servovalve and motor with load becomes

$$\frac{\Delta \theta_m}{\Delta e_g} = \frac{rK_0 \dfrac{K_q}{D_m}}{s\left(\dfrac{s^2}{\omega_h{}^2} + \dfrac{2\delta_h}{\omega_h}s + 1\right)\left(\dfrac{s^2}{\omega_0{}^2} + \dfrac{2\delta_0}{\omega_0}s + 1\right)\left(\dfrac{s}{\omega_r} + 1\right)} \qquad (7\text{-}64)$$

The transfer function of the servovalve alone can be defined as

$$\frac{\Delta x_v}{\Delta e_\nu} = \frac{rK_0}{(s/\omega_r + 1)[s^2/\omega_0{}^2 + (2\delta_0/\omega_0)s + 1]} \qquad (7\text{-}65)$$

However, usual practice is to use no load (i.e., $P_L = 0$) flow $\Delta Q_L = K_q \Delta x_v$ as an output parameter rather than spool position.

Let us now turn to the static performance. During steady-state operation the torque generated by the torque motor is consumed by the armature and flow force springs. Therefore

$$K_t \Delta i + K_m\left(\frac{x_v}{r}\right) = K_a\left(\frac{x_v}{r}\right) + 0.43rw(P_s - P_L)x_v \qquad (7\text{-}66)$$

Solving for x_v and substituting into (7-46) yields

$$\frac{Q_L}{Q_0} = \left[\frac{\sqrt{1 - \dfrac{P_L}{P_s}}}{1 + K_R\left(1 - \dfrac{P_L}{P_s}\right)}\right]\frac{\Delta i}{\Delta i_{max}} \qquad (7\text{-}67)$$

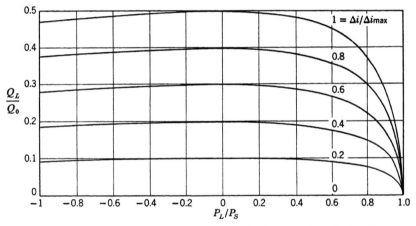

Figure 7-13 Plot of pressure-flow curves for a single-stage electrohydraulic servovalve with $K_R = 1$.

where $Q_0 \equiv rK_tC_dw\sqrt{P_s/\rho}\,\Delta i_{max}/(K_a - K_m) =$ load flow with $K_R =$ $P_L = 0$ and maximum input current, in.^{3}/sec

Equation 7-67 gives the pressure-flow curves for the single-stage servovalve. These curves are plotted in Fig. 7-13 for $K_R = 1$ and positive input current. Note that these curves are relatively flat compared with the usual square root type curves which occur when $K_R = 0$. Thus, flow forces tend to compensate the single-stage servovalve for load pressure variations so that the valve behaves more nearly like a true flow control valve. For $P_L/P_s < 0$ these curves have a slight positive slope. Differentiating (7-67) it can be shown that Q_L/Q_0 reaches a maximum when $P_L/P_s = 1 - (K_R)^{-1}$. Although the slope of the pressure-flow curves is near zero, this should not be interpreted as also making the damping ratio δ_h zero. The damping ratio δ_h is affected by $\partial Q_L/\partial P_L|_{x_v}$ and not by $\partial Q_L/\partial P_L|_{\Delta i}$. Thus, the load compensating effect of the flow force does not alter the motor-load damping ratio.

Referring to (7-67) and letting $P_L = 0$, we note that the maximum flow is reduced by a factor $(1 + K_R)^{-1}$ when flow forces are present. Because K_R should be about unity, a 50% reduction in flow can be expected. However, the upper limit on K_R of unity also imposes an upper limit on the area gradient of single-stage servovalves. This limit is given by

$$w \le \frac{(1/r^2)(K_a - K_m)}{0.43 P_s} \qquad (7\text{-}68)$$

for critical center spool valves. A very large commercial torque motor might have a spring constant of 500 lb/in. so that a servovalve designed for system pressures to 3000 psi would have an upper limit of

$$\frac{500}{0.43(3000)} \approx 0.4 \text{ in.}^2/\text{in.}$$

on the area gradient. This in turn limits the servovalve flow capacity for a given stroke.

One other point warrants discussion before leaving the single-stage servovalve. The steady state gain characteristic of an element is quite important, especially when the element is placed in a servo loop. Changes in gain of an element causes changes in the loop gain and stability and/or response problems may result. Thus, it is quite essential that gain changes be minimized (vary less than say 1.5 or 2 to 1). Because P_L is often near zero, the steady state no load gain of the single-stage servovalve is

$$\left.\frac{Q_L/Q_0}{\Delta i/\Delta i_{\max}}\right|_{\substack{\text{no} \\ \text{load}}} = \frac{1}{1 + K_R} \qquad (7\text{-}69)$$

If K_R is constant, the gain is relatively constant, which is the case for uncompensated valve spools. However, K_R varies considerably with spools having flow force compensation because such compensation is not effective at small spool displacements. As the compensation becomes effective at larger strokes, K_R decreases and causes an increase in the servovalve gain. This increase in gain can cause stability problems at larger flows. Therefore, single-stage servovalves with flow force compensation often have an increasing gain characteristic. For this reason, as well as the additional cost involved, flow force compensation is generally not desirable.

7-4   TWO-STAGE ELECTROHYDRAULIC SERVOVALVE WITH DIRECT FEEDBACK

Two-stage servovalves overcome the disadvantages of limited flow capacity and instability which are inherent in single-stage servovalves. Two-stage servovalves with position feedback are most common and have pressure-flow curves, as illustrated in Fig. 7-1a. Position feedback may be achieved in three basic ways: direct position feedback, as shown in Fig. 7-14; using a spring to convert position to a force signal which is

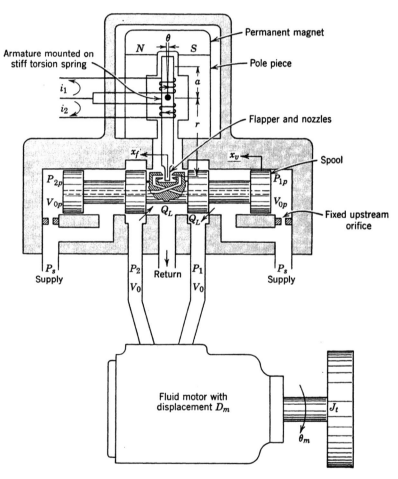

Figure 7-14   Schematic of a two-stage electrohydraulic servovalve with direct feedback controlling a motor with inertia load.

fed back to the torque motor, as illustrated in Fig. 7-17; and by placing stiff springs at the spool ends that act to center the spool against the pressure differential caused by the pilot stage. Direct and force feedback schemes are most widely used for two-stage servovalves and will be treated in this and the next section.

A two-stage servovalve with flapper-nozzle pilot stage and direct position feedback is illustrated in Fig. 7-14. With a positive current differential (i.e., $i_1 > i_2$) the flapper deflects to the left, increasing the pilot pressure P_{1p} (P_{2p} simultaneously decreases), which in turn causes the spool to move to the left. Spool motion ceases when the flapper is again centered between the nozzles. Thus, the main spool valve follows the pilot flapper valve directly because the spool carries the nozzles of the flapper valve. The purpose of this section is to discuss the static and dynamic characteristics of this type of servovalve and indicate design principles when possible.

Let us assume that a linearized analysis about the operating point where both flapper and spool valve load pressure differences (i.e., P_{Lp} and P_L) are zero gives meaningful results. This assumption is supported by a substantial amount of experimental evidence. The basic equations required in this analysis have been derived in other sections so that we need only collect and arrange them. Let us assume a four-way critical center spool valve with a linear area gradient so that the following relations derived in Section 5-3 are applicable: The load flow is given by

$$Q_L = C_d w x_v \left[\frac{1}{\rho} \left(P_s - \frac{x_v}{|x_v|} P_L \right) \right]^{1/2} \qquad (7\text{-}70)$$

Assuming no net damping length, no flow force compensation, and that the reaction of the nozzle flow forces ($F_1 - F_2$) on the spool can be neglected, spool forces consist only of mass and steady state flow force so that

$$P_{Lp} A_v = M_v \frac{d^2 x_v}{dt^2} + 0.43 w (P_s - P_L) x_v \qquad (7\text{-}71)$$

where $P_{Lp} = P_{1p} - P_{2p}$ = pilot (flapper) valve load pressure, psi
  P_{1p}, P_{2p} = pilot valve line pressures, psi
    A_v = end area of spool, in.^{2}
    M_v = mass of spool, lb-sec^2/in.

Linearizing about $P_{L0} = 0$ and Laplace transforming yields

$$A_v \Delta P_{Lp} = M_v s^2 \Delta x_v + 0.43 w P_s \Delta x_v - 0.43 w x_{v0} \Delta P_L \qquad (7\text{-}72)$$

From Fig. 7-14 it is apparent that the spool valve acts as a piston load on the flapper valve. Flow forces on the spool will result in a spring type of load in addition to spool inertia. The transfer function for such a valve-piston combination was derived in Section 6-2 as (6-41). Thus we can write directly that

$$\frac{x_v}{x_\epsilon} = \frac{\dfrac{K_{qp}}{\omega_f A_v}}{\left(\dfrac{s}{\omega_f} + 1\right)\left(\dfrac{s^2}{\omega_{hp}^2} + \dfrac{2\delta_{hp}}{\omega_{hp}} s + 1\right)} \qquad (7\text{-}73)$$

where $x_\epsilon = x_f - x_v$ = variation in flapper clearance about the null value of x_{f0}, in.

$\omega_{hp} = \sqrt{\dfrac{2\beta_e A_v^2}{V_{0p} M_v}}$ = hydraulic natural frequency of pilot stage,* rad/sec

$\delta_{hp} = \dfrac{\omega_{hp} K_{cp} M_v}{2A_v^2}$ = damping ratio of pilot stage, dimensionless

$\omega_f = \dfrac{0.43 w P_s K_{cp}}{A_v^2}$ = break frequency due to flow force on spool, rad/sec

V_{0p} = contained volume at each end of spool, in.[3]

K_{qp} = flow gain of pilot (flapper) valve, in.[3]/sec/in.

K_{cp} = flow-pressure coefficient of pilot valve, in.[3]/sec/psi

Let us now establish the torque motor equations. The torque motor analysis which resulted in (7-40) did not include any forces due to flow impingement on the armature (flapper) but did include inertia and torsion spring forces. The force on the flapper due to flow impingement

---

* Neglecting fluid mass sometimes causes considerable error in computed hydraulic natural frequencies. This is especially so in servovalves because the fluid mass is often larger than the spool mass. It should be apparent that the pressures P_{1p} and P_{2p} in Figs. 7-14 and 7-17 are generated at the nozzles. These pressures must accelerate (a) the mass of fluid in the lines connecting the nozzles to the spool ends, (b) the total mass of fluid in both spool end chambers ρV_f, and (c) the mass of the spool M_v. The connecting lines may be composed of line segments with differing cross-sectional areas and volumes. Let a typical line segment have an area of A_i and a volume of V_i. The total mass can be shown to be

$$M_t = M_v + \rho V_f + \sum_{i=1}^{n} \rho V_i \left(\frac{A_v}{A_i}\right)^2$$

where n is the number of line segments, ρ is the fluid mass density, and A_v is the spool end area. Because A_i is quite small compared with A_v, it is clear that the reflected mass of the fluid in a line segment, that is, $\rho V_i (A_v/A_i)^2$, can be substantial. The fluid mass has been omitted from equations in this chapter to simplify the presentation. However, M_t rather than M_v should be used in computing the hydraulic natural frequency ω_{hp}.

was derived as (5-134). Therefore,

$$F_1 - F_2 = A_N P_{Lp} - (8\pi C_{df}{}^2 P_s x_{f0})(x_f - x_v) \tag{7-74}$$

where x_v is included because the nozzles move with the spool. These three force terms can be included in (7-40) by considering the load on the torque motor armature to be

$$T_L = r A_N P_{Lp} + r(8\pi C_{df}{}^2 P_s x_{f0}) x_v \tag{7-75}$$

and redefining the armature spring rate to also include the negative spring rate due to flow impingement; that is,

$$K_{ae} = K_a - r^2(8\pi C_{df}{}^2 P_s x_{f0}) \tag{7-76}$$

where K_{ae} is the effective armature spring rate. Thus the mechanical spring rate is decreased slightly due to flow forces on the flapper. Therefore, the torque motor is described by

$$\theta = \frac{K_0 e_g - \dfrac{1}{K_{ae}\left(1 - \dfrac{K_m}{K_{ae}}\right)}\left(1 + \dfrac{s}{\omega_a}\right) T_L}{\left(\dfrac{s}{\omega_r} + 1\right)\left(\dfrac{s^2}{\omega_0{}^2} + \dfrac{2\delta_0}{\omega_0}s + 1\right)} \tag{7-77}$$

where $A_N = \dfrac{\pi D_N{}^2}{4} = $ nozzle area, in.^{2}

$x_{f0} = $ clearance between flapper and nozzle at null, in.

$\omega_m = \sqrt{\dfrac{K_{ae}}{J_t}} = $ natural frequency of armature, rad/sec

$r = $ working radius of torque motor, in.

$K_0 = \dfrac{2K_t \mu}{(R_c + r_p)K_{ae}(1 - K_m/K_{ae})} = $ gain constant, rad/volt

The quantities K_t, μ, R_c, r_p, K_a, K_m, ω_a, and J_a are as defined in Section 7-2. The quantities ω_r, ω_0, and δ_0 are found from Figs. 7-6, 7-7, and 7-8 by using K_m/K_{ae} for K_m/K_a.

The final equations required are those that describe the hydraulic motor and load. These relations are needed to establish P_L in the flow force terms. Let us assume simple inertia load so that

$$P_L D_m = J_t s^2 \theta_m \tag{7-78}$$

206

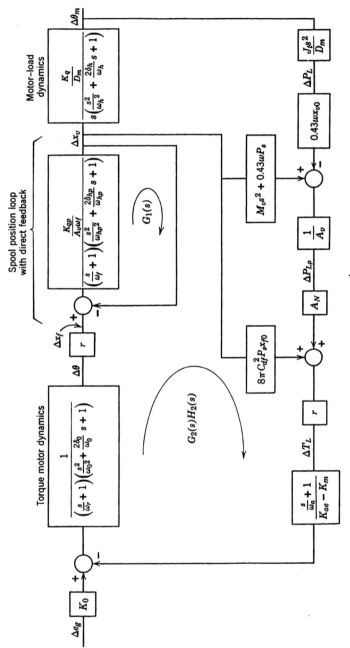

Figure 7-15 Block diagram of a two-stage servovalve with direct feedback and controlling a motor with inertia load.

and the valve-motor transfer function, derived in Section 6-1, becomes

$$\frac{\Delta\theta_m}{\Delta x_v} = \frac{\dfrac{K_q}{D_m}}{s\left(\dfrac{s^2}{\omega_h{}^2} + \dfrac{2\delta_h}{\omega_h}s + 1\right)} \tag{7-79}$$

where $\omega_h = \sqrt{\dfrac{2\beta_e D_m{}^2}{V_0 J_t}}$ = hydraulic natural frequency of load, rad/sec

$\delta_h = \dfrac{\omega_h K_c J_t}{2D_m{}^2}$ = damping ratio, dimensionless

K_q = flow gain of spool valve, in.^{3}/sec/in.
K_c = flow-pressure coefficient of spool valve, in.^{3}/sec/psi
Other symbols are defined in Section 6-1.

Servovalve Block Diagram and Analysis

Equations 7-72, 7-73, 7-77, 7-78, 7-79 and the fact that $x_f = r\theta$ can be combined to form the block diagram in Fig. 7-15. Note that two feedback loops are involved, and these are due to spool position feedback and to pressure feedback on the flapper. Because either loop could pose stability problems, both loops require design attention. Because these loops establish servovalve dynamic performance, let us examine the stability of these loops and determine parameter relationships useful in design.

The principal loop is the spool positioning loop with direct feedback. Because ω_f is usually small, the loop gain function, (7-73), can be approximated by

$$G_1(s) = \frac{\dfrac{K_{qp}}{A_v}}{s\left(\dfrac{s^2}{\omega_{hp}{}^2} + \dfrac{2\delta_{hp}}{\omega_{hp}}s + 1\right)} \tag{7-80}$$

Applying Routh's criterion to this loop, we find that it is stable if

$$\frac{K_{vp}}{\omega_{hp}} < 2\delta_{hp} \tag{7-81}$$

where $K_{vp} = \dfrac{K_{qp}}{A_v}$ = open loop gain (or velocity) constant, sec^{-1} (7-82)

The Bode diagram of $G_1(s)$ is sketched in Fig. 7-16, and we see that the asymptotic crossover frequency is equal to the velocity constant, that is,

$\omega_c \approx K_{vp}$. We also note that ω_c must be considerably less than ω_{hp} when δ_{hp} is small, to prevent the resonant peak from reaching unity gain level and the subsequent instability that would occur. Thus stability requires that

$$\frac{\omega_c}{\omega_{hp}} < 2\delta_{hp} \qquad (7\text{-}83)$$

Because δ_{hp} is on the order of 0.1 for typical cases, the crossover frequency

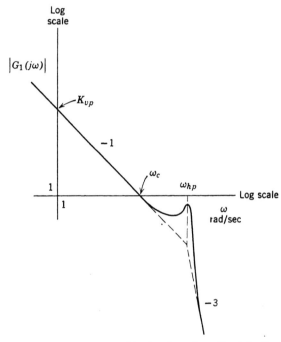

Figure 7-16   Bode diagram of spool positioning loop in a direct feedback two-stage servovalve.

is limited to about 20% of the hydraulic natural frequency of the pilot stage ω_{hp}.

The other loop shown in Fig. 7-15 is the result of the hydraulic pressure feeding back onto the flapper of the torque motor. This feedback loop is undesirable in a direct position feedback servovalve for at least three reasons:

1. Because the loop gain depends on flow induced forces which are not easily predicted and controlled in design, instabilities could result in some servovalve applications.

2. This pressure feedback loop has no proven benefits which outweigh the stability problem. This loop should not be confused with the loop formed when load pressure difference is used as an intentionally designed feedback.

3. Such a feedback loop, if it plays some dominant role, limits the dynamic response of the servovalve. In general, feedback around any element reduces its response compared with its open loop response.

Therefore it is desirable that design criteria be established to ensure that this loop will be stable and will have negligible influence on servovalve response. This result may be achieved by requiring that the loop gain function be less than unity. Let us now determine this loop gain function and its maximum value. As we shall see, the algebraic complexity is such that many approximations must be made to yield a useful result.

Because the closed loop response of the position loop is a block in the pressure feedback loop, see Fig. 7-15, we must first obtain this transfer function. The closed loop response is

$$\frac{\Delta x_v}{\Delta x_f} = \frac{G_1(s)}{1 + G_1(s)} \qquad (7\text{-}84)$$

The characteristic equation $1 + G_1(s) = 0$ is a cubic and difficult to factor. However, it is shown in Section 8-2 that approximate factors consist of a lag at the crossover frequency ω_c and a quadratic at the frequency ω_{hp}. Because $\omega_c \approx K_{vp}$, with approximate factors (7-84) becomes

$$\frac{\Delta x_v}{\Delta x_f} \approx \frac{1}{\left(\dfrac{s}{K_{vp}} + 1\right)\left[\dfrac{s^2}{\omega_{hp}{}^2} + \dfrac{\left(2\delta_{hp} - \dfrac{K_{vp}}{\omega_{hp}}\right)}{\omega_{hp}} s + 1\right]} \qquad (7\text{-}85)$$

Let us now determine the transfer from Δx_v to ΔP_{Lp} in Fig. 7-15. Assuming that $\sqrt{0.43 w P_s / M_v} \gg \omega_h$, M_v can be neglected. Now, because $K_q/K_c = 2P_s/x_{v0}$ and $2\delta_h/\omega_h = K_c J_t/D_m{}^2$, the two paths from Δx_v can be added to give

$$\frac{\Delta P_{Lp}}{\Delta x_v} = \frac{\dfrac{0.43 w P_s}{A_v}\left(\dfrac{s^2}{\omega_h{}^2} - \dfrac{2\delta_h}{\omega_h} s + 1\right)}{\left(\dfrac{s^2}{\omega_h{}^2} + \dfrac{2\delta_h}{\omega_h} s + 1\right)} \qquad (7\text{-}86)$$

Assuming the lead at ω_a effectively cancels the lag at ω_r and neglecting the high frequency break at ω_{hp} in (7-85), the open loop gain function of the pressure feedback loop is approximately given by

$$G_2(s)H_2(s) \approx \frac{\dfrac{r^2(8\pi C_{df}^2 P_s x_{f0}) + r^2 A_N \dfrac{\Delta P_{L,p}}{\Delta x_v}}{K_{ae} - K_m}}{\left(\dfrac{s}{K_{vp}} + 1\right)\left(\dfrac{s^2}{\omega_0^2} + \dfrac{2\delta_0}{\omega_0}s + 1\right)} \tag{7-87}$$

This loop gain function is quite complicated in spite of the many approximations made, and therefore it would be difficult to control the stability of this loop by design. About the best that can be done is to restrict the maximum value of $|G_2 H_2|$ to be less than unity. In this manner the pressure feedback loop can be rendered ineffective for any purpose. If δ_0 is large enough to minimize the resonant peak at ω_0, the maximum value of $|G_2 H_2|$ is approximately

$$|G_2 H_2|_{\max} \approx \frac{r^2(8\pi C_{df}^2 P_s x_{f0}) + (A_N/A_v)(0.43 r^2 w P_s)}{K_{ae} - K_m} \tag{7-88}$$

If this value is restricted well below unity, the Nyquist diagram is confined to the unit circle and stability is achieved. Now, by definition

$$K_{R1} \equiv \frac{r^2(8\pi C_{df}^2 P_s x_{f0})}{K_a - K_m} \tag{7-89}$$

which is the ratio of the negative spring rate due to flapper flow forces to the net spring rate of the torque motor. Combining with (7-76) yields the stability criterion for the pressure feedback loop.

$$|G_2 H_2|_{\max} = \frac{K_{R1} + (A_N/A_v)K_R}{1 - K_{R1}} < 1 \tag{7-90}$$

where K_R is defined by (7-58). In practice, A_N/A_v is usually very small so that $(A_N/A_v)K_R$ is often negligible. In this event the criterion simply states that $K_{R1} < \frac{1}{2}$. It should be emphasized again that rigor has been sacrificed to arrive at a useful result. This criterion is probably conservative because most approximations made were conservative.

Assuming that (7-90) is satisfied by design, we can neglect the pressure feedback path in Fig. 7-15, and the over-all dynamic response of the servovalve becomes

$$\frac{\Delta Q_L}{\Delta e_g} = \frac{r K_0 K_q}{\left(\dfrac{s}{\omega_r} + 1\right)\left(\dfrac{s^2}{\omega_0^2} + \dfrac{2\delta_0}{\omega_0}s + 1\right)\left(\dfrac{s}{K_{vp}} + 1\right)\left[\dfrac{s^2}{\omega_{hp}^2} + \dfrac{\left(2\delta_{hp} - \dfrac{K_{vp}}{\omega_{hp}}\right)}{\omega_{hp}}s + 1\right]} \tag{7-91}$$

where $\Delta Q_L = K_q \, \Delta x_v$ is the no load (i.e., $P_L = 0$) flow. It is apparent that the servovalve transfer function is quite complex because the denominator is eighth order. However, not all of the factors contribute materially to the servovalve response. Usually, K_{vp} is lowest and ω_{hp} is highest of the break frequencies involved. Often, ω_a is large and may be neglected. Thus, an approximate transfer function for a direct feedback two-stage servovalve is

$$\frac{\Delta Q_L}{\Delta e_g} \approx \frac{r K_0 K_q}{\left(\dfrac{s}{K_{vp}} + 1\right)\left(\dfrac{s^2}{\omega_0{}^2} + \dfrac{2\delta_0}{\omega_0} s + 1\right)} \qquad (7\text{-}92)$$

This expression provides a description of servovalve dynamics that is useful in the design of electrohydraulic servos. Because K_{vp} is sometimes less than ω_0, a simple lag at K_{vp} is often used to describe the servovalve; however, this approximation is justified only in cases where the servovalve is much faster than other elements so that a crude description of servovalve dynamics is satisfactory.

For steady state operation $x_v \approx x_f$ and spool position is related to differential current by

$$x_v = \left(\frac{r K_t}{K_a - K_m}\right) \Delta i \qquad (7\text{-}93)$$

which can be combined with Equation 7-70 to yield

$$Q_L = \frac{C_d w r K_t \Delta i}{K_a - K_m}\left[\frac{1}{\rho}\left(P_s - \frac{x_v}{|x_v|} P_L\right)\right]^{\frac{1}{2}} \qquad (7\text{-}94)$$

This is the equation of the pressure-flow curves illustrated in Fig. 7-1a.

From the servovalve analysis just completed it is possible to establish guidelines for design. Servovalve design must begin with given requirements for flow capacity and pressure range. Then, in sizing the servovalve parameters, it is probably best to start with the spool and work back to the torque motor. This procedure is repeated until a compatible set of parameter values is obtained.

Assuming a value for the valve stroke, which may have to be altered later, we can determine the area gradient and major dimensions of the spool as outlined in Section 5-8. Next comes design of the spool positioning loop which is relatively straightforward. From a rough layout of the servovalve, values for spool mass M_v and contained volume V_{op} can be obtained to compute the natural frequency ω_{hp}. The crossover frequency is then selected to be perhaps 10% of ω_{hp} to allow sufficient stability margin. It is usually quite easy to get crossover frequencies in excess of 100 cps for direct feedback servovalves because ω_{hp} is usually in excess of

1000 cps. With ω_c thus established, K_{vp} is also determined because $K_{vp} \approx \omega_c$. With K_{vp} now known and selecting a suitable value for the spool actuation area A_v, we may use (7-82) to determine the required flow gain of the pilot (flapper in this case) valve. Usually some juggling of the values for A_v and K_{qp} is required to yield a jet nozzle diameter that is not so small that it will cause clogging problems nor so large that it will result in a large quiescent flow loss. Once the nozzle diameter is established, then well-known rules discussed in Section 5-9 can be applied to determine flapper clearance and upstream orifice diameter.

The criterion given by (7-90) assures that servovalve performance is independent of load and of flow forces on the spool. This criterion can be used as an aid in selecting torque motor stiffness values. It was shown that the net torque motor spring rate should be at least twice the spring rate due to flow striking the flapper, that is, $(K_a - K_m) > 2r^2(8\pi C_{df}^2 P_s x_{f0})$, to insure stability of the pressure feedback loop. This establishes a minimum for $(K_a - K_m)$ because x_{f0}, P_s, and r are determined from other considerations. At the other extreme K_a should be large enough so that the torque motor resonance ω_m is not so low as to limit servovalve bandwidth. However, if K_a is too large, torque motor size and electrical power requirements become undesirable. Some weighing of these factors and the desired stroke, making sure that (7-90) is adequately satisfied, are required to arrive at a satisfactory value for K_a. Once K_a is established, K_m is selected so that the static curve of the torque motor is reasonably linear because the torque motor appears open loop in direct feedback servovalves. The entire design process is now repeated, if necessary, to obtain final parameter values.

In summary, an analysis of direct feedback two stage servovalves has been given. The analysis included flow forces on the spool and pressure feedback to the torque motor flapper, and two main feedback loops were identified: the principal position feedback loop and a pressure feedback loop. Potential stability problems make the pressure feedback loop undesirable. Criteria for servovalve design were developed, and a design procedure was suggested.

7-5 TWO-STAGE ELECTROHYDRAULIC SERVOVALVE WITH FORCE FEEDBACK

As described in the introductory paragraphs of Section 7-4, another common method of obtaining spool position feedback is to use a feedback spring that connects between spool and flapper to provide a force balance. Referring to Fig. 7-17, a positive current difference (i.e., $i_1 > i_2$) causes a torque on the flapper which moves it to the left, increasing pressure P_{p1}

and decreasing P_{p2}. The spool then moves to the right and continues to move until the torque on the flapper (armature) due to the feedback spring balances the torque due to the input current. At this point the flapper is approximately centered between the nozzles, but the spool has taken a new position directly proportional to the input current.

This type of servovalve will now be analyzed to determine static and dynamic performance and design criteria. This discussion will be somewhat abbreviated because many points applicable here were discussed in detail in Section 7-4. Let us now collect the appropriate equations and

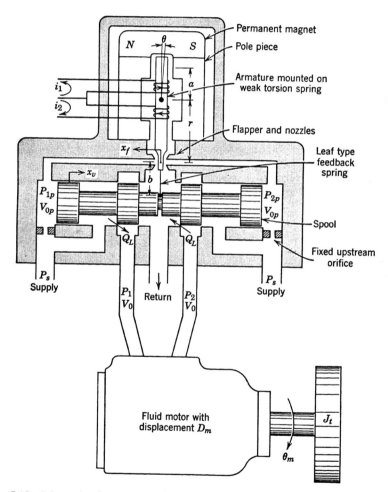

Figure 7-17   Schematic of a two-stage electrohydraulic servovalve with force feedback controlling a motor with inertia load.

construct the block diagram of the servovalve. Equation 7-21 describes the voltage drops around the coil circuits and can be written

$$\mu e_g = \tfrac{1}{2}(R_c + r_p)\left(1 + \frac{s}{\omega_a}\right)\Delta i + K_b s\theta \qquad (7\text{-}95)$$

where $\omega_a = \dfrac{(R_c + r_p)}{2L_c}$ is the armature circuit break frequency.

Referring to Fig. 7-17, the total deflection of the cantilevered feedback spring at its free end is $[(r + b)\theta + x_v]$. Now, by combining the torques defined in Equations 5-134 and 7-31 with that due to the feedback spring, the armature torque equation becomes

$$K_t \Delta i = J_a s^2 \theta + K_{an}\theta + rP_{Lp}A_N + (r + b) K_f[(r + b)\theta + x_v] \qquad (7\text{-}96)$$

where symbols not previously defined are:

$$K_{an} = K_a - K_m - r^2(8\pi C_{df}{}^2 P_s x_{f0}) = \text{net spring rate, in.-lb/rad}$$
$$x_f = r\theta = \text{flapper displacement from null, in.}$$

$$K_f = \frac{3EI}{l^3} = \begin{array}{l}\text{spring constant of the cantilevered feedback spring at}\\ \text{the free (i.e., spool) end, lb/in.}\end{array}$$

The transfer function of the flapper valve controlling the spool is given by (7-73). Neglecting the flow force on the spool, $\omega_f \rightarrow 0$, and this equation reduces to

$$\frac{\Delta x_v}{\Delta x_f} = \frac{\dfrac{K_{qp}}{A_v}}{s\left(\dfrac{s^2}{\omega_{hp}{}^2} + \dfrac{2\delta_{hp}}{\omega_{hp}}s + 1\right)} \qquad (7\text{-}97)$$

These equations can now be combined to form the block diagram in Fig. 7-18. As in the case with direct feedback, two main feedback paths can be identified. The principal loop is the spool positioning loop which is formed by the force feedback through the spring connection between the spool and the flapper. The other loop is due to pressure feedback on the flapper and is of secondary importance.

The spool position loop is a Type 1 servo loop and has a velocity constant of

$$K_{vf} = \frac{r(r + b)K_f K_{qp}}{A_v[K_{an} + K_f(r + b)^2]} \quad \text{rad/sec} \qquad (7\text{-}98)$$

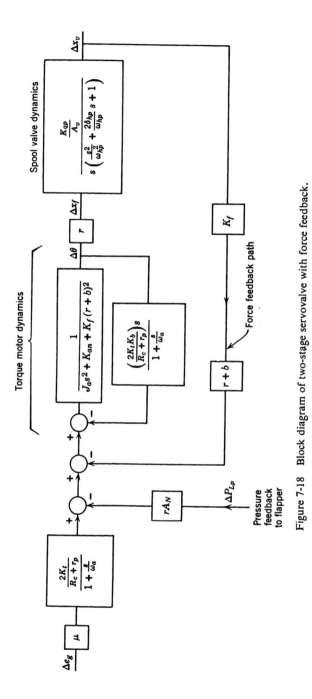

Figure 7-18 Block diagram of two-stage servovalve with force feedback.

215

To maximize this gain constant, these servovalves are usually designed so that $K_{an} = 0$; that is,

$$K_a = K_m + r^2(8\pi C_{df}{}^2 P_s x_{f0}) \tag{7-99}$$

Thus the mechanical spring rate on the torque motor armature just balances the negative spring rates due to magnetics and to flow forces on the flapper. Therefore, the torque motor by itself is just on the borderline of static stability. However, K_f becomes the major spring rate when the torque motor is installed in the servovalve and static stability is assured. With this design criterion, (7-98) reduces to

$$K_{vf} = \left(\frac{r}{r + b}\right)\frac{K_{qp}}{A_v} \tag{7-100}$$

The dimension b is dictated by the feedback spring design but should be kept short to maximize gain. The maximum value of K_{vf} is limited by stability considerations to less than about 20% of the lowest natural frequency in the loop. Referring to Fig. 7-18, two such natural frequencies occur in the force feedback loop. Although not readily apparent, the inner loop containing the torque motor dynamics has a cubic characteristic equation which has a quadratic factor at around the frequency

$$\omega_{mf} = \left[\frac{K_{an} + K_f(r + b)^2}{J_a}\right]^{1/2} \tag{7-101}$$

The other natural frequency is ω_{hp}. Because ω_{mf} is nearly always the lower of these two frequencies, the stability criterion for the force feedback loop becomes

$$K_{vf} < 0.2\omega_{mf} \tag{7-102}$$

With ample margin, this relation may be used for design purposes.

As pointed out in Section 7-4, the pressure feedback loop serves no useful purpose but could be the source of stability problems. This loop can be rendered ineffective by keeping the open loop gain below unity at all frequencies. Assuming a rotary motor with inertia load, (7-86) approximately describes the feedback connection between Δx_v and ΔP_{Lp} in Fig. 7-18. The maximum gain for this portion of the loop is $0.43wP_s/A_v$ and the maximum value for the remainder of the loop is $rA_N/(r + b)K_f$. The product of these quantities is the approximate maximum value for the open loop gain of the pressure feedback loop and should be less than unity to ensure stability. Therefore

$$\left(\frac{r}{r + b}\right)\left(\frac{A_N}{A_v}\right)\frac{0.43wP_s}{K_f} < 1 \tag{7-103}$$

This criterion is usually met with no difficulty.

Neglecting the pressure feedback loop, the dynamic response of the force feedback servovalve is approximately given by

$$\frac{\Delta x_v}{\Delta e_g} = \frac{\dfrac{2\mu K_t}{(R_c + r_p)(r + b)K_f}}{\left(1 + \dfrac{s}{K_{vf}}\right)\left(1 + \dfrac{s}{\omega_a}\right)\left(\dfrac{s^2}{\omega_{mf}^2} + \dfrac{2\delta_{mf}}{\omega_{mf}}s + 1\right)} \qquad (7\text{-}104)$$

where the closed loop response of the position loop has been approximated with a lag at K_{vf} and a quadratic at ω_{mf}. Usual practice is to use the no load flow (i.e., $\Delta Q_L = K_q \Delta x_v$) as an output parameter rather than Δx_v. Because K_{vf} is lowest of the break frequencies in (7-104), the lag at this frequency dominates servovalve response and is often used as an approximation of servovalve dynamics.

For steady-state operation $x_f \approx 0$ and spool position is related to differential current by

$$x_v = \frac{K_t}{(r + b)K_f} \Delta i \qquad (7\text{-}105)$$

which can be substituted into Equation 7-70 to give the pressure-flow curves illustrated in Fig. 7-1a.

Design of these servovalves proceeds as outlined in Section 7-4. Spool valve dimensions are sized to satisfy load flow and pressure requirements. The flapper valve gain is determined from (7-100), and the other flapper valve dimensions follow from Section 5-9. The armature spring rate is sized in accordance with (7-99); K_f should be large enough to satisfy (7-103) and to keep ω_{mf} high so that K_{vf} is not limited to a low value. However, K_f should be small to minimize the input current required for full stroke and to minimize torque motor size. In accordance with (7-105), K_t must be large enough to obtain the desired spool stroke. The spool stroke should be as large as is practical to improve resolution and silting performance and to reduce over-all servovalve size. This is more possible with force feedback servovalves because torque motor linearity is not as critical since the armature operates about its neutral position. This is the basic advantage of the force feedback arrangement. The major disadvantage is that the torque motor dynamics appear inside the spool position loop and, consequently, limit the dynamic response of the servovalve. This disadvantage is more acute with valves of higher flow capacity.

7-6 SPECIFICATION, SELECTION, AND USE OF SERVOVALVES

Servovalves convert electric signals into a hydraulic output which controls the acceleration, velocity, and position of an actuator. Because servovalves occupy a unique position at the electric-hydraulic interface of systems and require some knowledge of both areas to understand their operation, they are often the focal point of attention in servo design and development and are often blamed when system troubles occur. This has resulted in an extraordinarily long list of parameters which specify servovalve performance and the recommendation of standards by subcommittees of SAE and of AIEE.* When compared with other hydraulic components, the specification of servovalve performance is way overdone. However, there are some fundamental static and dynamic performance characteristics which are useful to both designer and user in selecting and evaluating servovalves.

Servovalves are rated by the load flow (Q_L), called *rated flow* and designated Q_R, obtainable with maximum input current and at a given valve pressure drop $P_v = P_s - P_L$. A valve drop of 1000 psi is virtually standard and stems from the fact that maximum power transfer to the load occurs with this valve drop when the supply pressure (P_s) is 3000 psi. Although servovalves are rated at 1000 psi drop, the flow at other valve drops can be computed from $Q_R\sqrt{P_v/1000}$ for the usual servovalve with square-root type pressure-flow curves.

The *pressure-flow curves*, illustrated in Fig. 7-19a, completely define the static performance of a servovalve but are troublesome to measure and are difficult to read near null, where most operation occurs. These curves are mainly useful in determining servovalve type (see Fig. 7-1) and in sizing the servovalve to match load flow and load pressure requirements. The principal method of sizing a servovalve is to select a valve with a flow rating large enough that the curve for Δi_{max} in Fig. 7-19a encompasses all load flow and load pressure points expected in the system duty cycle, making sure that $P_L < \frac{2}{3}P_s$. This ensures that all loads are within the servovalve range but may require full current with the associated system error. If loads must be handled with less static error, then load points should be confined within a curve for less current. The servovalve flow rating can also be used as a basis for selection by choosing a rating such that $Q_R\sqrt{P_v/1000}$ is greater than the desired load flow at all valve drops, making sure that $P_L < \frac{2}{3}P_s$, as explained in Section 8-1.

The *flow gain curve* (Fig. 7-19b) is a plot of load flow (Q_L) versus input current (Δi) for a given pressure drop, usually 1000 psi, across the servovalve. It is desirable that the flow gain be measured by continuously plotting the variables on an x-y recorder with input current going through

---

* Servovalve terminology and specification standards have also been proposed by manufacturers such as Moog Servocontrols, Inc., East Aurora, New York and Vickers, Inc., Detroit, Michigan, and are available from these companies.

a complete cycle between the extreme plus and minus values. The flow gain is useful because it indicates static hysteresis width, linearity, symmetry, and most important it shows the type of null characteristic. As illustrated in Fig. 5-2, the gain at null depends on the fit between the spool

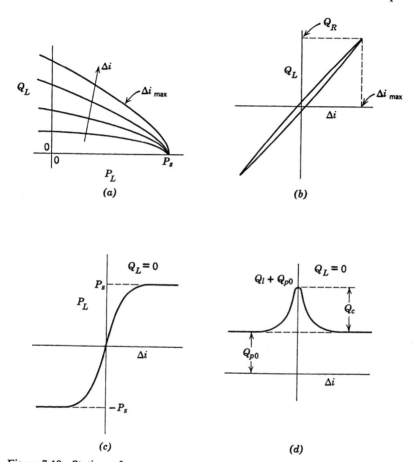

Figure 7-19   Static performance curves for a servovalve: (*a*) pressure-flow curves; (*b*) flow gain curve;   (*c*) pressure sensitivity curve;   (*d*) leakage curve.

and the sleeve at neutral.   A critical center valve is usually desirable, as discussed in Sections 5-1 and 5-8, because of its linear gain. The gain of the flow curve at null is also an indicator of quality of manufacture because it reveals the calibration of spool and sleeve.

The *pressure sensitivity curve* (Fig. 7-19*c*) is obtained by blocking the load lines and measuring the load pressure as the servovalve is stroked through a complete current cycle for a particular supply pressure (usually 1000 psi). This curve should have a steep slope and snap to saturation in a small portion of full stroke current. A large pressure sensitivity is desirable, especially to overcome stiction loads and constant loads with minimum error. A low pressure sensitivity indicates large center flow, poor valve at fit neutral, and may result in slow, sluggish servo performance.

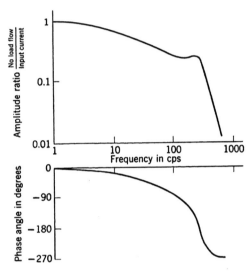

Figure 7-20   Frequency response of servovalve from input current to no load flow output.

The *leakage curve* (Fig. 7-19*d*) is a plot of flow to (or from) the servovalve with load lines blocked and current cycled over the full range. For a two-stage servovalve, the leakage flow consists of the pilot or first stage valve leakage flow Q_{po} and the spool leakage flow Q_l. As pointed out in Section 5-3, the ratio Q_c/P_s can be used as the flow-pressure coefficient for the spool. The total leakage flow should only be a few percent of rated flow except on small servovalves where a higher proportion is to be expected. Large null leakage results from a poor valve fit or from valve wear and causes low pressure sensitivity, excess flow drain on the hydraulic supply, and unnecessary standby power loss. The center flow Q_c is probably the best indicator of valve wear and can be used to signal a worn out servovalve.

The dynamic characteristics of servovalves are usually specified by giving the *frequency response* from input current to the flow supplied to a test actuator (Fig. 7-20). The flow is obtained by measuring actuator velocity and multiplying by area; hence this flow is actually the ideal flow (or no load flow) to the actuator. The dynamics of the actuator, specifically the resonant frequency ω_h, and connecting lines must be higher than that of the servovalve for the data obtained to be valid. The frequency at which the phase shift is $-90°$ is usually taken as the servovalve bandwidth. Bandwidth is a measure of the servovalve speed of response. It is well known that servos are limited by the lowest natural frequency that occurs in the loop. Usually the motor-load natural frequency ω_h is lowest and is the limiting factor on over-all servo response. There is no justifiable reason, except perhaps in some special application, for the servovalve bandwidth being low enough to limit servo performance. Hence, the servovalve should be selected to have a bandwidth greater than other natural frequencies in the loop. However, a very large servovalve bandwidth may allow electric noise and dither signals to be transmitted to the load. Hence, it is desirable that the servovalve be fast enough not to limit servo response but slow enough to attenuate unwanted high frequency signals.

Parameters which define servovalve *threshold* and *null shift* with temperature and pressure are often useful in establishing servo error due to these effects. However, these parameters are usually specified with a single numerical parameter rather than a curve.

Servovalves have *hysteresis* in the flow gain curve because of the magnetic paths in the torque motor. Width of the hysteresis loop varies directly with input signal amplitude, and it is usually expressed as a percentage of full stroke current, for example, $\pm 2\%$. However, the same percentage would approximately apply to smaller input signals as well. Because width of the loop shrinks as the input signal decreases, hysteresis does not cause problems in most servo systems. However, all servovalves have a certain amount of backlash in which the loop width is a constant. This backlash is due to slipping of the mechanical anchorages in the torque motor, but it is mainly due to friction between spool and sleeve. This backlash can be greatly increased if the oil is dirty [7] and may cause servo system instability.

Dither is often used to improve the performance of servovalves. Dither is a high frequency oscillation which is added to the signal at the servovalve input. The action of dither is to reduce or remove the backlash in a servovalve. Dither is especially effective in two stage servovalves that have a spool type pilot stage. In these cases, dither provides a source of relative motion between pilot spool and its sleeve and, thereby, effectively removes the backlash due to friction between these elements. Dither is not effective in removing the servovalve hysteresis due to the magnetic circuit of the torque motor. Because dither keeps the main spool in motion (at very low amplitudes), it is also effective in preventing silting because small particles in the fluid at the metering orifices are in agitation and are swept through as the orifice opens during a portion of the dither cycle.

Neither the frequency nor the waveform of the dither signal seem to be critical to its action. The dither frequency should be considerably in excess of the expected signal frequencies and should not coincide with resonances in the servovalve or in the actuator and load. Excitation of these resonances could cause fatigue failures or saturate the elements involved. There does not seem to be an upper limit to the dither frequency, but higher frequencies would require a larger dither amplitude because of attenuation in the torque motor. Sine, triangular, and square waves seem to be equally effective as a dither waveform. The amplitude of the dither should be large enough so that its peak-to-peak value just fills the backlash width. This may amount to 0.0001 in. or so of main spool motion. The dither amplitude should not be so large that it is transmitted to the load by the servovalve flow. In this event, dither could result in excessive wear or fatigue failure of hydraulic motors, gears, and other mechanical elements in the drive system.

There are some practical matters which require attention in the use of servovalves. Radial clearance between spool and sleeve may vary from 0.000050 in. ($1.25\ \mu$) to perhaps 0.0004 in. ($10\ \mu$). Thus, particles in the fluid with dimensions in this range can lodge in these clearances and cause the spool to bind. The friction forces thus generated cause backlash to appear in the servovalve flow gain curve, which in turn may cause servo system instability. Contaminants in the fluid can also cause erosive wear of the orifice metering edges, which increases center flow and shortens valve life, clogs internal servovalve orifices and filters, and leads to silting. Silting is the buildup of fine particles at the metering ports and is most noticeable at valve neutral. When the spool is at neutral, silting can seal the lines to the load so effectively that line pressures, which are normally around $P_s/2$, will drop to zero due to leakage, and the system becomes dead. For these reasons servovalve manufacturers recommend at least 10 micron filtration of the fluid entering the servovalve. Filtration is more art than science at the present time, but the care to be taken in this area cannot be over emphasized and the rewards are long component life and system reliability.

One further area should be discussed. Occasionally, oscillation problems arise which are associated with servovalves and require troubleshooting. Limit cycle oscillations are almost always a feedback instability (as opposed to forced excitation) and, as such, a complete loop having both forward and feedback paths is required. An attempt should be made to identify both of these paths. Generally, the feedback path for these oscillations, assuming that the basic spool positioning loop is stable, is through the steady-state flow force on the spool, which in turn reflects back onto the torque motor armature. The forward path involves the lines connecting the servovalve to the load and the actuator. When such oscillations occur, it is usually helpful to make some physical changes in line lengths, mounting manifolds, and/or hydraulic actuators to determine contributors to the instability. Quite often line resonances are the major path contributor to oscillations [8, 9].

Another clue in oscillation problems is the frequency which may be traceable to some particular element or feedback loop. For example, the frequency of oscillation may coincide with resonant frequencies of structural elements internal to the servovalve, one quarter wave length frequency of connecting lines, the pilot natural frequency ω_{h_p}, or the motor-load natural frequency ω_h. Over-all we should be certain that the instability is due to the servovalve and load and not to other effects, such as an unstable relief valve in the hydraulic supply, resonances of auxiliary lines, pump pulsations, or forced vibrations from some source. The comments given in Section 10-10 might also prove helpful in this problem.

REFERENCES

[1] White, D. C., and H. H. Woodson, *Electromechanical Energy Conversion*. New York: Wiley, 1959, p. 396.

[2] Dunn, J. F., "A Study of Permanent-Magnet Torque Motors," *D.A.C.L. Research Memorandum No. R.M.* 6387-5, Dynamic Analysis and Control Laboratory, M.I.T., 1954.

[3] Noton, G. J., and D. E. Turnbull, "Some Factors Influencing the Stability of Piston-Type Control Valves," *Proc. Inst. Mech. Eng.* (London), **172**, No. 40, 1958, 1065–1081.

[4] Williams, H., "Effect of Oil Momentum Forces on the Performance of Electro-Hydraulic Servomechanisms," *Proc. Conf. Recent Mech. Eng. Dev. Automatic Control* (Institute of Mechanical Engineers, London), No. 31, 1960.

[5] Eynon, G. T., "Developments in High-Performance Electromechanical Servo-Mechanisms at the Royal Aircraft Establishment, Farnborough," *Proc. Conf. Recent Mech. Eng. Dev. Automatic Control* (Institute of Mechanical Engineers, London), No. 93, 1960.

[6] McCloy, D., and H. R. Martin, "Some Effects of Cavitation and Flow Forces in the Electro-Hydraulic Servomechanism," *Proc. Inst. Mech. Eng.* (London), **178**, Pt. 1, No. 21, 1963–64.

[7] Wheeler, Jr., H. L., *Filtration in Modern Fluid Systems*. Bendix Corporation, 434 W. 12 Mile Road, Madison Heights, Michigan, 1964.

[8] Ainsworth, F. W., "The Effect of Oil-Column Acoustic Resonance on Hydraulic Valve Squeal," *Trans. ASME*, **78**, 1956, 773–778.

[9] Blackburn, J. F., G. Reethof, and J. L. Shearer, *Fluid Power Control*. New York: Technology Press of M.I.T. and Wiley, 1960, Chapter 12.

8

Electrohydraulic Servomechanisms

Electrohydraulic servos are capable of performance superior to that of any other type of servo. Large inertia and torque loads can be handled with high accuracy and very rapid response. The power portion of these servos consist of an electrohydraulic servovalve, hydraulic actuator, and hydraulic power supply. The error portion of the servo loop consists of feedback transducers which measure the quantity being controlled and electronic amplifiers, discriminators, and compensation networks appropriate to furnish a signal to the servovalve to close the loop. Electrohydraulic servos derive their flexibility from the electronic portion and their power handling ability from the hydraulic portion.

The electronic portion of the loop adds flexibility to the electrohydraulic servo in many respects. The wide variety of electrical transducers permit control of many quantities. The versatility of electronic amplifiers allow gain changes to be made easily and permit use of compensation networks to correct an unsatisfactory servo loop gain. Such networks consist of resistors and capacitors which are simpler and much less susceptible to parameter variations compared with equivalent hydraulic and/or mechanical networks. These corrective networks manipulate the loop gain and allow the designer a choice of low frequency (essentially static) performance characteristics such as steady-state error and compliance due to static loads. Resistance adders and difference amplifiers permit other quantities to be fedback in subsidiary (minor) loops to alter dynamic characteristics of particular elements in the over-all loop. A variety of feedback transducers, ease of signal correction, and addition of minor loops make electrohydraulic servos very versatile and the best choice for many applications.

Feedback measuring devices having an electrical output can be obtained to sense almost any physical quantity such as position, velocity, acceleration, force, pressure, temperature, humidity, light intensity, etc., and electrohydraulic servos may be designed to control any of these quantities. Position is probably most often controlled, and typical position sensing

224

devices are synchros, variable reluctance pickups, and potentiometers. Tachometers and accelerometers are used to obtain velocity and acceleration as electrical signals. Electrical signals proportional to force or pressure can be obtained from many types of commercial sensors. The waveform of the electrical output depends on the particular transducer. Some devices have an electrical output which varies continuously with the quantity being measured. This type output is usually referred to as d-c, but the name is not particularly appropriate because the output waveform may have any shape. A second type of output uses the signal to amplitude modulate a carrier frequency of 60 or 400 cps or sometimes higher. Because the carrier frequency occurs in the output only for d-c inputs signals, the result is called suppressed-carrier modulation. This type output, typical of synchros and many other common transducers, requires a discriminator to recover the phase and amplitude information of the signal. A third type of output consists of a pulse train whose amplitudes (or widths) are modulated proportional to the quantity being measured. Hence, the quantity being measured is sampled at discrete intervals. This type of waveform is found in systems having digital inputs. Systems using these three types of electrical signals are referred to as being d-c data, a-c data, and sampled data, respectively.

Electrohydraulic servos are usually classified by the name of the quantity being controlled. Other design features such as type of data in the electronic section, type of series compensation, quantities fedback in minor loops, rated horsepower, etc. are also often used for descriptive purposes.

In this chapter we discuss analysis and design of the principal types of electrohydraulic servos.

8-1 SUPPLY PRESSURE AND POWER ELEMENT SELECTION

In this section we will discuss the factors involved in selecting system supply pressure and determining the size of servovalve and of hydraulic actuator for servos.

Supply Pressure Selection

One of the first steps in the design of a hydraulic control system operating from a constant pressure supply is to select the value for the supply pressure. Factors influencing this selection are:

1. Many considerations favor a large supply pressure. Horsepower is the product of pressure and flow. As supply pressure is increased, less flow is required to achieve a given horsepower. Smaller pumps, lines, valves, oil supply, etc., are then possible. As a result, over-all weight is

decreased. Beyond 4000 psi, however, strength considerations dictate an increase in weight [1]. Faster response is often possible because of smaller oil volumes and higher bulk modulus, but these are not major considerations because a low pressure system can be designed with comparable response.

2. Several factors suggest lower supply pressures. Leakages increase with supply pressure and results in higher oil temperatures. As a result, tolerances must be tightened and system cost increased. Objectionable noise also increases with pressure. Component life, especially that of rotary motors with antifriction bearings, decreases substantially because of higher bearing loads.

These considerations have resulted in the use of 500 to 2000 psi for industrial systems and 3000 to 4000 psi for aircraft systems because of the weight reduction benefit. In general, lower supply pressures are always desirable when choice permits because they are more conducive to long component and system life, produce lower leakages which minimize power loss, need less maintenance, and permit larger contaminants without failure. Of course, the final choice of supply pressure must be made in conjunction with the hydraulic actuator sizing to accommodate expected loads and the pressure rating of the components involved.

Maximum Power Transfer to Load with a Servovalve

It will prove helpful to determine the conditions for maximum power transfer to the load with a servovalve controlled actuator. For simplicity, let us assume a critical center valve for which the flow with a positive spool displacement has been shown to be

$$Q_L = C_d w x_v \left[\frac{1}{\rho} (P_s - P_L) \right]^{\frac{1}{2}} \tag{8-1}$$

The horsepower supplied to the load is then

$$\text{hp} \big|_{\text{load}} = P_L Q_L = C_d w x_v \left(\frac{P_s}{\rho} \right)^{\frac{1}{2}} P_s \left(\frac{P_L}{P_s} \right) \left(1 - \frac{P_L}{P_s} \right)^{\frac{1}{2}} \tag{8-2}$$

which is plotted in Fig. 8-1. The load horsepower is zero when $P_L = P_s$ because no pressure drop remains across the servovalve orifices to yield a flow to cause actuator motion; the actuator is stalled. When $P_L = 0$ no horsepower is required by the load. Maximum horsepower occurs between these extremes and is found by forming $d\text{hp} \big|_{\text{load}} / dP_L = 0$ and simplifying to give the result which is

$$P_L = \tfrac{2}{3} P_s \tag{8-3}$$

Thus, maximum power is transferred to the load when the pressure across the load is two thirds of supply pressure. However, because P_L is usually changing during servo operation, it will be at $\frac{2}{3}P_s$ only a small portion of the operating time. Hence power is usually transferred to the load in a less optimum manner. Therefore this relationship is not very useful in design unless the load is relatively constant over a duty cycle.

However, as a general rule power elements are sized such that P_L does not exceed $\frac{2}{3}P_s$ for the maximum loads normally expected. Although this

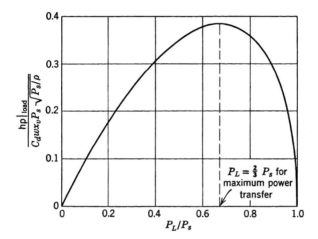

Figure 8-1   Normalized plot of power at load versus load pressure.

rule promotes efficiency by handling the largest expected load at the point of maximum power transfer, this is not the most important reason for the rule. It should be apparent that as load is increased, P_L approaches P_s and the hydraulic actuator tends to stall. Because the pressure drop across the servovalve approaches zero, the flow gain is decreased and the servo tends to loose control of the load. Hence, it is desirable to limit P_L so that the flow gain is high enough for the servo to maintain control. If it is limited to $\frac{2}{3}P_s$, then the flow gain is reduced to not more than

$$\frac{\sqrt{P_s - \frac{2}{3}P_s}}{\sqrt{P_s}} = 0.577 = 57.7\%$$

of the no load flow gain. Because servos are usually designed to provide sufficient control if the loop gain drops by half to allow margin for gain changes in components due to tolerances, aging, temperature, loads, etc., then the $\frac{2}{3}P_s$ rule seems adequate and should be disregarded with caution.

Although it is quite possible that high acceleration rates during rapid transients may cause P_L to exceed $\frac{2}{3}P_s$, such conditions are usually of short duration and need not be taken into account.

Maximum Power Transfer and Efficiency

Combining (8-2) and (8-3), the maximum power that can be transmitted to the load is

$$\text{hp}_{\text{max}}\big|_{\text{load}} = \frac{2C_d w x_{v\text{max}}}{\sqrt{\rho}}\left(\frac{P_s}{3}\right)^{\!\!3/2} \tag{8-4}$$

and occurs when $P_L = \frac{2}{3}P_s$. The operating efficiency of a servovalve controlled actuator may be stated as

$$\eta_{op} = \frac{\text{hp}\big|_{\text{load}}}{\text{hp}\big|_{\text{gen}}} \times 100 = \frac{P_L Q_L}{P_s Q_L} \times 100 = \frac{P_L}{P_s} \times 100 \tag{8-5}$$

This equation assumes that the hydraulic power supply uses a variable delivery pump which adjusts its output flow to just meet the required load flow. Some additional pump flow, of course, would be required to meet leakage losses. The operating efficiency is zero at no load (for the obvious reason that no horsepower is required by the load) and increases to 100% with a stall load. Because P_L/P_s is normally limited to $\frac{2}{3}$, the maximum operating efficiency is 67% for servovalve controlled actuators. The power generated at no load ($P_L = 0$), which is dissipated to heat oil across the servovalve orifices, may vary widely from zero to $P_s Q_{L_{\text{max}}}$, depending on the flow required to achieve maximum velocity ($Q_{L_{\text{max}}}$).

If the hydraulic power supply uses a constant displacement pump, the operating efficiency would be

$$\eta_{op} = \frac{P_L Q_L}{P_s Q_{L_{\text{max}}}} \times 100 \tag{8-6}$$

which is always less than that with a variable displacement pump. The power loss at no load and zero actuator velocity is $P_s Q_{L_{\text{max}}}$, which is dissipated to heat the fluid across the relief valve and requires the supply to have an oil cooler.

Although hydraulic power losses must be reckoned with because of higher oil temperatures, efficiency is not a major consideration in hydraulic servo design and operation. A very desirable attribute of a servo is its ability to hold a given position against disturbing influences. Because no motion takes place, no work is done and the efficiency of this operation is zero. It is clear then that factors such as accuracy, response, and compliance are considerations that outweigh efficiency in importance. This fact is typical of control as a discipline. Generally, the power involved in the process under control is much larger than the power required for control purposes. Therefore, improvements in efficiency of the process under control would net greater return in over-all cost reduction.

Selection of Hydraulic Actuator

Two basic considerations govern the size (i.e., piston area or motor displacement) selected for the power output device of a hydraulic system:

1. The size should be large enough to handle the loads expected during a duty cycle. This is often the prime purpose of an actuator and requires an analysis of typical duty cycles to obtain horsepower and force or torque load requirements.

2. Closed loop response of a servo is limited by the lowest open loop resonance (see Section 8-2). Because this is usually the motor-load hydraulic natural frequency, it must be large enough to permit acceptable servo response. Hence, the actuator size should be large enough so that the associated hydraulic natural frequency is adequate.

Both these factors are important and warrant attention. An actuator that handles the load but is slow in response will be as unsatisfactory as one undersized for loads. However, it is also important not to oversize actuators so that the flow required for maximum velocity is kept to a minimum. Otherwise, the hydraulic power supply becomes bulky with large no load power losses. This is the basic disadvantage to selecting large size actuators. In some cases where extremely large loads and horsepowers are involved, actuator size is chosen mainly to meet the loads, and response characteristics are secondary. Simultaneous requirements for fast response and high loads are exceptional.

As an example, suppose the output device is a piston. The required piston area may be found by solving (6-34). Therefore

$$A_p = \frac{M_t(d^2x_p/dt^2) + B_p(dx_p/dt) + Kx_p + F_L}{P_L} \tag{8-7}$$

Analysis of the most stringent duty cycle will give values for maximum acceleration and velocity which may be assumed to occur simultaneously as a first approximation. Using the $\frac{2}{3}P_s$ value for P_L, there is adequate flow gain control and the power required for maximum load is transferred in an optimum manner. If the dynamic specifications for the system are stringent, the allowable valve position error can be multiplied by the pressure sensitivity of the servovalve to determine a value for P_L. If this value is used, maximum load would be handled within the specified error. However, the piston area would be large and would require large flows for maximum velocity but may be warranted in extreme cases.

Referring to (6-37), we find that the hydraulic natural frequency for a servovalve-piston combination is

$$\omega_h = \left(\frac{4\beta_e A_p{}^2}{V_t M_t}\right)^{\frac{1}{2}} \tag{8-8}$$

and must be large enough for adequate dynamic response. Theoretically, an analysis of the most demanding input signal and the desired response will give the significant frequency spectrum which the servo must pass and establish the required bandwidth frequency. As indicated in Section 8-2, the required hydraulic natural frequency can be related to the bandwidth needed. However, as with most other specifications as well, acceptable natural frequency values usually evolve from experience and/or past performance of particular servo applications. For hydraulic servos which perform computational functions or which have light loads, the area is chosen mainly to give a large enough hydraulic natural frequency.

Special mention of pistons with long strokes is in order. As discussed in Section 6-2, the hydraulic natural frequency varies with stroke and is usually lowest when the piston is at midstroke for double-acting pistons and at full stroke for single-acting pistons. Care must be taken to ensure adequate natural frequency with the large contained volumes of oil. If this frequency is too low, the piston area should be increased with the attendent disadvantage of high oil flow required for maximum velocity, or a rotary motor might be used in the application with a ball screw or rack and pinion arrangement used to convert rotary to linear motion.

The two factors mentioned thus far are by no means the only considerations in actuator sizing. When large friction loads are present it may be desirable to choose actuator size so that the error required to overcome friction loads is tolerable; that is, the tolerable error times the pressure sensitivity of the servovalve produces a pressure drop across the motor which is sufficient to overcome the stiction load [2]. Other actuator performance characteristics which might be important are speed range, stiffness, smoothness (i.e., absence of velocity variations at operating speeds), reliability, life, backlash, pressure rating, and cost.

Gear Ratio Selection

The selection of the displacement of a rotary actuator is made on a basis similar to that outlined for pistons. However, additional considerations are necessary to establish gear ratios. For example, suppose that torque and horsepower loads are satisfied with a motor displacement of 10 in.³/rev at the final output shaft. There are several combinations that can be used to yield this displacement. A 10 in.³/rev motor could be directly coupled to the load. Alternate possibilities are the use of a 5 in.³/rev motor with a 2:1 gear reduction, a 2 in.³/rev motor with a 5:1 gear reduction, a 1 in.³/rev motor reduced 10:1, etc. The product of motor displacement and gear ratio is held constant so that the same speed and torque is delivered to the load. This provides a realistic basis for comparison. Of course, the speeds of the smaller displacement motors are

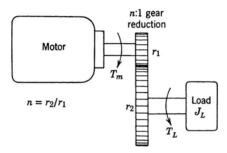

Figure 8-2   Motor and single mesh gear train.

proportionally larger to provide the same horsepower and speeds at the load. However, the same flow rate to each motor and gear combination will achieve the same output speed. Under these circumstances, that is, constant torque at the load, which combination should be selected? Several considerations favor as direct a drive as possible.

1. The most important reason is that the torque to inertia ratio at the load is maximum. Thus, maximum load acceleration is achieved with the least torque. This fact may be demonstrated by assuming a single mesh gear train as shown in Fig. 8-2 and writing the torque equation at the load.

$$T_g = nT_m = (n^2 J_m + J_L)\alpha_L + T_L \qquad (8\text{-}9)$$

where T_m = torque generated at motor, in.-lb
$T_g = nT_m$ = torque from motor referred to the output, in.-lb
T_L = constant load torque at output, in.-lb
J_m = inertia of motor and pinion, in.-lb-sec²
J_L = inertia of driven gear and load, in.-lb-sec²
α_L = acceleration at the load (output), rad/sec²
n = gear reduction from motor to output shaft
We are assuming that J_L, α_L, and T_L are given load conditions.

The torque to inertia ratio at the load is

$$\left.\frac{\text{torque}}{\text{inertia}}\right|_{\text{load}} = \frac{T_g}{n^2 J_m + J_L} = \frac{T_g/J_L}{n^2(J_m/J_L) + 1} \tag{8-10}$$

This function must be maximized with respect to gear ratio. However, this in turn hinges on finding the minimum value for reflected motor inertia $(n^2 J_m)$ because smaller motors used with larger gear ratios have less inertia. Empirically, motor inertia is proportional to the 1.5 power of displacement. Hence, the inertia of a direct coupled motor can be written

$$J_m \big|_D = k D^{1.5} \tag{8-11}$$

where k is a constant and D is the displacement (10 in.^{3}/rev for the example under discussion). The inertia of a geared motor of displacement D_G would be

$$J_m \big|_G = k D_G^{1.5} \tag{8-12}$$

Because $D = n D_G$ by assumption, we can form the ratio of these inertias at the load to determine the larger. Therefore,

$$\frac{n^2 J_m \big|_G}{J_m \big|_D} = \frac{n^2 k (D/n)^{1.5}}{k D^{1.5}} = n^{1/2} \tag{8-13}$$

A sketch of the function $n^{1/2}$ shows that the reflected inertia of the geared motor is always greater than that of direct drive if $n > 1$. Thus, we see that the motor inertia at the load increases faster due to gear ratio than it is reduced because of smaller motor inertias. Since n is greater than unity and minimum inertia is desirable, we can conclude that (8-10) is maximized and highest acceleration is obtained with as direct a drive as possible. If the reflected motor inertia is small compared with load inertia, this distinction becomes academic.

Other factors favoring direct drive are the following:

2. Minimum of backlash since gears are eliminated. Backlash is most severe in lag compensated type 1 servos and type 2 servos because of their tendency toward limit cycle oscillation. Gear backlash will appear outside the loop and not affect stability if feedback is taken from the motor shaft rather than from the output shaft.

3. High coupling stiffness is achieved by eliminating the resiliency of the gear train.

4. Lower operating speeds promote reliability and long life but require

motors with good low speed performance, that is, minimum torque and velocity ripple.

There are many considerations which encourage a geared arrangement:

1. The most important consideration is hydraulic natural frequency that increases with gearing and is maximum for the largest possible gear ratio. As many gear meshes are added, the inertia of the load and driven gear meshes near the load become progressively negligible when reflected to the motor. If torque at the load is held constant as a realistic basis for comparison, smaller motors with less inertia can be used as gear ratio is increased. Hence, in the limit the hydraulic natural frequency would be that of a small motor with inertia of the first pinion. This would be quite high for small motors because, empirically, natural frequency is inversely proportional to the 0.25 power of displacement (i.e., $f_n \propto (1/D^{0.25})$).

2. Gear ratios permit higher operating speeds which can improve smoothness and increase piston pulse frequencies. The piston cycling frequency in rotary motors may coincide with structural modes of vibrations and cause undesirable forced oscillations.

3. Smaller motors are more readily available and are less costly.

In general, some gearing is usually necessary to ensure adequate hydraulic natural frequency. Weighing the acceleration capability of direct drive and the higher natural frequency capability of geared drive, *the best gear ratio is the smallest ratio which will give an adequate hydraulic natural frequency.* However, there is usually a range of gear ratios that will provide acceptable performance. Some of the other factors mentioned in actuator selection should be evaluated to arrive at a final choice. The gear train should be designed for minimum inertia consistent with the power to be transmitted [3].

Servovalve Selection

Electrohydraulic servovalves are universally rated or sized by the load flow obtainable at full stroke and with a valve drop of 1000 psi (500 psi across each orifice of the common critical center four way spool valve). The important considerations in determining servovalve size were discussed in Sections 7-6 and 5-8. Hence, we need only summarize these factors:

1. As illustrated in Fig. 8-3, the pressure-flow curve for maximum stroke should encompass all load flow and load pressure points such that $P_L < \frac{2}{3}P_s$. This assures that adequate flow and horsepower is delivered to the hydraulic actuator and is the basic factor in servovalve sizing.

2. As explained in Section 5-8, the flow gain should be reasonably linear. This fact makes rectangular ports preferable.

3. The pressure sensitivity should be large. A value of 10^6 psi/in. per 1000 psi of system supply pressure is often specified. This figure is independent of valve size as discussed in Section 5-3.

4. Leakage flow should be limited to a reasonable percentage of rated flow to prevent unnecessary power loss.

5. Threshold and null shifts with temperatures and pressures should be minimum.

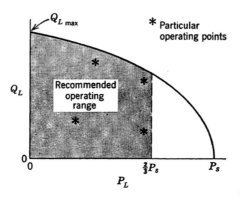

Figure 8-3   Pressure-flow curve at maximum valve stroke.

6. The servovalve bandwidth should be higher than the hydraulic natural frequency of the actuator and load. Otherwise the servovalve dynamics will limit the response of the servo.

Other factors such as servovalve stability (the servovalve itself may oscillate and cause undesirable flow fluctuations), sensitivity to contamination, dither requirements, electrical power requirements, weight, reliability, and cost may contribute to the final selection.

8-2 ELECTROHYDRAULIC POSITION CONTROL SERVOS

The most common type of electrohydraulic servos are those used for position control. An analysis of such a servo, omitting any form of compensation such as minor loops or lag networks, will establish significant performance criteria and suggest a design procedure. A thorough understanding of the basic parameters and problems associated with this type of servo is fundamental to the design of other electrohydraulic servos as well including those employing compensation schemes and those used to control other physical quantities.

Block Diagram of Servo Loop

A simple, uncompensated electrohydraulic rotary position control servo is illustrated in Fig. 8-4. A pair of synchros is used to measure the error between input (reference) and output (controlled) shaft positions. These devices give a suppressed carrier amplitude modulated electrical signal output whose envelope is proportional to the instantaneous error [3, 4]. The envelope or signal portion of the output is given by

$$e_s = K_e \sin (\theta_r - \theta_c) \qquad (8\text{-}14)$$

where K_e is the output voltage at maximum coupling between rotors and stators and is a constant which depends on the reference voltage and transformation ratios between rotors and stators of the synchro units.

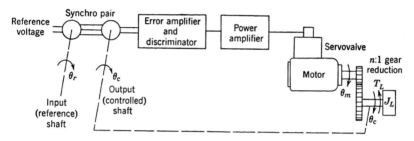

Figure 8-4   Schematic of an electrohydraulic position control servo. Electronic and hydraulic power supplies are not shown.

To improve system accuracy it is often desirable to introduce gearing between the synchro shafts and the input and output shafts so that many revolutions, say 10, of the synchro shafts correspond to one revolution of the input and output shafts. However, the system will seek a false null when the error at the synchro shafts exceeds 180° (half a revolution) because the synchro output voltage is a sinusoidal function of error. To prevent the problem of having false nulls it is necessary to use a coarse synchro system which is geared to the reference and controlled shafts so that the coarse synchro error can never exceed 180° and false nulls do not occur. For large errors the coarse synchros are in use. When the coarse error reduces to the point where it corresponds to less than 180° error of the accurate or fine synchro system from true null, then the fine synchros may be switched into control. However, in practice it is usually undesirable to switch to the fine synchros when the fine synchro error is about 90°. This allows some latitude in the switching zone as a hedge against parameter variations and gives a smooth transition to the fine system because the fine error at this point is usually large enough to saturate the system, as does the coarse error signal. A good rule to follow in synchro gearing and switching is that the error between the shafts of the synchro pair in control should be less than 90° (and must never exceed 180°) from true null under any conditions. The most stringent conditions causing large errors are those associated with start-up when the servo is synchronizing, large transients, and large constant velocity inputs.

The servo loop gain in fine operation is higher than in coarse operation by the speed ratio of the fine to coarse synchros. However, during coarse operation the servo loop is usually saturated and stability is not a problem. For this reason the decreased loop gain during coarse operation need not be compensated. Thus, for design purposes in establishing gains and time constants for stability, it is sufficient to assume the fine synchros are in control. Because the largest synchro gain occurs for small errors where $\sin(\theta_r - \theta_c) \approx \theta_r - \theta_c$ in radians, the synchro gain can be represented as a constant K_e volts/rad.

Referring to Fig. 8-4, the synchro error signal e_s is amplified and fed to a discriminator which removes the d-c signal portion of the carrier modulated wave. This d-c signal e_g is then amplified in a power amplifier which provides a differential current Δi to control the servovalve spool position. In series compensated servos, compensation networks are inserted between the discriminator and the power amplifier.

The design of amplifiers and discriminators in the electronic portion of the loop are discussed elsewhere [3]. Because these components have negligible dynamics compared with the hydromechanical components, they can be described by their steady-state gain constants. Let the gain of the error amplifier and discriminator be

$$\frac{e_g}{e_s} = K_d \quad \text{volts/volt} \tag{8-15}$$

The transfer function from power amplifier input voltage to servovalve spool position was shown in Chapter 7 to be of the form

$$\frac{x_v}{e_g} = \frac{K_s}{\left(\dfrac{s}{\omega_1} + 1\right)\left(\dfrac{s}{\omega_2} + 1\right)\left(\dfrac{s^2}{\omega_0^{\,2}} + \dfrac{2\delta_0}{\omega_0} s + 1\right)} \tag{8-16}$$

where K_s in./volt is the servovalve and amplifier gain constant. The lags at ω_1 and ω_2 rad/sec stem from the inductive time constant (L/R) of the torque motor armature and from the crossover frequency of the spool position loop. The natural frequency ω_0 is due to the spring-mass resonance of the torque motor.

The remaining element in the loop is the valve-motor combination. The transfer function of the particular combination can be selected from Chapter 6, in which these elements are treated in detail. Let us assume simple inertia load and no structural dynamics so that the power element is described by (6-15). This case is by far the most common and is often used as a model in system design. Therefore,

$$\theta_m = \frac{\dfrac{K_q}{D_m} x_v - \dfrac{K_{ce}}{D_m{}^2}\left(1 + \dfrac{V_t}{4\beta_e K_{ce}}s\right)\dfrac{T_L}{n}}{s\left(\dfrac{s^2}{\omega_h{}^2} + \dfrac{2\delta_h}{\omega_h}s + 1\right)} \tag{8-17}$$

where T_L is the load torque at the output shaft. The load inertia must be properly reflected in computing the hydraulic natural frequency ω_h. The motor and load speeds are related by the gear ratio. Thus

$$\frac{\theta_c}{\theta_m} = \frac{1}{n} \tag{8-18}$$

These mathematical descriptions of the elements in the loop can be used to form the block diagram (Fig. 8-5) of the servo. Stability and other performance characteristics can be computed from this diagram.

Stability Analysis

Stability is probably the most important performance characteristic of a servo and often requires some sacrifice in the speed of response. The design of the loop dynamics is usually centered around the requirements for stability. Compared with root locus and Nyquist diagrams, the Bode diagram method of stability analysis is the most useful in hydraulic control because parameters tend to vary. Because Bode diagrams are sketched rapidly, stability can be investigated for a wide range of parameter variations, and this should always be done.

The Bode diagram is a plot of the open loop gain function which, see Fig. 8-5, is given by

$$A_u(s) = \frac{K_v}{s\left(\dfrac{s}{\omega_1} + 1\right)\left(\dfrac{s}{\omega_2} + 1\right)\left(\dfrac{s^2}{\omega_0{}^2} + \dfrac{2\delta_0}{\omega_0}s + 1\right)\left(\dfrac{s^2}{\omega_h{}^2} + \dfrac{2\delta_h}{\omega_h}s + 1\right)} \tag{8-19}$$

where $K_v = K_e K_d K_s(K_q/D_m)(1/n)$ sec^{-1} is the open loop gain constant (also called the velocity constant). The free s in the denominator indicates an integration so that this servo loop is type 1 and has zero position error. The steady-state error for constant velocity inputs is simply the input velocity divided by K_v.

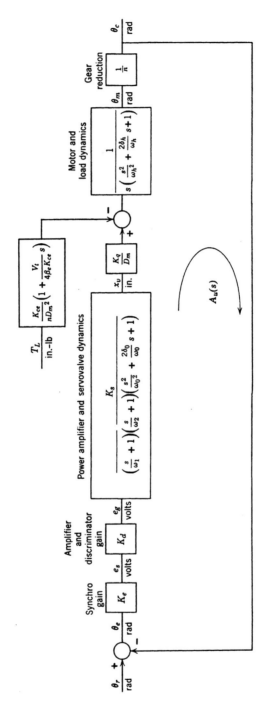

Figure 8-5 Block diagram of electrohydraulic position control servo.

This loop gain function is very complicated and a simpler expression, which still retains information essential to stability, is desirable. Because servovalves have fast response, the hydraulic natural frequency ω_h is usually the lowest break frequency in the loop and dominates dynamic

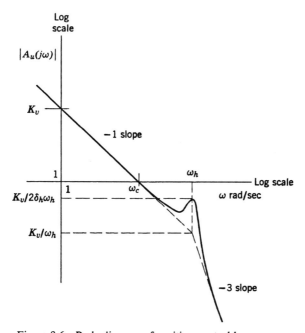

Figure 8-6   Bode diagram of position control loop.

performance. Hence the loop gain can be approximated by

$$A_u(s) = \frac{K_v}{s\left(\dfrac{s^2}{\omega_h^2} + \dfrac{2\delta_h}{\omega_h}s + 1\right)} \tag{8-20}$$

This approximation should be valid in all but exceptional cases because it is difficult to conceive of the servovalve being slower than the power actuator and load. In any case, ω_h should be interpreted as being the lowest natural frequency in the loop.

Figure 8-6 shows the Bode diagram of (8-20). If the resonant peak of the quadratic rises above unity gain, then the system becomes unstable because the critical point of the Nyquist diagram would be encircled. From the geometry of the asymptotic diagram, note that the crossover frequency is approximately equal to the velocity constant (actually, somewhat higher), that is,

$$\omega_c \approx K_v \qquad (8\text{-}21)$$

The gain level of the asymptotic diagram is K_v/ω_h at the frequency ω_h. This level is amplified by the factor $1/2\delta_h$, which is the amplification factor of the quadratic, at resonance. Thus, the gain level at the resonant peak is $K_v/2\delta_h\omega_h$ and must be less than unity for stability. Hence, the stability criterion is

$$\frac{K_v}{\omega_h} < 2\delta_h \qquad (8\text{-}22)$$

This result is fundamental to all hydraulic servo design and can be rigorously derived by forming the characteristic equation $1 + A_u(s) = 0$, using (8-20), and applying Routh's stability criterion. Because damping ratios of 0.1 and 0.2 are characteristic of systems, the velocity constant is limited to 20 to 40% of the hydraulic natural frequency. This fact provides a rule of thumb useful for design purposes. Because K_v and ω_c are approximately equal, this rule also gives the permissible crossover frequency.

From this stability criterion we note that large velocity constants require large hydraulic natural frequencies and large damping ratios. Both of these quantities are fixed once the power element is selected. As discussed in Section 6-1, ω_h can be computed with fair confidence but δ_h is much more variable and nebulous. Computed damping ratios are lower than measured and the following reasons are offered for this:

1. Servovalve underlap and friction forces in the actuator and load act to increase damping. Internal friction torque in rotary motors contribute considerable damping especially at null (i.e., near zero velocity); however, this effect diminishes with velocity increase because a perturbation in velocity will not cause a reversal in velocity and the associated reversal in friction torque.

2. The tendency toward pressure saturation, that is, P_L approaching P_s, increases in the vicinity of resonance for sinusoidal inputs. Hence, the flow gain decreases and causes the resonant peak to be attenuated. This gives the impression of higher damping ratios when analyzing test data.

3. Damping ratio is proportional to $|x_v|$ (see Section 6-7). Hence, an increased damping ratio is to be expected at higher velocities of the output actuator.

This area has not been fully explored and needs study to improve the prediction of damping ratios. It should be clear, however, that (8-22) has a sound physical and mathematical basis even though computed damping ratios cannot be wholly relied upon.

In some applications specifications might require the uncompensated velocity constant and/or crossover frequency to be near or in excess of ω_h. Although claims of servo bandwidths (crossover frequency is a measure of bandwidth) greater than ω_h have been made, these must be viewed with skepticism and such a design objective should be thoroughly reviewed. It may well be that a servo might have *acceptable response* at frequencies higher than ω_h, but this is not the same as bandwidth. However, the following techniques have been or are being used to increase damping and permit larger ratios of K_v/ω_h to be obtained.

1. A capillary or fixed orifice is placed across the actuator lines. This leakage path acts to increase damping especially at spool neutral where it is often lowest. However, such paths decrease static stiffness of the actuator and contribute to power losses.

2. Feedback of actuator pressure difference in subsidiary loops acts to increase damping ratio and extend servo bandwidth. This is because P_L tends to become large near resonance and feedback of this quantity will reduce the gain which reduces the resonant peak. This effect can also be achieved by using dynamic pressure feedback (DPF) servovalves.

3. Electrical networks, such as antiresonant circuits, having characteristics which cancel or extend the motor-load dynamics, are often considered. However, the resonant peaks must be closely matched to be effective, and this is difficult to achieve because the hydraulic damping is low and varies with temperature and spool valve position. This technique is unreliable because frequent adjustment would be necessary.

Leakage shorts and pressure feedback are the most common methods. The third technique must be viewed as an extreme measure because the added elements reduce reliability and make the servo difficult to adjust and service. In either case, outstanding results should not be expected.

Response to Reference and Load Torque Inputs

Two important characteristics of servo loops are the closed loop response to reference inputs, that is, $\Delta\theta_c/\Delta\theta_r$, and the compliance due to load torque inputs, that is, $\Delta\theta_c/\Delta T_L$. Closed loop stiffness, which is the reciprocal relation $\Delta T_L/\Delta\theta_c$, is often preferred over compliance. Let us compute these characteristics from the simplified block diagram of Fig. 8-7, which is based on the approximate loop gain function.

The closed loop response is given by

$$\frac{\Delta\theta_c}{\Delta\theta_r} = \frac{1}{\dfrac{\omega_h}{K_v}\left(\dfrac{s}{\omega_h}\right)^3 + 2\delta_h\left(\dfrac{\omega_h}{K_v}\right)\left(\dfrac{s}{\omega_h}\right)^2 + \dfrac{\omega_h}{K_v}\left(\dfrac{s}{\omega_h}\right) + 1} \qquad (8\text{-}23)$$

The denominator of this expression is the system characteristic equation, which is a cubic. Because this cubic can be represented by a linear and a quadratic factor, we can write

$$\frac{\Delta\theta_c}{\Delta\theta_r} = \frac{1}{\left(\dfrac{s}{\omega_b} + 1\right)\left(\dfrac{s^2}{\omega_{nc}^2} + \dfrac{2\delta_{nc}}{\omega_{nc}}s + 1\right)} \tag{8-24}$$

where ω_b is the break frequency of the linear factor, ω_{nc} is the natural frequency of the quadratic factor (sometimes called the closed loop natural

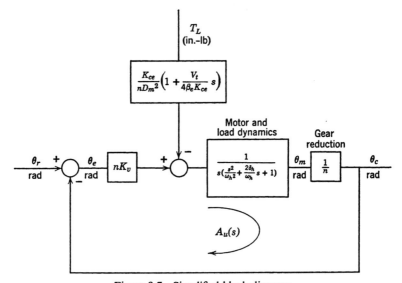

Figure 8-7   Simplified block diagram.

frequency), and δ_{nc} is the damping ratio of the quadratic factor. Normally, these quantities can be found only if the cubic has numerical coefficients because literal solution is cumbersome. However, we are fortunate in this case because the coefficients of the cubic bear such a relationship to one another that it is possible to solve the cubic and plot the results in a normalized manner, as shown in Figs. 8-8, 8-9, and 8-10. Given the open loop parameters K_v, ω_h, and δ_h, the closed loop parameters ω_b, ω_{nc}, and δ_{nc} can be obtained from these charts. For low values of δ_h and K_v/ω_h, the following approximations, useful in preliminary design calculations, can be made: $\omega_{nc} \approx \omega_h$, $\omega_b \approx K_v$, and $2\delta_{nc} \approx 2\delta_h - K_v/\omega_h$. Hence, the closed loop break frequencies are roughly equal to the hydraulic natural frequency and to the crossover frequency. However, exact values can always be obtained from the cubic solution charts.

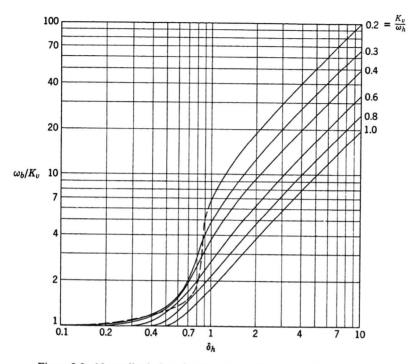

Figure 8-8   Normalized plot of real root of cubic characteristic equation.

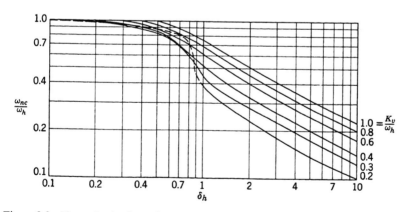

Figure 8-9   Normalized plot of natural frequency of quadratic factor of cubic characteristic equation.

243

The closed loop frequency response function (8-24) is plotted in Fig. 8-11. This curve is a measure of servo response capability. The concept of bandwidth has been developed in an effort to reduce the meaningful information in this curve to a single numerical value. Several measures of bandwidth can be used. These are:

1. The most common measure of bandwidth is the frequency at which the amplitude ratio falls to 0.707 (3 db down) of its low frequency value.

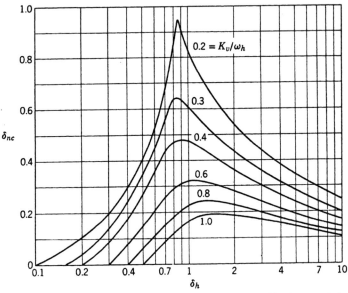

Figure 8-10   Plot of damping ratio of quadratic factor of cubic characteristic equation.

From Fig. 8-11 this occurs approximately at ω_b. Because Fig. 8-8 shows that ω_b/K_v is slightly greater than unity and since $K_v \approx \omega_c$, we can conclude that this measure of bandwidth (ω_b) is somewhat greater (perhaps 1.5 times) than the crossover frequency of the loop.

2. Another frequently used measure of bandwidth is the frequency at which the phase lag is 90°. This frequency will also be somewhat larger than ω_c.

3. The closed loop natural frequency ω_{nc} is sometimes used as a bandwidth index, but this measure has not found broad acceptance.

From any of these measures we can conclude that system response is fundamentally limited by the actuator dynamics, that is, by ω_h and δ_h. If a certain system bandwidth is required, one can work back to establish the needed crossover frequency.

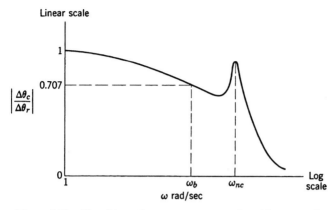

Figure 8-11   Closed loop frequency response of position control servo.

Another system characteristic often useful is the closed loop compliance. Using the cubic factors discussed and referring to Fig. 8-7, this function can be written

$$\frac{\Delta\theta_c}{\Delta T_L} = \frac{-\dfrac{K_{ce}}{K_v n^2 D_m^2}\left(1 + \dfrac{V_t}{4\beta_e K_{ce}}s\right)}{\left(\dfrac{s}{\omega_b} + 1\right)\left(\dfrac{s^2}{\omega_{nc}^2} + \dfrac{2\delta_{nc}}{\omega_{nc}}s + 1\right)} \tag{8-25}$$

Inverting this function and combining with (6-22), the dynamic stiffness is obtained. Therefore,

$$\frac{\Delta T_L}{\Delta\theta_c} = \frac{-\dfrac{K_v(n D_m)^2}{K_{ce}}\left(\dfrac{s}{\omega_b} + 1\right)\left(\dfrac{s^2}{\omega_{nc}^2} + \dfrac{2\delta_{nc}}{\omega_{nc}}s + 1\right)}{\left(1 + \dfrac{s}{2\delta_h\omega_h}\right)} \tag{8-26}$$

Because the break frequencies ω_b and $2\delta_h\omega_h$ are about the same value (but both are normally lower than ω_{nc}), the factors at these frequencies effectively cancel. Hence, the plot of the closed loop stiffness is as shown in Fig. 8-12 and is considerably higher than the open loop stiffness shown in Fig. 6-4. The servo loop acts to increase the open loop stiffness by a factor equal to the open loop gain. Therefore, high loop gains are desirable for stiffer servos. The closed loop stiffness is lowest, see Fig. 8-12, near resonance where the value is approximately given by

$$\left.\frac{\Delta T_L}{\Delta\theta_c}\right|_{min} = \frac{2\delta_{nc}K_v(n D_m)^2}{K_{ce}} \tag{8-27}$$

It should be apparent that the output shaft deflection due to a load torque represents a position error in the servo loop. Therefore, it is important to compute the closed loop stiffness and determine this error for expected loads. If the stiffness must be improved over that of the simple uncompensated loop discussed thus far, perhaps lag compensation in the loop or velocity feedback in a minor loop should be incorporated.

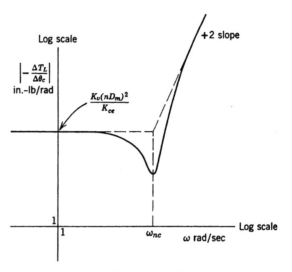

Figure 8-12   Closed loop stiffness of position control servo.

These techniques will be the subject of the next two sections. However, it often happens that structural stiffnesses of drive members such as gears, shafts, motor mountings, etc. are less than that of the servo. In such cases improving servo stiffness will not increase over-all stiffness. Structural improvements are then necessary.

8-3 LAG COMPENSATED ELECTROHYDRAULIC POSITION CONTROL SERVOS

Lag compensation is often used in electrohydraulic servos to boost low frequency gain and still retain stability. The network shown in Fig. 8-13 is inserted in the d-c signal portion of the loop between the discriminator and the power amplifier driving the servovalve. Because α (the lag to lead ratio) is greater than unity, that is, the lag time constant is larger, the network has a net lagging phase and for this reason is called a *lag network* or *lag-lead network*. With the lag network time constants properly chosen, the uncompensated Bode diagram of Fig. 8-6 can be altered to the compensated diagram of Fig. 8-14. Now it is well known that desirable servo performance results only when the loop gain is greater than unity. Note that the area under the curve where $|A| > 1$ is much greater for the compensated servo. Hence, better performance should be expected. In particular, the advantages of lag compensation result from the increased low frequency (i.e., below crossover frequency) gain and are:

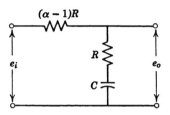

$$\frac{e_o}{e_i} = \frac{1 + s/\omega_{rc}}{1 + (\alpha/\omega_{rc})s}$$

$\omega_{rc} = 1/RC = $ lead break frequency

$\alpha = $ lag to lead ratio

Figure 8-13   Single stage lag compensation network.

1. Steady-state error is decreased by a factor of $1/\alpha$ for constant velocity inputs because the velocity constant is increased by α. This is the major reason for using lag compensation.

2. Control accuracy is improved in the frequency range below crossover.

3. Steady-state and low frequency closed loop stiffness is increased.

The closed loop response and bandwidth are not improved and in fact are reduced slightly. Hence, lag compensation offers improvements in low frequency accuracy and stiffness but not in response. Other disadvantages of lag compensation will be discussed later in this section.

The selection and design of lag compensation is based on linear servo theory and the procedure is well documented in the literature (see Section 6-4 of Reference 5 for an excellent discussion). However, this procedure will be summarized here for convenience. Referring to Fig. 8-14, the compensated loop gain function is given by

$$A_c(s) = \frac{K_{vc}\left(1 + \dfrac{s}{\omega_{rc}}\right)}{s\left(1 + \dfrac{\alpha}{\omega_{rc}}s\right)\left(\dfrac{s^2}{\omega_h^2} + \dfrac{2\delta_h}{\omega_h}s + 1\right)} \tag{8-28}$$

where $K_{vc} = \alpha K_v$ is the compensated velocity constant. The quantities ω_h and δ_h are fixed by the power element selection described in Section 8-1. Therefore, quantities to be determined are α, ω_c, K_{vc}, and ω_{rc}. The values for ω_{rc} and ω_c will be found in the following manner:

1. Determine the frequency between ω_{rc} and ω_h in Fig. 8-14 where the phase lag is a minimum. Because this frequency will have the highest

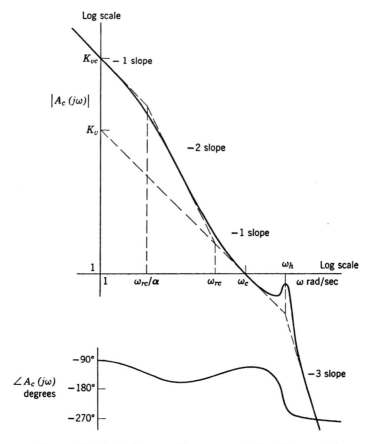

Figure 8-14   Bode diagram of lag compensated position control servo.

phase margin, ω_c will be located at this frequency. In this sense ω_c is the optimum crossover frequency.

2. Adjust ω_{rc} to obtain adequate margin ϕ_M. A phase margin of about 60° is usually desirable.

Step 1 requires some approximations to simplify the algebra involved. Let us assume that $\delta_h = 1$ so that the phase at crossover of the quadratic in (8-28) is $-2 \tan^{-1} (\omega_c/\omega_h)$ rad. While this phase lag is greater than with less damping and is, therefore, more destabilizing, this offsets to some extent the destabilizing effect of the high resonant peak when damping is low. Let us also assume that the lag at ω_{rc}/α contributes $-(\pi/2)$ rad ($-90°$) at crossover. With these assumptions, the phase lag of (8-28) at crossover is given by

$$\beta_c = -\pi + \tan^{-1}\frac{\omega_c}{\omega_{rc}} - 2\tan^{-1}\frac{\omega_c}{\omega_h} \qquad (8\text{-}29)$$

Using the series approximations of $\tan^{-1} x \approx x$ for $x \ll 1$ and $\tan^{-1} x \approx \pi/2 - 1/x$ for $x \gg 1$ where x is in radians, this relation simplifies to

$$\beta_c = -\pi + \frac{\pi}{2} - \frac{\omega_{rc}}{\omega_c} - 2\frac{\omega_c}{\omega_h} \quad \text{rad} \qquad (8\text{-}30)$$

Differentiating and equating to zero (i.e., $d\beta_c/d\omega_c = 0$) gives the optimum crossover frequency.

$$\omega_c = \sqrt{\tfrac{1}{2}\omega_{rc}\omega_h} \qquad (8\text{-}31)$$

If the phase margin is to be ϕ_M rad, the permissible phase at crossover is $\beta_c = -\pi + \phi_M$ rad and combining with (8-30) yields

$$\phi_M = \frac{\pi}{2} - \frac{\omega_{rc}}{\omega_c} - 2\frac{\omega_c}{\omega_h}$$

Solving (8-31) and (8-32) simultaneously we obtain the required values.

$$\omega_{rc} = \frac{\omega_h}{32}(\pi - 2\phi_M)^2 \quad \text{rad/sec} \qquad (8\text{-}32)$$

$$\omega_c = \frac{\omega_h}{8}(\pi - 2\phi_M) \quad \text{rad/sec} \qquad (8\text{-}33)$$

The remaining quantities to be determined are K_{vc} and α. From the geometry of Fig. 8-14, we note that $K_v = \omega_c$. Because $K_{vc} = \alpha K_v = \alpha\omega_c$, establishing α will define K_{vc}; α should be selected to meet system requirements, but there are three limiting factors.

1. The lag network resistors become widely separated in value and it becomes difficult to maintain impedance separation of the network in the circuit.

2. The lag network capacitor value becomes large and the long charge and discharge times may cause undesirable transient performance.

3. The loop approaches conditional stability with the attendant problems of synchronization and limit cycle oscillation.

These considerations limit α to a practical maximum of 10 to 20 or so.

Because of the many approximations made, especially in the derivation of (8-32) and (8-33), the values found thus far must be treated as preliminary. An exact Bode diagram should be plotted using these values, and a check should be made to see if specifications are met. Particular attention should be given the resonant peak at ω_h to make sure that it does not increase above unity gain. Some adjustments in ω_{rc} and ω_c may be necessary before final values are established.

Values for the elements in the lag network can now be obtained. The value for one element is arbitrarily chosen to establish the over-all impedance level of the network so that it neither loads nor is loaded by surrounding circuitry. The remaining resistor and capacitor values are selected to give the required α and ω_{rc}.

The closed loop response and closed loop stiffness, discussed in Section 8-2, can be computed using (8-28) with the final break frequency values.* The characteristic equation will be a quartic and can be conveniently solved only for a numerical case. However, an approximate expression for the closed loop response can be found as follows: For frequencies less than ω_h, the quadratic factor $[s^2/\omega_h{}^2 + (2\delta_h/\omega_h)s + 1]$ may be considered unity and the closed loop response computed on this basis will show a quadratic denominator factor with a natural frequency of $\omega_0 = \sqrt{K_{vc}\omega_{rc}/\alpha}$. Also, it is well known that factors in the open loop transfer function having break frequencies well beyond ω_c also appear as approximate factors of the closed loop transfer function. Hence, it is reasonable to assume that the quartic characteristic equation has two quadratic factors with natural frequencies of ω_0 and ω_h. Multiplying out two such quadratic factors and equating coefficients with those of the exact characteristic equation will yield expressions for the two damping ratios. Performing these manipulations, the approximate closed loop response is

$$\frac{C}{R} = \frac{\left(\dfrac{s}{\omega_{rc}} + 1\right)}{\left(\dfrac{s^2}{\omega_0{}^2} + \dfrac{2\delta_0}{\omega_0}s + 1\right)\left(\dfrac{s^2}{\omega_h{}^2} + \dfrac{2\delta_1}{\omega_h}s + 1\right)} \tag{8-34}$$

where $\omega_0 = \sqrt{\dfrac{K_{vc}\omega_{rc}}{\alpha}}$ = closed loop natural frequency, rad/sec

$$\delta_0 = \left(\frac{1}{K_{vc}} + \frac{1}{\omega_{rc}}\right)\frac{\omega_0}{2} = \text{damping ratio, dimensionless}$$

$$\delta_1 = \delta_h - \frac{\delta_0\omega_0}{\omega_h} = \text{damping ratio, dimensionless}$$

* We are talking about the response computed from linearized equations. A nonlinear analysis is very difficult using hand calculation (see Reference 6). Machine computation using an analogue computer is probably the best way.

Because $K_{vc} = \alpha\omega_c$ and using (8-32) and (8-33) these relations can be simplified to

$$\omega_0 = \frac{\omega_c}{2}\sqrt{\pi - 2\phi_M}$$

$$\delta_0 = \left[1 + \frac{(\pi - 2\phi_M)}{4\alpha}\right](\pi - 2\phi_M)^{-\frac{1}{2}}$$

$$\delta_1 = \delta_h - \frac{\delta_0}{16}(\pi - 2\phi_M)^{\frac{3}{2}}$$

If $\delta_0 \ll 1$, stability suffers but the response is fast. If $\delta_0 \gg 1$, the system is very stable but slow in response. A value of $\delta_0 \approx 1$ offers a good compromise and may be considered a design objective. If α is fairly large, say greater than 5, a phase margin of $\phi_M = 1.07$ rad $= 61.3°$ is required to make $\delta_0 \approx 1$. In this event, the system values become $\omega_{rc} = \omega_h/32$, $\omega_c = \omega_h/8$, $\omega_0 = \omega_c/2$, $\delta_0 = 1$, and $\delta_1 = \delta_h - \frac{1}{16}$. These values are quite typical. It is important to note that the crossover frequency is limited to only 12.5% of the hydraulic natural frequency. Hence, if the system bandwidth must be large, it is essential that the design begin by making ω_h as large as possible.

We can now compute a good approximation to the step response of the system. If $\delta_0 \approx 1$ and the quadratic in ω_h is neglected in (8-34), then the unit step response can be found as

$$C(t) = 1 - \left[1 - \left(\frac{\omega_c}{2\omega_{rc}} - 1\right)\frac{\omega_c}{2}t\right]e^{-(\omega_c/2)t}$$

By differentiation the maximum percentage overshoot is

$$\text{max percent OS} = 100\left(\frac{\omega_c}{2\omega_{rc}} - 1\right)e^{(2\omega_{rc}/\omega_c - 1)^{-1}}$$

and occurs at time t_0 where

$$t_0 = \frac{2}{\omega_c - 2\omega_{rc}}$$

For the typical case where $\phi_M = 61.3°$, $\omega_c = 4\omega_{rc}$ and max percent OS $= 100(2 - 1)e^{-2} = 12.5\%$ which occurs at $t_0 = 4/\omega_c = 32/\omega_h$. This rather large overshoot is undesirable, but it is typical of systems using lag compensation.

The major drawback to lag compensation is the fact that the servo loop becomes much more prone to limit cycle oscillation. Several multivalued nonlinearities, such as hysteresis, backlash, and friction have describing functions which would be intersected by the linear part of the loop gain function of a lag compensated servo. Backlash is the most serious non-linearity in this regard and will always cause a limit cycle if α is much beyond two. If α is very large, the loop may become conditionally stable. It is best to avoid use of two-stage lag networks because the loop will always be conditionally stable. Conditionally stable loops are undesirable because:

1. A shift in gain either up or down may cause instability.

2. The loop becomes prone to limit cycle oscillation with almost any nonlinearity. Even simple single valued nonlinearities such as saturation and deadband will cause limit cycles.

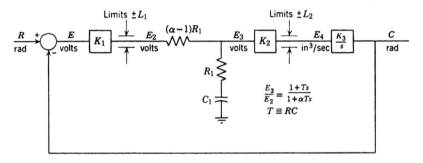

Figure 8-15   Simplified block diagram of electrohydraulic servo.

3. Such loops have difficulty in synchronizing to the desired position when power is first applied (see Section 10-3).

4. Because these loops are sensitive to parameter variations, it becomes difficult to maintain proper adjustments. Troubleshooting and servicing are more involved.

In practice, type 2 servos are often conditionally stable and one must contend with the above disadvantages to reap the benefit of zero steady-state error to constant velocity inputs.

Another difficulty with lag compensated servos is an extremely large overshoot which can occur in the step response [7]. This large overshoot, which cannot be explained by linear servo theory, is a very disconcerting observation when the loop has been designed with 60° of phase margin and should be stable with normal overshoot. It makes one doubt the usefulness of linear theory in design. However, this phenomenon can be explained using a simple mathematical model and has a relatively simple physical correction.

Let us simplify the electrohydraulic positioning servo to that of Fig. 8-15: R and C are the reference and controlled positions, respectively, in radians; E is the error $(R-C)$ signal; and K_1 is the error path gain in volts/rad and physically represents the lumped gains of synchros, amplifier, and discriminator. The output E_2 is limited at $\pm L_1$ volts due to saturation in the error path elements. The lag network has a lead time constant of T seconds and a time constant ratio of α. It is assumed that the lag network is driven from a low impedance source and is not loaded by elements which follow in the loop. The voltage output of the network, E_3, feeds a d-c power amplifier and servovalve with lumped gain of K_2. The flow E_4 in.^{3}/sec is limited by the hydraulic supply to $\pm L_2$ in.^{3}/sec. The hydraulic actuator is approximated by a simple integration. Because of the limit L_2, the output velocity has a maximum value of $\dot{C}_M = L_2 K_3$ rad/sec.

Let us now determine the step response of a lag network. For a step input $E_2(t) = E_{2o}u(t)$ so that $E_2(s) = E_{2o}/s$ and the transformed output is

$$E_3(s) = \frac{E_{2o}(1 + Ts)}{s(1 + \alpha Ts)} \tag{8-35}$$

Inverse transforming, we obtain

$$E_3(t) = E_{2o}\left[1 - \left(1 - \frac{1}{\alpha}\right)e^{-t/\alpha T}\right] \tag{8-36}$$

Thus at $t = 0$ the output jumps immediately to E_{2o}/α and then approaches a value of E_{2o} exponentially with a time constant of αT. This result will prove useful later.

Let us now put a step input of $R_0 u(t)$ into the system of Fig. 8-15 and sketch the transients that occur at all points in the loop. These are shown on Fig. 8-16 and may be explained as follows: When the step is first applied, the error E is large and begins to decrease as the output C starts to increase. The quantity E_2 goes instantly to the limit value L_1, providing that the step input is large enough, that is, $R_0 > L_1/K_1$. The input to the lag network is, therefore, a step $L_1 u(t)$. The output of the lag network, which is the key transient, rises to L_1/α when the step is applied and further increases as the capacitor in the network is charged, approaching a value of L_1 exponentially. If $L_2 < (L_1/\alpha)K_2$, then E_4 goes to its limit value of L_2. Thus, the output moves at its maximum rate of $\dot{C}_M$ immediately when the step is applied. This action continues until time $t_2 = R_0/\dot{C}_M$, when the output reaches its final value for the first time, is reached. At this instant the error switches polarity and, if the overshoot is large enough (much greater than L_1/K_1), the error path saturates in the opposite direction to $-L_2$. At $t = t_2$ the output of the lag compensation immediately steps $-(2L_1/\alpha)$ volts in the negative direction and decreases, further approaching a value of $-L_1$ volts. When the output of the lag network goes through

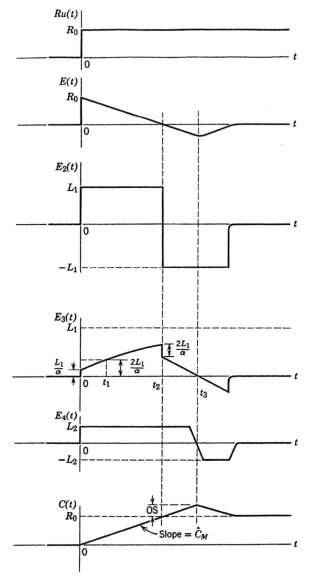

Figure 8-16   Step response wave forms.

zero in the negative direction (at time $t = t_3$), the flow to the motor is reversed and the output velocity reverses.

During the interval $t_3 - t_2$, the output moved at maximum velocity $\dot{C}_M$ and, therefore, overshot its final position. The reason for the overshoot is clearly the saturations L_1 and L_2 in the loop. The saturation L_1 in the error path allows the capacitor in the lag network to charge up to such a large value that it takes quite some time ($t_3 - t_2$) for it to decay to zero once the error signal is reversed. The flow saturation L_2 limited the output velocity, thereby allowing more time for the capacitor to assume the charge. If the system were linear, the input to the network would be a rapidly decreasing function of error because $\dot{C}$ would not be limited. The large lag time constant of the network then prevents following of such a fast signal and there would be negligible overshoot as predicted by linear theory.

For the somewhat idealized case where the input to the lag network in Fig. 8-15 can be represented by step functions, the overshoot can be easily computed. The time t_1 where $E_3(t) = 2L_1/\alpha$ can be computed from (8-36). Therefore

$$t_1 = \alpha T \log_e \left(\frac{\alpha - 1}{\alpha - 2} \right) \tag{8-37}$$

The time t_2 is by definition

$$t_2 = \frac{R_0}{\dot{C}_M} \tag{8-38}$$

Up to time t_3 the input E_2 can be written

$$E_2(t) = L_1 u(t) - 2L_1 u(t - t_2) \tag{8-39}$$

Referring to (8-36), the output of the network for this input can be written

$$E_3(t) = L_1 \left[1 - \left(1 - \frac{1}{\alpha} \right) e^{-t/\alpha T} \right] u(t)$$

$$- 2L_1 \left[1 - \left(1 - \frac{1}{\alpha} \right) e^{-(t-t_2)/\alpha T} \right] u(t - t_2) \tag{8-40}$$

$E_3(t) = 0$, $u(t) = 1$, and $u(t - t_2) = 1$ at the time t_3. Substituting these conditions into (8-40) and solving for t_3 yields

$$t_3 = \alpha T \log_e \left[\left(1 - \frac{1}{\alpha} \right) (2 e^{t_2/\alpha T} - 1) \right] \tag{8-41}$$

If $t_2 \leq t_1$, the negative lead jump of $2L_1/\alpha$ at time t_2 will be enough to reverse the flow to the actuator and a large overshoot will not occur.

Combining (8-37) and 8-38), this condition may be written

$$\frac{R_0}{\dot{C}_M \alpha T} \le \log_e \left(\frac{\alpha - 1}{\alpha - 2} \right) \tag{8-42}$$

Thus, if $\alpha \le 2$ there will be no overshoot. If $\alpha > 2$ there may or may not be overshoot, depending on whether the step input is large enough to exceed the critical value obtained from (8-42).

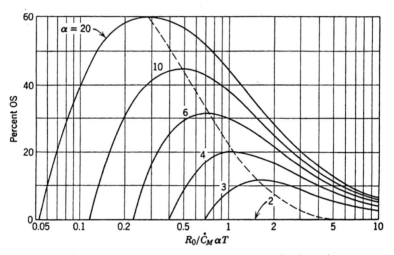

Figure 8-17   Percent overshoot versus normalized step input.

If $t_2 > t_1$, overshoot will occur and is given by

$$OS = (t_3 - t_2)\dot{C}_M \tag{8-43}$$

if the output moves at maximum velocity in the interval $t_2 \le t \le t_3$. The percentage overshoot is

$$\text{percent } OS = 100 \left(\frac{OS}{R_0} \right)$$

Substituting in t_2 and t_3, the final result is

$$\text{percent } OS = 100 \left\{ \frac{\alpha T \dot{C}_M}{R_0} \log_e \left[\left(1 - \frac{1}{\alpha} \right)(2e^{R_0/\dot{C}_M \alpha T} - 1) \right] - 1 \right\} \tag{8-44}$$

This relation, plotted in Fig. 8-17, is valid if the step input is large enough that waveforms in Fig. 8-16 provide reasonable approximations and the output moves at maximum velocity in the interval between t_2 and t_3. For given values of α, T, $\dot{C}_M$, and R_0, the overshoot can be obtained from Fig. 8-17. It is apparent that very large percentage overshoots are possible.

For small step inputs that do not saturate the system, the overshoot may be computed from linear theory.

Let us now consider the cure for this phenomenon. Either increasing the saturation levels L_1 and L_2 to beyond the range of interest (i.e., make the system linear) or decreasing the lag network ratio α will reduce the overshoot. However, the first is not generally possible and the second is not attractive because α is chosen to meet design requirements. Referring to the trace of $E_3(t)$ in Fig. 8-16, there will be no overshoot if $E_3(t)$ is limited to a value

$$E_3(t)\,|_{\text{limit}} \leq \frac{2L_1}{\alpha} \qquad (8\text{-}45)$$

because t_1 and t_2 will then be coincident and the negative lead jump of $-(2L_1/\alpha)$ volts at t_2 is potentially enough to instantly reverse the direction of flow to the actuator. Thus, a method of physically limiting E_3 is needed.

A very simple solution to this problem is to place biased diodes or zener diodes in parallel with the capacitor in the lag network to limit the charge that can be stored.* Very often the diode conduction voltage provides adequate bias so that a diode strapped around the capacitor is sufficient. Another diode in parallel but with polarity reversed is required to limit the capacitor voltage of the other polarity as the lag network has input voltages of both polarities. Analysis of such a diode corrected lag network will show that (8-45) is satisfied if the voltage across the capacitor is limited to a value of

$$E_c \leq \frac{L_1}{\alpha - 1} \qquad (8\text{-}46)$$

It should be apparent that the lag network behaves linearly and gives the desired increase in open loop gain when $E_3(t)$ is less than $E_3(t)\,|_{\text{limit}}$. However, for large transients the capacitor is effectively shorted and the lag network becomes a simple voltage divider with gain of $1/\alpha$ which reduces the open loop gain from K_{vc} to K_v. Thus, for large transients the loop reverts back to the uncompensated form and the necessary steady-state gain reduction is automatically made.

To ensure that the lag network is operating and gives reduced error for constant velocity inputs, a check should be made in the design stage to make sure that $E_3(t)\,|_{\text{limit}}$ is not reached for such inputs. Referring to Fig. 8-15, it should be apparent that

$$K_2 E_3(t)\,|_{\text{limit}} > L_2 \qquad (8\text{-}47)$$

* To the author's knowledge, this ingenious solution was first proposed by Norman D. Neal of the Cincinnati Milling Machine Company.

must be satisfied to ensure that maximum output velocity is achieved. This condition favors a large K_2. However, from the standpoint of error due to drift in the d-c amplifier, K_2 should be small. In general, the gain ahead of the point where the unwanted disturbance is injected into the loop must be large enough to reduce the resulting error to within specified limits. Hence, some juggling of parameter values and gain distribution in the loop may be necessary to arrive at a satisfactory final design.

8-4 ELECTROHYDRAULIC VELOCITY CONTROL SERVOS

For two basic reasons it is often desirable to control the velocity of a hydraulic actuator by sensing velocity and feeding it back to form a servo loop (Fig. 8-18).

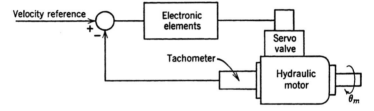

Figure 8-18   Schematic of velocity control loop.

1. Velocity may be the principal output of the system. Examples are rate tables, tracking modes of antennas, radars, gun turrets, and machine tool feed drives. In these applications accuracy demands feedback control of velocity.

2. Velocity may be fedback in a subsidiary or minor loop as a means of compensating the major control loop. The advantages of this type compensation are increased major loop stiffness, gain variations in servovalve are minimized in the major loop, and amplifier gain in minor loop can be made large to reduce errors due to drift and load friction. However, the disadvantages are decreased bandwidth in the major loop and increased system cost and complexity. Because these minor loops limit response capability of the major loop, they should not be used unless some clear-cut objective is served.* It is often the case that these minor loops serve no useful purpose whatever.

---

* Velocity feedback minor loops actually increase over-all response in servos using electric servomotors as power elements. This is because electric motors respond like a simple lag. Hence, there is no stability problem and the crossover frequency of the velocity loop can be made greater than the electric motor break frequency. In contrast, stability of hydraulic velocity control loops demands that the crossover frequency be lower than the hydraulic natural frequency.

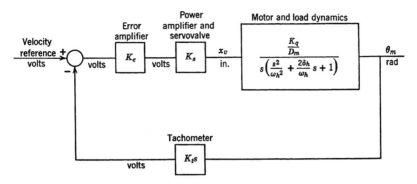

Figure 8-19   Block diagram of velocity control servo.

As discussed in Section 8-2, the servovalve break frequencies are usually lower than the hydraulic natural frequency. Therefore, the transfer function of the power element provides an adequate description of the hydraulic portion of the loop. We assume simple inertia load on a rotary motor; (6-18) is applicable and the block diagram of the velocity control loop becomes that in Fig. 8-19. This is a type 0 servo and the loop gain function is

$$A_{vu}(s) = \frac{K_0}{\dfrac{s^2}{\omega_h^2} + \dfrac{2\delta_h}{\omega_h} s + 1} \tag{8-48}$$

where $K_0 = K_e K_s K_t (K_q/D_m)$ is the open loop gain constant. The Bode diagram of this function is sketched in Fig. 8-20. The crossover frequency occurs on a -2 slope and there is very little phase margin, especially for

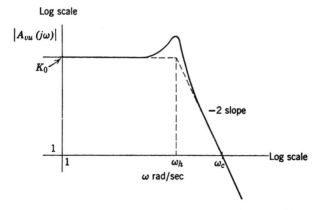

Figure 8-20   Bode diagram of uncompensated velocity control loop.

the usual case of low δ_h. The loop is stable only because the loop dynamics are so simply represented. However, if other lags, such as those associated with the servovalve, occur in the frequency range between ω_h and ω_c, the loop will become unstable. In practice this always seems to be the case and even loops with gain constants like $K_0 \approx$ 1 will be unstable. Because feedback control is effective only for gain values in excess of unity, the usefulness of a low gain, say $K_0 < 2$, velocity loop is questionable. As reasonable gain values, say $K_0 > 5$, are desirable, then *electrohydraulic velocity control loops must always be compensated to achieve stability.* There are few exceptions to this rule. An example might be a loop which controls high velocities. The servovalve would then be open and increase damping (δ_h) to well above unity. The motor-load quadratic would then factor into a dominant lag and stability could be achieved.

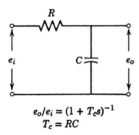

$$e_o/e_i = (1 + T_c s)^{-1}$$
$$T_c = RC$$

Figure 8-21   Simple lag network.

The simplest method of compensation is to place an *RC* lag network (Fig. 8-21) in the electronic portion of the loop ahead of the servovalve. The corrected loop gain function (Fig. 8-22) is then stable with adequate phase margin. The break frequencies of the hydraulic elements occur well

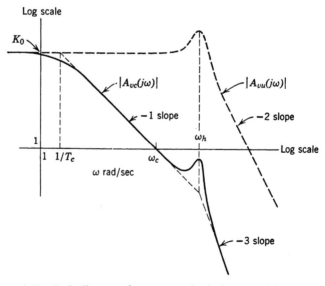

Figure 8-22   Bode diagram of compensated velocity control loop.

beyond the crossover frequency and, consequently, do not influence stability. The loop is sized by selecting the power element and choosing a value for K_0 which meets specifications. ω_c is then limited to a value of 0.2 to 0.4 of ω_h to keep the resonant peak below unity gain as explained in Section 8-2. From the geometry of Fig. 8-22, the required network time constant is

$$T_c = \frac{K_0}{\omega_c} \qquad (8\text{-}49)$$

for which values of R and C can be selected. Because ω_c must be fixed to preserve stability, large T_c values are necessary for large gains. Similar compensation can be achieved by using a d-c operational amplifier as an integrator.* However, large T_c values have some drawbacks. If the velocity servo is a minor loop, then stability of the major loop can be adversely affected. Also, if T_c is large, the network capacitor is slow in charging and discharging. This may cause synchronizing problems when the servo is turned on or slow transient response under certain conditions. However, this behavior may be eliminated by placing zener diodes around the capacitor to limit the voltage to no more than that necessary to cause maximum output velocity. In this manner excess charge cannot be stored.

If velocity feedback is used as a minor loop, the closed loop response of the loop must be computed. Approximately, the result will be a linear lag at the corrected crossover frequency and a quadratic at ω_h. The lag at ω_c, because it is lower than ω_h, becomes a limiting factor on the response of the major loop. This fact should be carefully weighed before such a minor loop is chosen in design.

The velocity control loop is a very good example of the need for compensation. Referring to Fig. 8-22, we note that the crossover frequency of the compensated loop gain is much lower than that of the uncompensated loop gain. Hence, speed of response must be sacrificed to achieve stability. The trade off of accuracy or response for stability is often necessary in control.

8-5 SERVO DESIGN CONSIDERATIONS

Design necessarily begins with a statement of specifications. If they are not given, and they often are not, the designer must first determine them. This is not an easy task; usually, a set of specifications evolve from experience and past history for a particular kind of control system. From

---

* Another compensation technique would be to insert a lead between ω_h and ω_c in Fig. 8-20. However, this scheme is impractical because variation in ω_h would jeopardize stability. Furthermore, the servovalve must be very fast because its break frequencies must be higher than the new crossover frequency.

an analysis of the input signal, desired output, and loads to be controlled, the following specifications can be obtained: (a) bandwidth, (b) disturbances, (c) accuracy or allowable error, and (d) degree of stability.

The bandwidth requirement may take the form of M-peak frequency, step response rise time, crossover frequency, or any other suitable measure. Noise and stability prevent the bandwidth from being arbitrarily large. In fact, it is desirable to keep the bandwidth at a minimum consistent with specifications. A reduced bandwidth usually simplifies compensation and, because peak power outputs are associated with high frequencies, relaxes requirements on individual elements, thereby producing a savings in cost.

Extraneous disturbances such as noise, loads, and drifts in components have an adverse effect on over-all accuracy. Load disturbances, in particular, may be negligible in instrument servos, but can seriously effect stability and accuracy of servos which control appreciable loads. This specification usually takes the form of a steady-state value for compliance, that is, the maximum steady-state error which can be tolerated for a given load.

Accuracy is a basic requirement for all servo systems. Indeed, their existence is based principally on this requirement. In general, servo error results from many factors: (a) errors due to input position, velocity, and acceleration, (b) errors due to drifts, tolerances, and changes in component parameters, (c) errors due to disturbances such as noise and force loads, and (d) errors due to nonlinearities such as backlash and friction. The accuracy specification may take the form of requirements for position, velocity, and acceleration error constants, final value of steady-state error, error resulting from sinusoidal inputs, etc.

The degree of stability can be specified by any of the conventional measures of relative stability: phase margin, gain margin, M-peak, percent overshoot, damping ratio, etc.

Thus, we can arrive at a realistic and consistent set of specifications such as bandwidth, velocity constant, total error, output stiffness, and phase margin. The problem now is to design the servo loop such that its loop gain function meets these specifications [5]. Components are then sized to obtain this loop gain, keeping in mind their dynamic limitations. Arbitrary selection of components with no regard for their gain and time constants can prove fatal with regard to stability.

There are some general criteria applicable to the "good" design of any servo system.

1. There must always be a range of frequencies where the loop gain is substantially greater than unity. All the desirable characteristics of feedback control, indeed their very existence, stem from this simple fact. When

the loop gain is less than unity, feedback is not effective and the loop is essentially open. The crossover frequency gives the borderline between open loop and closed loop control.

2. There is always an accuracy requirement and this necessitates some loop gain greater than unity.

3. For reasons outlined in Section 8-3, a system should never be designed conditionally stable unless it cannot be avoided.

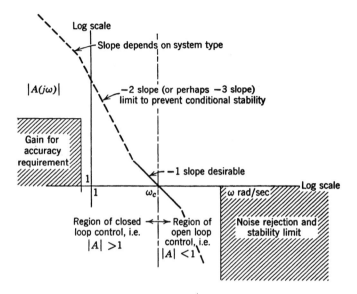

Figure 8-23   Bode diagram of a "good" servo loop.

4. For satisfactory stability, the crossover frequency should occur on an asymptotic −1 slope and it must be "controlled." That is, establishment of the crossover frequency must be an explicit part of design, and its value and variation must be computed to assure stability under all operating conditions.

5. Noise rejection and stability always limit the system bandwidth.

6. Accuracy requirements usually dictate the slope of the Bode diagram at low frequencies, that is, zero for type 0 systems, −1 for type 1 systems, −2 for type 2 systems, etc.

These constraints or good design features are represented in the Bode diagram of Fig. 8-23. Because loop gain functions fall off with increasing frequency, the range of permissible Bode diagrams is limited. Therefore, shaping of the Bode diagram for a particular design consists of adjusting the various break frequencies and slopes to meet the specifications. Conflicting or inconsistent specifications are usually spotted without particular difficulty.

Let us now list some general remarks that are applicable to the particular area of hydraulic servo design.

1. Selection of supply pressure, servovalve, and actuator are based on factors discussed in Section 8-1. This step fixes the power element parameters such as D_m, n, δ_h, ω_h, and K_q in Fig. 8-5.

2. The error detector is chosen to fit the application and give the specified accuracy. However, some portion of the allowable error must be alloted to other sources of error such as amplifier and servovalve drift, component tolerances, parameter variations, loads, and component nonlinearities such as hysteresis, deadband, and friction.

3. The remaining portion of the loop consists of electronic a-c error amplifiers, discriminators, compensations networks, and d-c power amplifier to drive the servovalve. The gains of these elements are selected to maintain stability of the loop. Some range of gain adjustment is desirable to facilitate final setting at manufacture.

4. When circumstances permit some flexibility in gain distribution, it is best to keep higher gain portions of the electronic section near the error detector in the a-c amplification stages. This minimizes the error due to drift in the d-c amplifier.

5. System specifications should always be checked to make sure they are met.

6. With component gains and time constants known and the block diagram complete, a check should be made to see if limit cycle oscillations will result from component nonlinearities. Describing function analysis is very useful in this regard.

7. The design should be checked to see that individual elements are not saturated for steady-state position and velocity inputs. This assures that linear design theory is applicable and the system, therefore, should perform as expected.

8. As described in Sections 3-7 and 6-6, a check is in order to see that destructive pressure transients are under control.

In a nutshell, hydraulic servo design proceeds by first selecting the power element to meet the prime considerations of load and response capability. The remainder of the loop is then sized to obtain the best possible servo performance (i.e., accuracy and response) under the limitations imposed by the power element dynamics. This amounts to selecting the highest possible gain constants and crossover frequency with adequate stability margin. Use of lag compensation and velocity feedback minor loops do permit some latitude in meeting accuracy specifications. However, speed of response cannot be altered except by choosing a faster power element. In a sense, then, detailed performance specifications are useless because there may not be sufficient design flexibility to meet them.

Although structural natural frequencies are occasionally lowest and become the limiting factor, it is the hydraulic natural frequency which is usually lowest and limits servo performance. The hydraulic natural frequency is highly dependent on the bulk modulus. As discussed in Section 2-5, the bulk modulus is lowered by entrained air and mechanical compliances. It also decreases substantially at higher temperatures. Hence, it could happen that a given system design might become unstable at elevated temperatures. This is offset somewhat by the increased leakage due to lower fluid viscosity at higher temperatures, which results in increased damping of the hydraulic natural frequency. In general, however, stability and over-all performance deteriorates at higher temperatures. There is less viscous drag on lubricated elements, and coulomb friction becomes more dominant in addition to the decreased hydraulic natural frequency.

REFERENCES

[1] Cooke, C., "Optimum Pressure for a Hydraulic System," *Product Engineering*, May 1956, 162–168.
[2] Merritt, H. E., and J. T. Gavin, "Friction Load on Hydraulic Servos," *Proc. Natl. Conf. Indl. Hydraulics*, 16, 1962, 174.
[3] Gibson, J. E., and F. B. Tuteur, *Control System Components.* New York: McGraw-Hill, 1958.
[4] Davis, S. A., and B. K. Ledgerwood, *Electromechanical Components for Servomechanisms.* New York: McGraw-Hill, 1961.
[5] Bower, J. L., and P. M. Schultheiss, *Introduction to the Design of Servomechanisms.* New York: Wiley, 1958.
[6] Nikiforuk, P. N., and D. R. Westlund, "The Large Signal Response of a Loaded High-Pressure Hydraulic Servomechanism," *Proc. Inst. Mech. Eng.* (London), 180, Pt. 1, 1966.
[7] Acker, R., "Designing Servos for Large-Signal Stability," *Control Eng.*, May 1965, 71–75.

9

Hydromechanical Servomechanisms

Hydromechanical servos, as the name implies, have no electronic elements in the loop. These servos are used almost exclusively for position control. A spool type servovalve ports fluid to a piston or rotary motor whose motion is fed back through some mechanical arrangement to stroke the servovalve and close the loop. Hydromechanical servos are used extensively in aircraft, missile, and jet engine controls. Other examples include the positioning loop of direct feedback two-stage servovalves, automobile power steering, and tracer controlled machine tools.

Simple examples of hydromechanical position control servos are illustrated in Fig. 9-1. In both cases input motion opens the servovalve and ports flow to a piston. Piston motion is fed back and subtracted from the input motion so as to null the servovalve. In the case of Fig. 9-1a, the feedback is direct and spool position from neutral is given by

$$x_v = x_i - x_p \qquad (9\text{-}1)$$

An error bar is used to sum the motions for the case in Fig. 9-1b. For small motions of the error bar, the spool position is given by

$$x_v = \frac{b}{a+b} x_i - \frac{a}{a+b} x_p \qquad (9\text{-}2)$$

In general, the feedback connection might consist of cams, linkages, gears, shafts, differential gears, and, if a rotary actuator is used, rotary to linear converters such as screw-and-nut and rack-and-pinion combinations. In any event the input, output, and spool positions can be related by an expression of the form

$$x_v = K_i x_i - K_f x_p \qquad (9\text{-}3)$$

Dynamics of the mechanical feedback elements are negligible compared with the hydraulic elements and are usually omitted. If the feedback path is nonlinear, as is often the case if cams are involved, (9-3) must be considered a linearized expression. The constants would have to be evaluated and the stability determined for every operating condition.

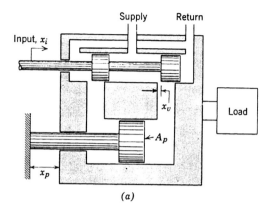

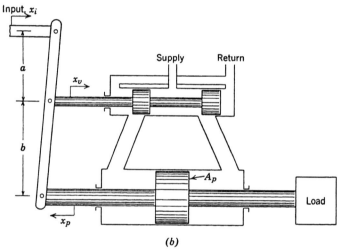

Figure 9-1   Schematic of simple hydromechanical position control servos.

The transfer function of the hydraulic portion of the loop is obtained from Chapter 6 for the appropriate case. Assuming a servovalve and piston combination with simple inertia load (6-36) is then applicable. This expression can be combined with (9-3) to form the block diagram in Fig. 9-2. The loop gain function is

$$A_u(s) = \frac{K_v}{s\left(\dfrac{s^2}{\omega_h{}^2} + \dfrac{2\delta_h}{\omega_h}s + 1\right)} \tag{9-4}$$

where $K_v = K_f K_q/A_p$ sec^{-1} is the velocity constant. This expression is identical to (8-20) and has the same Bode diagram (Fig. 8-6). The reader is referred to the discussion of this equation in Section 8-2 for a detailed treatment of stability.

The sizing of hydromechanical positioning servos is relatively straightforward. The actuator is sized to handle expected loads and have a hydraulic natural frequency large enough to meet over-all response requirements, as explained in Section 8-1. The feedback gain K_f depends on the

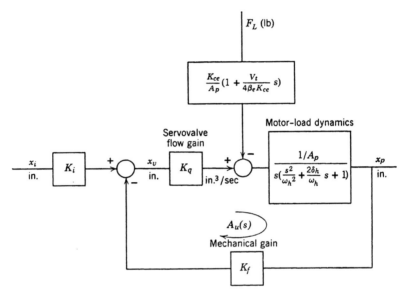

Figure 9-2   Block diagram of hydromechanical position control servo.

particular mechanical arrangement used. The remaining parameter, the servovalve flow gain K_q, is sized so that the velocity constant does not exceed the value 0.2 to 0.4 of ω_h required for stability. Invariably, the permissible flow gain is so low that full periphery valve ports cannot be used. Small width rectangular ports, beveled or notched metering edges on the spool, and round ports are common methods of obtaining a low area gradient and flow gain. Other considerations in spool design are discussed in Section 5-8.

For several reasons the bandwidth requirements for these servos are comparatively low. Many of these servos, such as those used for aircraft control surfaces and vehicle power steering, are used to "power boost" an input from a human operator and the bandwidth need not be much greater than human response capability of perhaps 2 to 5 cps. In other cases, such as jet engines and aircraft flight data computers, these servos perform subsidiary functions and are inherently much faster than the over-all process being controlled. Because it is quite easy in practice to get crossover frequencies of 25 to 50 rad/sec, these applications can be met with no particular difficulty.

In general, the accuracy requirement for these servos is also modest. Hence, there is no need for a lag-lead type of compensation (such as that discussed in Section 8-3) to boost low frequency gain. It would, of course, be possible to construct equivalent mechanical compensation networks from springs, dashpots, and masses. However, placement in the loop, parameter variations, and reliability are major drawbacks to such struc-tures. Fortunately, such schemes are not usually required in position control servos. However, mechanical compensation is often used in the hydromechanical control of other quantities. For example, a jet engine speed control may use a flyweight tachometer to sense speed and feedback a signal to control fuel flow to the engine. In this case it might be desirable to make a hydromechanical lead network to compensate the large engine lag and improve over-all response. However, the parameters in such a network vary with temperature and other factors and may cause stability and response problems. Hence these networks should be used sparingly and due regard given their peculiarities in the design stage.

There are several difficulties which can be encountered with hydro-mechanical servos. With no pretense to completeness, some of these are:

1. Very often the servovalve area gradient is chosen for manufacturing convenience or to obtain a certain flow at full stroke. The resulting flow gain is usually too high and, if so, the servo will become unstable.

2. A servovalve with low flow gain may require a very long stroke to achieve the flow required for maximum output velocity. Because valve strokes are physically limited, this condition requires an increased flow gain. If this flow gain exceeds that for stability, then the area gradient is increased with valve stroke. In this manner stability is maintained at null, and maximum velocity is achieved at full stroke. Round ports achieve this effect of low gain at null and increased gain away from null. However, servos which have an increasing gain characteristic usually oscillate above a certain velocity. If this performance is undesirable, then linear ports should be used and the specification on maximum velocity should be re-examined. As explained in Sections 5-1 and 5-8, linear ports are the best all around choice.

3. In some cases the structural mounting of hydraulic piston or cylinder may not be sufficiently rigid, and the dynamic characteristics of the com-pliant mounting may appreciably alter the power element dynamics and cause servo instability.

4. In the design of any hydraulic system, gain adjustments should be provided. In hydromechanical servos this is difficult. Flow gains of servovalves are usually easiest to alter without major modifications in the mechanical structure of the system.

5. The mechanical feedback elements should be of sufficient strength and the path should be as direct and simple as possible. Loose fitting linkages, pins, and gears are a source of backlash and may cause instability. Friction loads on flexible mechanical members is also a source of backlash. Therefore, tight as well as loose fits should be avoided.

6. If hydromechanical compensation networks are used, stability should be checked for all operating conditions, taking into account the variation in parameters values of the network.

7. Hydraulic lines, especially that of interdrillings, should be large enough to ensure that adequate pressure is supplied at all flow conditions.

Because many of the points discussed in Chapter 8, especially in Sections 8-1 and 8-5, are also applicable to hydromechanical servos, a reading of that chapter is recommended to gain a broader understanding of these servos.

10

Nonlinearities in Control Systems

Why a chapter on nonlinearities in a book on hydraulic control? Most of the peculiar phenomena observed in the test of a system can be traced to certain nonlinearities. A knowledge of the effect of nonlinearities on performance is required to make sense of these experiences and fill the gap between theory and design on one side and actual practice on the other. The purpose of this chapter is to help fill this gap and to cover in detail nonlinearities such as shaped gain, backlash, and friction which are of particular importance in hydraulic control.

Linear systems, by definition, have elements described by linear differential equations. Some attributes of linear systems are:

1. Because a mathematical theory exists for the solution of these equations, all aspects of performance, such as stability and response, are predictable.

2. Using the principle of superposition, we obtain the system response to several inputs by adding individual responses.

3. They are completely described by transfer functions. It should be apparent that the transfer function is obtained from a sinusoidal test and the governing differential equation can then be obtained from the transfer function. Because the differential equation can be solved for any input, the *sinusoidal response is indicative of system response to any input.* The usefulness of transfer functions and sinusoidal analysis stems from this basic fact. Other advantages are simplicity of analysis and availability of many design techniques. The unique status of sinusoidal analysis, that is, of implying response to any input, is valid only for linear systems.

Because a theory exists for linear systems, many systematic design and test procedures have been developed. Performance criteria such as bandwidth, phase margin, M-peak, rise time, and settling time, to name a few, have been defined and serve as means of comparing systems. These criteria do not bear the same relationship to one another in nonlinear systems. It is important to realize that linear theory is fundamental to system understanding, analysis, and design and provides a backdrop for

271

system comparisons. Therefore, even if nonlinear solutions were easily obtained for particular inputs, there would still be need for indices to measure and compare performance.

Although all real systems are nonlinear to some extent because of saturations, there is usually some linear region of operation. However, some elements such as relays are nonlinear to such an extent that no linear region exists. In most nonlinear systems the problem is one of determining the effect on performance of certain nonlinear elements. No general nonlinear theory exists, but there are several techniques which are applicable to certain classes of problems. The major techniques are:

1. Linearization or small signal analysis. Assuming small excursions in the system variables, the system can be treated as linear. Performance is computed for every operating point. This technique is not a method of solving nonlinear differential equations but in many cases it is all that can be done. It is most often used.

2. Piecewise linear analysis. This technique is useful when the system is describable by linear equations valid over certain ranges. By inserting appropriate initial conditions into the differential equations, an over-all solution can be patched together.

3. Describing function analysis. This technique is an approximate method of determining sinusoidal response, jump resonance phenomenon, and possible limit cycles. It is the most fruitful of the paper and pencil nonlinear analysis techniques. The describing function concept is also useful in system design, test, and troubleshooting.

4. Phase plane analysis. This is a graphical technique for solving nonlinear differential equations and is useful in determining limit cycles and step function response. It becomes tedious for third- and higher-order systems.

5. Analog computer analysis. This technique consists of making an electronic model of the differential equations and recording the solutions on an oscillograph. It is very general in application and, especially for complex control problems, gives a very convenient method for analysis and design.

6. Digital computer analysis. In recent years many programs have been developed to solve nonlinear differential equations [1]. Presently, these programs are not available to all and, furthermore, digital simulation is very expensive for many dynamic problems.

10-1 TYPICAL NONLINEAR PHENOMENA AND INPUT-OUTPUT CHARACTERISTICS

Several performance phenomena are directly attributable to system nonlinearities. In test, these phenomena identify the nonlinear system as such.

Some of these are:

1. With a fixed input, the output may oscillate with a fixed amplitude and frequency. This condition is referred to as limit cycle oscillation. In an unstable linear system, the output increases without bound. However, in any physical system the oscillation amplitude is constrained by saturations.

2. With a sinusoidal input, discontinuous jumps in the output amplitude may occur as the input amplitude or frequency is changed continuously. This phenomenon is known as skewed or jump resonance.

3. With a sinusoidal input of frequency ω_1, the output may oscillate at frequencies of ω_1, $2\omega_1$, $3\omega_1$, . . . , $n\omega_1$. This phenomenon is known as the generation of harmonics or superharmonics of order n.

4. With a sinusoidal input of frequency ω_1, the output may oscillate at frequencies of ω_1, $\omega_1/2$, $\omega_1/3$, . . . , ω_1/n. This phenomenon is known as the generation of subharmonics of order $1/n$.

5. With a sinusoidal input, the output may oscillate at a combination of the input frequency and some other frequencies not present in the input. Little is known concerning this kind of response.

These phenomena are not usually observed under ordinary circumstances where systems are quite stable. Rapid input changes or proper input manipulations to a lowly damped system are necessary to cause many of these effects. These phenomena, which do not appear in linear systems, cannot be viewed as good or bad except, perhaps, limit cycle oscillations. The chief aim of a nonlinear analysis is to predict such "limit cycles."

Input-Output Characteristics of Nonlinear Elements

The output of a nonlinear element depends on the past history of the input as well as its present values. However, most of the nonlinearities of concern are completely described by an input-output characteristic. Typical characteristics are shown in Fig. 10-1. Many of these nonlinearities, especially those which are multivalued, such as backlash, pose serious threats to stability. In the next several sections describing function analysis will be used to investigate servo loops containing common nonlinearities.

10-2 DESCRIBING FUNCTION ANALYSIS

Describing function analysis is an approximate method of finding the sinusoidal response of certain nonlinear elements. The phenomena of skewed resonance and limit cycles can be predicted. However, the major

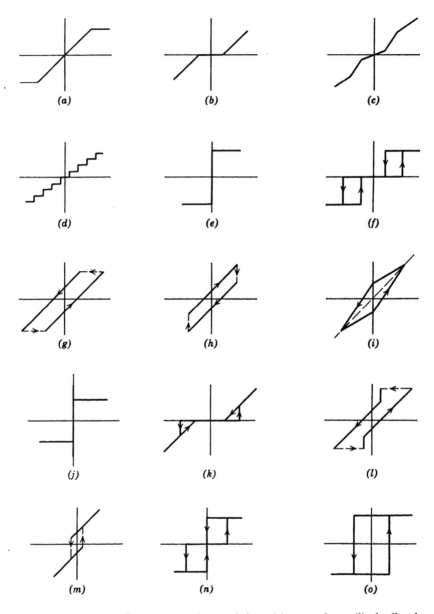

Figure 10-1 Typical input-output characteristics: (*a*) saturation; (*b*) deadband; (*c*) nonlinear gain; (*d*) granularity; (*e*) ideal relay or friction; (*f*) relay with deadband and hysteresis; (*g*) backlash; (*h*) inverse backlash; (*i*) hysteresis (electrical); (*j*) static and coulomb friction; (*k*) viscous and friction forces; (*l*) spring and friction forces; (*m*) negative deficiency; (*n*) clutch; (*o*) relay-actuator.

274

use is predicting limit cycles and the method gives the approximate amplitude, frequency, and waveform. Although the method is approximate, the results compare very well with test data and analog models. Of prime interest in any system is whether limit cycles will exist. A secondary interest is placed on the exact amplitude and frequency of such oscillations.

Limit cycles are undesirable in servo systems because:

1. It is usually difficult to reduce the amplitude and frequency to within limits considered adequate.

2. Depending on gain distribution, such oscillations may saturate a particular element, thereby reducing loop gain and causing a drop in accuracy and response.

3. The wear and strain caused by oscillations pose a threat of failure and may shorten system life.

4. Such oscillations constitute a power loss.

5. In certain cases, such oscillations may cause large errors.

Therefore, the detection of limit cycles should be an important consideration in the design, analysis, and test of a control system.

Describing function analysis is based on the following assumptions:

1. The input to the nonlinear element is sinusoidal.

2. The output waveform of the nonlinear element is adequately described by the fundamental component of a Fourier series, that is, harmonics are negligible.

These assumptions are justified by the following observations: (1) linear elements in the loop filter the higher frequency harmonics and emphasize the dominance of the fundamental; (2) the fundamental component of the output waveform is usually larger than that of the harmonics. Describing function analysis is restricted to cases in which all the nonlinearities in a loop can be lumped at a single point. However, in some cases little accuracy is lost if describing functions of individual nonlinearities are combined in a manner similar to that of linear elements (see the example in Section 10-6).

The theory of describing function analysis will be outlined. Consider the loop shown in Fig. 10-2 with frequency dependent (i.e., linear) elements $G_1(j\omega)$ and $G_2(j\omega)$. The nonlinear element is characterized by the describing function $G_d(\alpha, \beta)$, where α and β are ratios relating dimensions of the nonlinearity to the input amplitude M at the nonlinear element. The describing function is defined as

$$G_d(\alpha, \beta) \equiv \frac{C_1}{M} \underline{/\phi_1} \qquad (10\text{-}1)$$

which is the complex ratio of fundamental output to input of the nonlinear element. Because the describing function is defined only for sinusoidal signals, $j\omega$ must be substituted for s in the linear transfer functions. The closed loop response is given by

$$\frac{C}{R} = \frac{G_1(j\omega)G_d(\alpha, \beta)G_2(j\omega)}{1 + G_1(j\omega)G_d(\alpha, \beta)G_2(j\omega)} \qquad (10\text{-}2)$$

Stability of the system is dependent on the roots of the characteristic equation, which is the denominator equated to zero.* The characteristic

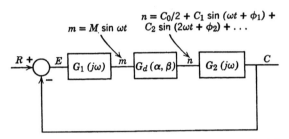

Figure 10-2   A general control loop containing a nonlinearity.

equation can be written

$$G_1(j\omega)G_2(j\omega) = -1/G_d(\alpha, \beta) \qquad (10\text{-}3)$$

Solutions are easily found by plotting both sides on the customary Nyquist diagram or gain-phase plane. If intersections of the two loci exist, a limit cycle will exist with frequency given by the left side of (10-3) and amplitude given by α and β. The quantity $-1/G_d(\alpha, \beta)$ can be considered as a generalized critical point in the Nyquist sense or simply as a locus of critical points. Therefore, stability is determined by separating the loop gain into two components, linear and nonlinear, and plotting them to see if intersections occur. Hence, there is the restriction that nonlinearities must be lumped.

The describing function is defined in terms of the fundamental output. This requires a Fourier analysis of the output waveform. If the nonlinearity is symmetrical, such as those in Fig. 10-1, there is no d-c component. Furthermore, if the nonlinearity is single-valued, there is no phase shift. Because multivalued nonlinearities have phase shift, usually phase lag,

---

* This can be shown as follows: Imagine the loop in Fig. 10-2 is broken at some point and a signal e_i is injected. Tracing around the loop, the signal which returns to this point is $e_0 = -G_1G_2G_de_i$. Now, when the loop is limit cycling it should be apparent that e_i and e_0 must be identically equal, that is, $e_0 = e_i$. Combining these two relations we obtain the condition for oscillation as: $G_1G_2G_d = -1 = 1\,\underline{/-180°}$.

these are most serious from the viewpoint of stability. A Fourier analysis is a tedious process in all but simple cases. Fortunately, however, this analysis can be performed for most commonly encountered nonlinearities, and the results can be plotted in a normalized manner [2, 3].

From a practical viewpoint, describing function analysis is rapid and the only method generally useful for higher order systems. It is applicable to a wide variety of nonlinearities, aids in system understanding, analysis, design, test, and troubleshooting. It indicates a design procedure—simply avoid intersections with the $-1/G_d$ locus. In many respects it is an extension of familiar frequency response concepts and techniques. The next several sections will be concerned with examples of describing function analysis.

10-3   SATURATION

The most common nonlinearity in control systems is saturation which is represented, somewhat idealized, in Fig. 10-3. The slope is unity because

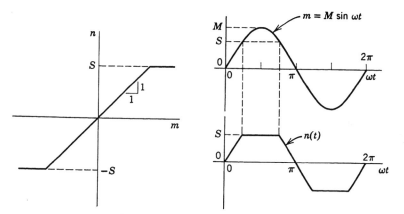

Figure 10-3   Saturation nonlinearity.

it is usual practice to treat any gain of the nonlinear element as part of the linear portion of the loop. A Fourier analysis of the output waveform for a sinusoidal input gives the following describing function

$$G_d \equiv \frac{C_1}{M} = \frac{2}{\pi}\left\{\sin^{-1}\frac{S}{M} + \frac{S}{M}\left[1 - \left(\frac{S}{M}\right)^2\right]^{\frac{1}{2}}\right\} \qquad (10\text{-}4)$$

The describing function gain G_d depends on the dimensionless ratio S/M and is plotted in Fig. 10-4. For a given saturation value (S), the gain is unity for small inputs (M) and approaches zero for infinite inputs.

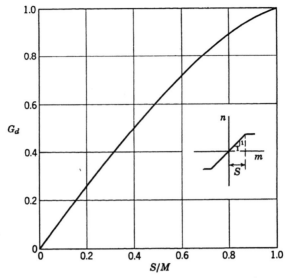

Figure 10-4 Describing function for saturation.

To determine possible limit cycles, we must plot both sides of (10-3). The left side of this equation is the conventional Nyquist diagram. For saturation, the right side, $-1/G_d$, is unity for small inputs and approaches $-\infty$ for large inputs. Suppose the two plots have the appearance of that in Fig. 10-5. Because the two curves intersect, sustained oscillations will occur at an amplitude given by the value of M on the $-1/G_d$ curve and at a frequency given ω on the $G_1(j\omega)G_2(j\omega)$ curve at the point of intersection. For this case the system is unstable if it is linear because the Nyquist

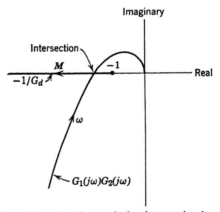

Figure 10-5 Describing function analysis of saturation in a control loop.

plot encircles the −1 point. Thus, a servo designed to be stable in the Nyquist sense will not become unstable due to saturation.

Suppose the Nyquist diagram represents a conditionally stable loop, Fig. 10-6. Two intersections will then occur. Once the system is synchronized, the system will be stable as long as the input amplitude M does not exceed that at point P_1. If the amplitude becomes greater, the Nyquist diagram encircles the critical point and oscillations will build up until a sustained limit cycle at point P_2 results. If M is greater than that at point P_2, the oscillation amplitude decreases because the critical point is no

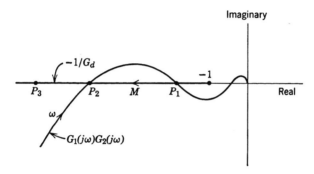

Figure 10-6 Describing function analysis of saturation in a conditionally stable loop.

longer encircled. Thus, the system is stable until the input amplitude places the critical point so that it is enclosed by the Nyquist diagram. Point P_1 is called a *divergent equilibrium point*, and point P_2 is called a *convergent equilibrium point*. At the divergent equilibrium point, the system becomes unstable and oscillations continue to grow until a sustained limit cycle is reached at the convergent equilibrium point. From these analyses we can conclude that saturation presents a stability problem if and only if the linear system is conditionally stable.

When a servo is first turned on or for large transients the error may be large enough to saturate some elements (especially error detectors). The critical point is then a large negative number (point P_3 in Fig. 10-6). Gradually, as the feedback quantity builds up, the large initial error in the system will be reduced. If the loop is conditionally stable, then, as the error is reduced, the critical point becomes that of point P_2 and sustained oscillations result. Thus, in a conditionally stable system, means must be provided to switch to a synchronizing system at an error small enough so that the critical point is still to the right of the divergent equilibrium point

P_1. Other disadvantages of conditionally stable loops stem from their sensitivity to parameter variations because a shift in gain either up or down may cause instability. Also, these loops are prone to oscillate with almost any nonlinearity present in the loop. Lag-lead compensated servos may be conditionally stable. Type 2 servos are usually conditionally stable.

10-4 DEADBAND

A deadband nonlinearity (Fig. 10-7) can result from coulomb friction and from overlap of valve ports in hydraulic systems. The linear gain of

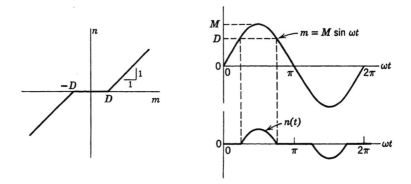

Figure 10-7   Deadband nonlinearity.

the deadband is normalized to unity and any gain present would be considered as part of the linear portion of the loop. Analysis of the output waveform gives the following describing function

$$G_d = \frac{2}{\pi}\left\{\frac{\pi}{2} - \sin^{-1}\frac{D}{M} - \frac{D}{M}\left[1 - \left(\frac{D}{M}\right)^2\right]^{1/2}\right\} \qquad (10\text{-}5)$$

which is plotted in Fig. 10-8.

We note that $-1/G_d$ is a large negative real number for small inputs to the deadband element and approaches -1 for large inputs. Suppose the plots of (10-3) are as shown in Fig. 10-9. The linear system with this Nyquist plot would be unstable. However, this system will be stable for small inputs to the deadband and unstable for large inputs. This is reasonable because for small inputs the system is operating with low gain due to the deadband. It may seem peculiar that a system which is linearly unstable may be stable if it contains a nonlinear element. In this case,

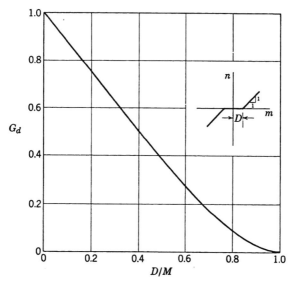

Figure 10-8   Describing function for deadband.

the point of intersection is a divergent equilibrium point and unbounded oscillations will occur if the input to the deadband is large enough.* In any practical system, however, saturation would be present and the oscillations would be bounded. This case is treated in the next section.

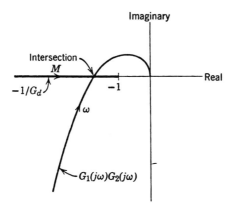

Figure 10-9   Describing function analysis of deadband in a control loop.

* If deadband were present in a conditionally stable system, the plot would be similar to Fig. 10-6, with the low frequency intersection being a divergent equilibrium point and the high frequency intersection being a convergent equilibrium point.

10-5 NONLINEAR GAIN CHARACTERISTICS

The describing function of the quite general nonlinear gain characteristic in Fig. 10-10 can be shown to be

$$G_d = k_3 + \frac{2}{\pi}(k_1 - k_2)\left\{\sin^{-1}\frac{D}{M} + \frac{D}{M}\left[1 - \left(\frac{D}{M}\right)^2\right]^{1/2}\right\}$$
$$+ \frac{2}{\pi}(k_2 - k_3)\left\{\sin^{-1}\frac{S}{M} + \frac{S}{M}\left[1 - \left(\frac{S}{M}\right)^2\right]^{1/2}\right\} \quad (10\text{-}6)$$

The describing functions for saturation and deadband can be obtained
from this expression by letting appropriate quantities be zero. With so

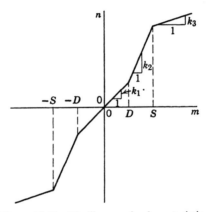

Figure 10-10 Nonlinear gain characteristic.

many parameters involved, normalized plots are impractical. It is preferable to compute a particular case. A characteristic of interest is the combination of saturation and deadband (Fig. 10-11). For this case $k_1 = k_3 =$
0 and $k_2 = 1$ and the describing function is plotted in Fig. 10-12. Note
that the "gain" is small for small inputs, increases to a maximum, then
decreases as the input amplitude M increases. Thus, the quantity $-1/G_d$
starts at $-\infty$ for small inputs, decreases to a minimum, then again approaches $-\infty$ as the input becomes very large. The $-1/G_d$ locus and a
Nyquist diagram of a linearly unstable system are shown in Fig. 10-13.
For the intersections shown, point P_1 is a divergent equilibrium point and
P_2 is a convergent equilibrium point. Note that this system is stable for
small inputs but once the input amplitude becomes greater than at point
P_2, oscillations will build up to a limit cycle at P_2. The $-1/G_d$ locus has
a minimum which approaches but never exceeds the -1 point. Thus, a

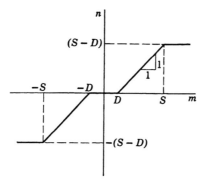

Figure 10-11   Saturation and deadband nonlinearity.

system having this characteristic and designed so that the frequency locus does not encircle the -1 point would be stable. However, it is possible for the system to be stable even if the -1 point is encircled because of the minimum of the $-1/G_d$ locus.

Another gain characteristic of interest (Fig. 10-14) is obtained by contouring the ports of hydraulic servovalves. Round ports and some types of flow force compensation of single-stage servovalves also give such a curve. The slope is normalized to unity near the origin and the second slope has a value greater than unity. The describing function for this characteristic can be obtained from (10-6), and it can be shown that the

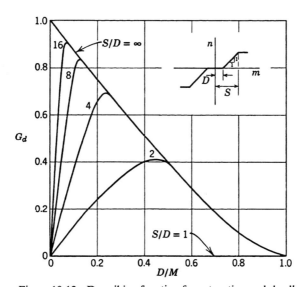

Figure 10-12   Describing function for saturation and deadband.

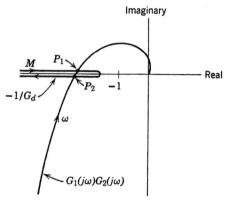

Figure 10-13 Describing function analysis of saturation and deadband nonlinearity in a control loop.

maximum value of G_d is always less than K. Therefore, the minimum critical point is greater than $-1/K$. Thus, if $K = 5$, for example, it is possible for the critical point to approach -0.2 as compared with -1. The $-1/G_d$ locus and a typical Nyquist diagram are shown in Fig. 10-15. The point P_1 is a divergent equilibrium point and P_2 is a convergent equilibrium point. The system is stable for small inputs, but if the input exceeds P_1 oscillations will build up to a limit cycle at P_2. Note that the frequency of oscillation will be higher than the crossover frequency. With most other nonlinearities, limit cycles occur at frequencies lower than the crossover frequency. Thus far this is the first nonlinearity to cause a system that was designed stable on a linear basis to be unstable.

In some cases this characteristic is intentionally inserted in the loop to provide increased gain for large signals with decreased gain near the origin

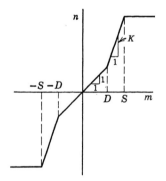

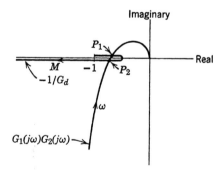

Figure 10-14 An upsweep gain nonlinearity.

Figure 10-15 Describing function analysis of an upsweep gain in a control loop.

for stability. However, because the system should be stable for the largest gain of the nonlinearity, it is debatable whether this technique is useful. For large signals the system would be oscillatory, and for small signals the system would be highly damped. It would seem better to compromise and use an intermediate and constant value of gain at the outset.

All the previous cases were concerned with single-valued nonlinearities. For such cases, an approximate describing function is the slope of a line drawn from the origin to a point on the nonlinear curve corresponding to the input amplitude chosen. Repeating this procedure for several values of input amplitude, an approximate plot of G_d is obtained rapidly and easily.

10-6 BACKLASH AND HYSTERESIS

These nonlinearities are multivalued. With backlash (Fig. 10-16), the input must be moved an amount $H/2$ before any motion of the output

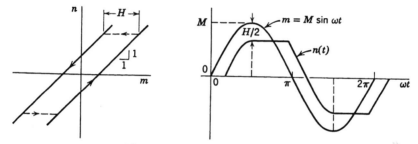

Figure 10-16   Backlash nonlinearity.

occurs. Upon reversal, the input must move a distance H before output motion again occurs. The backlash width H is a constant. Backlash is a nuisance in control systems and some of the physical situations which cause it are:

1. Loose gear meshes from which the term "backlash" stems.
2. Loose fitting pins in lever arrangements.
3. Flexure in lever arrangements which are loaded by friction. Thus, a tight fitting pin as well as a loose fit can cause backlash. This problem is treated in the next section.
4. Coulomb friction load on a hydraulic servo causes backlash in the closed loop response (see Section 10-7).
5. Deadband in the forward path of a type 1 servo loop causes backlash in the closed loop response. For example, a hydraulic servovalve with overlap will cause backlash in the closed loop response.

The describing function is given by

$$G_d = \left[\left(\frac{a_1}{M}\right)^2 + \left(\frac{b_1}{M}\right)^2\right]^{1/2} \bigg/ \tan^{-1}\frac{a_1}{b_1} \qquad (10\text{-}7)$$

where $\dfrac{a_1}{M} = \dfrac{H}{\pi M}\left(\dfrac{H}{M} - 2\right)$

$$\frac{b_1}{M} = \frac{1}{\pi}\left\{\frac{\pi}{2} + \sin^{-1}\left(1 - \frac{H}{M}\right) + \left(1 - \frac{H}{M}\right)\left[2\frac{H}{M} - \left(\frac{H}{M}\right)^2\right]^{1/2}\right\}$$

and is plotted in Fig. 10-17. Depending on the amplitude óf the input to the backlash, this nonlinearity can contribute up to 90° of phase lag. For this reason, backlash poses a very serious threat to the stability of a loop.

The plot of $-1/G_d$ for backlash and a Nyquist diagram for a type 1 system are sketched in Fig. 10-18. There are three intersections.* The low frequency intersection P_1 and the high frequency intersection P_3 are convergent equilibrium points. The intermediate point P_2 is a divergent equilibrium point. Note that the points of intersection are not as clear as in previous cases. Limit cycles predicted by describing function analysis are most accurate when the two curves intersect nearly perpendicular. Therefore, we can infer that accuracy of prediction may be poor in this case. The low frequency intersection corresponds to a value of $H/M \approx 2$. That is, the peak-to-peak amplitude of the oscillation at the input has a value equal to the total backlash width. The frequency of oscillation is very small and appears as a "wandering" within the backlash zone. The backlash width can be referred to the output and may properly be regarded as an error because control cannot be maintained within the backlash width.

Lag compensated servos have the higher frequency intersections in Fig. 10-18. If the amplitude to the backlash exceeds that at point P_2, oscillation will build up to a limit cycle at point P_3. Removal of the high frequency intersection and the associated limit cycle is very difficult. The following techniques may be useful:

1. Reduction of the loop gain constant until the linear locus is well within the $-1/G_d$ locus. This is a sacrifice of accuracy for stability.

2. Lead compensation could be inserted in the frequency range between points P_2 and P_3 to bend the linear locus so as to avoid an intersection. Describing functions are very useful in design because, as in this case, corrections to avoid limit cycles are suggested.

* For type 2 and 3 systems, a single intersection will always occur and will be a convergent equilibrium point.

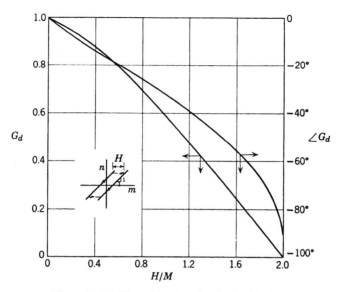

Figure 10-17   Describing function for backlash.

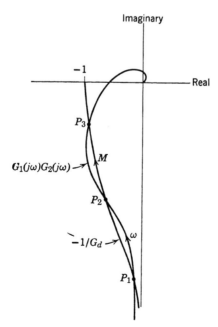

Figure 10-18   Describing function analysis of backlash in a type 1 control loop.

3. A gain shaping nonlinearity could be inserted to lower loop gain for signals greater than at point P_2 so as to shift the linear locus to the right and avoid intersections P_2 and P_3.

4. Dither is a widely used method of removing backlash. It is very effective where the backlash is caused by friction. Dither is a high frequency signal of constant amplitude and frequency which is added to the control signal at the input to the nonlinearity and has the effect of making the element appear linear. Dither cannot be used in cases such as gear backlash because it is difficult to inject, causes wear, and appears in the output.

5. If the cause of backlash is known, a reduction in the amount will reduce the amplitude of oscillation. The frequency, however, will not be altered.

6. A redistribution of gain in the loop can be effective in reducing limit cycle amplitude. If the backlash is in the forward path of the loop, increasing the gain ahead of the backlash (with a corresponding gain reduction after the backlash to maintain the same over-all loop gain) will decrease the oscillation amplitude at the output.

For type 2 and 3 systems, the limit cycle is almost impossible to avoid. Some of the techniques outlined may be useful at least in reducing the oscillation amplitude.

In all cases, limit cycles caused by backlash occur at a frequency lower than the crossover frequency of the loop that is oscillating. The frequency of oscillation as well as the distinctive flat-topped appearance of the output waveform usually makes for easy identification of backlash in component tests. Knowledge of the physical situations which cause backlash and abnormal phase lag across an element are also aids in detecting backlash.

In some cases there is more than one backlash in a loop. If dynamics between the elements with backlash are negligible, the individual backlash widths, properly referred to the same point, can be added to obtain a single backlash. If dynamics are not negligible, individual widths cannot be added and the situation is much more complicated. An extreme case can be considered where the filtering is so thorough that only the fundamental component of the output of the first backlash remains as an input to the second backlash. The over-all describing function can then be obtained using Fig. 10-17 for each backlash and combining the result. For example, a given input amplitude M_1 determines the gain and phase of the first backlash. The amplitude M_2 at the input to the second backlash is the product of the gain of the first backlash and M_1. Because M_2 is found, the gain and phase of the second backlash can be determined. The total gain across the two backlashes is the product of the individual gains, and the total phase shift is the sum of the individual phase shifts. Thus, if two elements having backlashes whose widths are of the same order referred to a point and if there is substantial filtering between the elements, it is conceivable that a phase lag approaching 180° could be obtained. Limit cycle oscillation therefore, would be extremely difficult to avoid. The procedure outlined to obtain the over-all describing function for several nonlinearities by simply combining individual functions must be used with caution. However, it affords a very approximate way of dealing with very complicated nonlinearities where most other techniques fail.

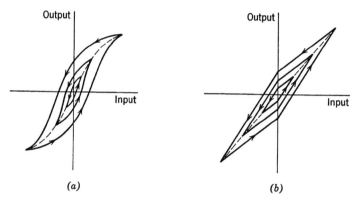

(a) (b)

Figure 10-19 (a) Actual hysteresis characteristic; (b) Analytical model of hysteresis.

Components with magnetic paths, such as electric servomotors and torque motors of electrohydraulic servovalves, have a multivalued non-linear characteristic commonly referred to as hysteresis (Fig. 10-19a). The width of the hysteresis loop varies directly with the input signal, whereas with backlash it is constant. Neal and Bunn [4] have proposed an approximation of hysteresis for analysis purposes. Assuming the ratio h/M to be constant, the describing function for the hysteresis model in Fig. 10-20 can be shown to be [4]

$$G_d = \left[\left(\frac{2}{\pi}\right)^2 \left(\frac{h}{M}\right)^2 + \left(1 - \frac{h}{M}\right)^2 \right]^{\frac{1}{2}} \underline{/\tan^{-1} \frac{-(2/\pi)(h/M)}{1 - h/M}}$$

Because h/M is a constant, the describing function locus $-1/G_d$ consists of a single point rather than an entire curve as with backlash. For example, if $h/M = 0.1$, that is, h is 10% of the input amplitude, then $G_d = 0.902$ $\angle -4°$ and $-1/G_d = 1.11 \angle -176°$. Thus, the critical point is shifted from $1 \angle -180°$ to $1.11 \angle -176°$.

Because the critical locus for hysteresis consists of a single point, this nonlinearity is not a serious threat to stability. The most noticeable attribute of elements with this nonlinearity is a fixed amount of phase lag at low frequencies. Frequency response measurements of electrohydraulic servovalves often show a fixed phase lag, that is, substantially independent of input amplitude, of a few degrees at low frequencies.

It is apparent that dither is not effective against electrical hysteresis because the width of the loop varies with input amplitude. Furthermore, it is questionable whether the shift in critical point location caused by

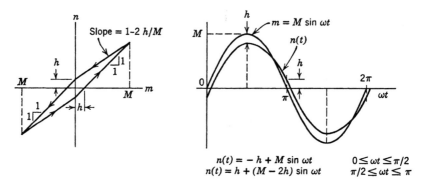

$$n(t) = -h + M \sin \omega t \qquad 0 \le \omega t \le \pi/2$$
$$n(t) = h + (M - 2h) \sin \omega t \qquad \pi/2 \le \omega t \le \pi$$

Figure 10-20   Hysteresis nonlinearity.

hysteresis is enough to warrant use of compensation networks. If h/M is quite large the shift in critical point may be significant. However, good components should have small values for h/M, say less than 0.1.

10-7   RELAY TYPE NONLINEARITIES

In many instances relay or relay type devices such as bistable hydraulic valves and clutches are used in the forward path of control loops. The major advantages offered by the use of on-off devices is low system cost and simplicity. The main disadvantage is that such elements are highly nonlinear and usually cause a limit cycle. The analysis and design of servos using such elements is more difficult. The technique of linearization cannot be used because no linear region exists. However, describing function analysis is particularly well suited to the treatment of these systems.*

Typical relay type input-output characteristics and their describing functions are tabulated in Fig. 10-21. Normalized plots are shown in Fig.

---

* As a matter of fact the technique was originally developed to analyze relay type servos and is credited to R. Kochenburger in this country.

(a) $D = H = 0$

$$G_d = \frac{4}{\pi}\frac{V}{M}$$

(a)

(b) $H = 0$

$$G_d = \frac{4}{\pi}\frac{V}{M}\sqrt{1 - \left(\frac{D}{2M}\right)^2} \quad \text{for} \quad M \geq \frac{D}{2}$$

$$G_d = 0 \quad \text{for} \quad M \leq \frac{D}{2}$$

(b)

(c) $D = 0$

$$G_d = \frac{4}{\pi}\frac{V}{M}\underline{/-\sin^{-1}\frac{H}{2M}} \quad \text{for} \quad M \geq \frac{H}{2}$$

$$G_d = 0 \quad \text{for} \quad M \leq \frac{H}{2}$$

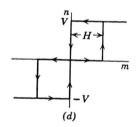

(c)

(d) $D = H$

$$G_d = \frac{2V}{\pi M}\left[\sqrt{1 + \frac{H}{M}} + \sqrt{1 - \frac{H}{M}}\right]\underline{/-\frac{\sin^{-1}\dfrac{H}{M}}{2}}$$

$$\text{for} \quad M \geq H$$

$$G_d = 0 \quad \text{for} \quad M \leq H$$

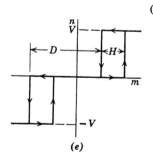

(d)

(e) General

$$G_d = \frac{4}{\pi}\frac{V}{M}\cos\left(\frac{\alpha + \beta}{2}\right)\underline{/\dfrac{\alpha - \beta}{2}} \quad \text{for} \quad M \geq \frac{D + H}{2}$$

$$G_d = 0 \quad \text{for} \quad M \leq \frac{D + H}{2}$$

$$\text{where} \quad \alpha = \sin^{-1}\frac{D}{2M}\left(1 - \frac{H}{D}\right)$$

$$\beta = \sin^{-1}\frac{D}{2M}\left(1 + \frac{H}{D}\right)$$

(e)

Figure 10-21   Relay describing functions. In all cases, input is $m = M\sin \omega t$.

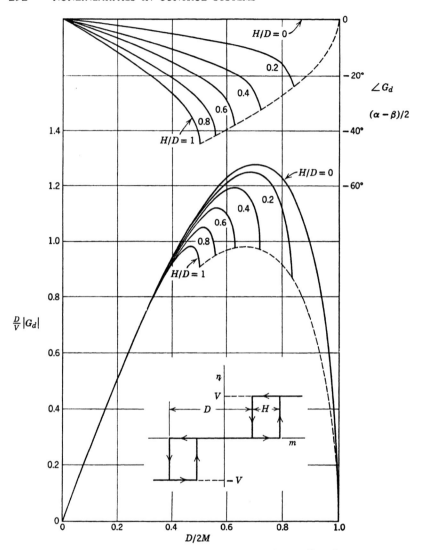

Figure 10-22   Describing function for a relay nonlinearity.

10-22. Let us consider the three cases shown in Fig. 10-21a, 10-21b, and 10-21e. The describing function analysis of servos with these nonlinearities are illustrated in Fig. 10-23.

The describing function analysis for ideal relay (Fig. 10-23a) gives a convergent equilibrium point at the intersection of the two curves. Thus, a limit cycle is predicted if the phase lag of the linear portion of the loop

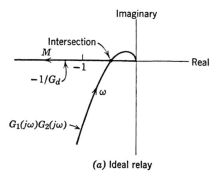

(a) Ideal relay

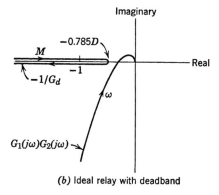

(b) Ideal relay with deadband

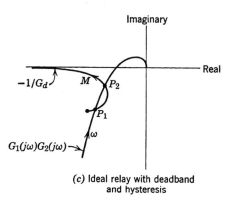

(c) Ideal relay with deadband
and hysteresis

Figure 10-23   Describing function analyses of relay type servos.

becomes 180° and, for physical systems, this will always occur. If the amplitude of the limit cycle is too large, it may be reduced by a redistribution of gain in the loop if this is permissible.

If the ideal relay has only deadband, it is possible to stabilize the system (Fig. 10-23b). The describing function for this case has a maximum of $G_d \big|_{\max} = 4/\pi D$, which occurs when $M = D/\sqrt{2}$. Hence the minimum value for $-1/G_d$ is $-D/1.275$, and if the linear locus crosses the real axis with less gain, the servo will be stable. Although deadband stabilizes the loop, it introduces static error. Again, we see that accuracy must be sacrificed to obtain stability.

If the relay has deadband and hysteresis, then the case shown in Fig. 10-23c results. The system is stable for small inputs, but if the input exceeds P_1, oscillation will build up to a limit cycle at the convergent equilibrium point P_2. For absolute stability, the linear locus must be shaped by proper compensation to avoid intersection with the describing function.

10-8   FRICTION NONLINEARITIES

This section will be devoted to the effects of static and coulomb friction in control loops. This phenomena is of such complexity that it is necessary to consider three describing functions, each having utility in different circumstances.

Newton's second law applied to any element, usually a power element,

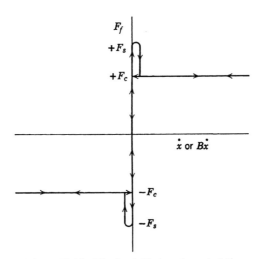

Figure 10-24   Idealized friction characteristic.

will result in an equation of the form

$$F_i = M_a \frac{d^2x}{dt^2} + B \frac{dx}{dt} + Kx + F_f + F_L \qquad (10\text{-}8)$$

where M_a is mass, B is a viscous damping coefficient, K is a spring constant, F_L is a constant or arbitrary load force, F_i is the driving force, and F_f is the friction force which is a sign dependent function of velocity with the somewhat idealized characteristic of Fig. 10-24. We can distinguish three cases depending on whether mass, viscous, or spring forces dominate. The block diagrams* of these cases are shown in Fig. 10-25. A fourth case where the driving force is consumed only by friction could be identified but this does not represent a situation of great interest. For each case in Fig. 10-25 the describing function must be found for the nonlinearity indicated and, by breaking the loop at point "p," a describing function analysis can be performed. Each analysis fails if the quantity on which the describing function is based is absent. For example, if M_a were small or negligible, a procedure based on Fig. 10-25a would fail because the open loop gain would be infinite. Similarly, a procedure based on Fig. 10-25b would fail if damping were zero. On the other hand, the case of Fig. 10-25c would survive if both M_a and B were zero, providing that K is not zero.

It is important to realize that one describing function cannot handle all nonlinearities due to friction. Because mass is present in all systems, Fig. 10-25a could be viewed as valid for all cases. However, in many systems mass can be neglected and a technique based on mass would be somewhat artificial with questionable accuracy. For example, the force required for acceleration during a limit cycle in a hydraulic servo is but a small fraction of that required for velocity dependent terms. Hence, a describing function based on the dominance of velocity forces would be in order.

Case a—Mass Force Dominant

The mathematical expression of the describing function for the nonlinearity in Fig. 10-25a is quite involved [5, 6]. However, it is plotted in Fig. 10-26. This describing function is independent of M_a and s (frequency)

* Strictly speaking, Eq. (10-8) cannot be transformed because it is nonlinear. However, it is customary to use s in the construction of block diagrams etc. with its meaning restricted to a shorthand notation for d/dt rather than as Laplace operator. Frequency response solutions are obtained by substituting $j\omega$ for s with suitable approximations of the frequency response of nonlinear elements. Transient response solutions must be found from the original differential equation.

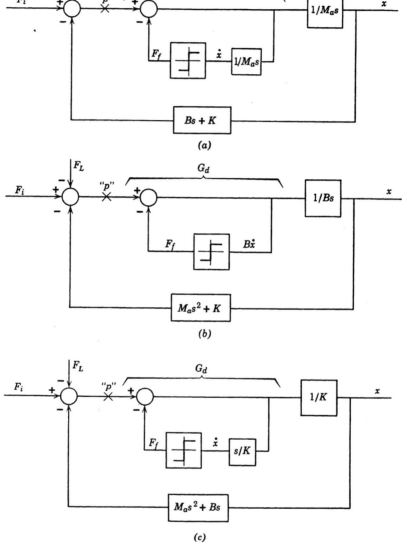

(a)

(b)

(c)

Figure 10-25 Block diagrams of Eq. (10-8).

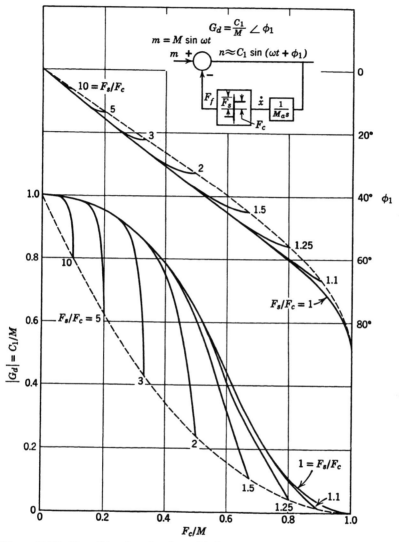

Figure 10-26 Describing function for a friction nonlinearity based on dominance of inertia forces.

297

because the friction force is dependent only on the sign of velocity. However, the value for M_a must be included in the linear part, as indicated in Fig. 10-25a. Motion ceases for $M \leq F_s$ and this relation in the form

$$\left.\frac{F_c}{M}\right|_{max} = \frac{1}{F_s/F_c} \tag{10-9}$$

is used to draw the dashed lines in Fig. 10-26.

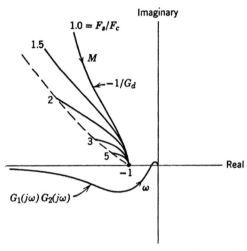

Figure 10-27   Describing function analysis of a control loop with a friction nonlinearity based on dominance of inertia term.

Note that this describing function can contribute up to 90° of *phase lead*. However, with the loop broken at point "p" in Fig. 10-25a, the linear locus usually has a double integration. Hence, the phase lead in G_d does not represent a stabilizing influence.

This describing function was originally derived for use with loops having electric motors as power elements [5]. Typical plots for such a case are shown in Fig. 10-27 and indicate no serious stability problem. Note that the linear locus does not have the usual appearance of a type 1 servo loop. This makes comparison of this nonlinearity with others very difficult because the linear loop gain must be in a special form.

Case b—Viscous Force Dominant

This case is applicable to power elements that are primarily integrating devices such as electric and hydraulic motors. For these devices the B

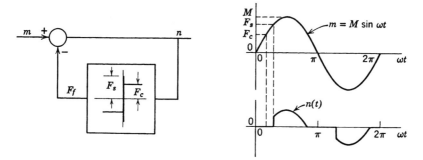

Figure 10-28   Friction nonlinearity based on dominance of viscous forces.

coefficient in Fig. 10-25b is quite large and always present because it mainly represents internal damping rather than load viscous friction.

The nonlinearity with waveforms for a sinusoidal input are shown in Fig. 10-28. An analysis of the output waveform yields the following describing function [7].

$$G_d = \frac{C_1}{M} \underline{/\phi_1} = \left[\left(\frac{a_1}{M}\right)^2 + \left(\frac{b_1}{M}\right)^2 \right]^{1/2} \underline{/\tan^{-1}\left(\frac{a_1}{b_1}\right)} \qquad (10\text{-}10)$$

where $\dfrac{\pi a_1}{M} = \left(\dfrac{F_c}{M}\right)^2 \left[2\,\dfrac{F_s}{F_c} - 1 - \left(\dfrac{F_s}{F_c}\right)^2 \right]$

$$\frac{\pi b_1}{M} = \pi - \sin^{-1}\left(\frac{F_s}{M}\right) - \sin^{-1}\left(\frac{F_c}{M}\right) - \frac{F_c}{M}\left[1 - \left(\frac{F_c}{M}\right)^2 \right]^{1/2}$$

$$+ \frac{F_c}{M}\left(\frac{F_s}{F_c} - 2\right)\left[1 - \left(\frac{F_s}{M}\right)^2 \right]^{1/2}$$

The magnitude and phase of this describing function is plotted in Fig. 10-29. There is no output motion for $M \leq F_s$ so that (10-9) can be used to draw the dashed lines through the tips of the loci. Note that for $F_s = F_c$ the describing function is identical to that of a deadband nonlinearity with a total deadband of $2F_c$. As a matter of fact the equation defining the nonlinearity is $m = B\dot{x} + F_f$ and is plotted in Fig. 10-30a. Because $n = B\dot{x}$, the ordinate and abscissa positions may be interchanged to yield Fig. 10-30b. This is the input-output characteristic of the loop in Fig. 10-28 and clearly shows deadband if $F_s = F_c$. Thus, we see how friction can result in deadband and, because it lowers gain, may tend to stabilize a system. However, we shall see in the next discussion—that is, case c— that friction can also be a very serious destabilizing influence.

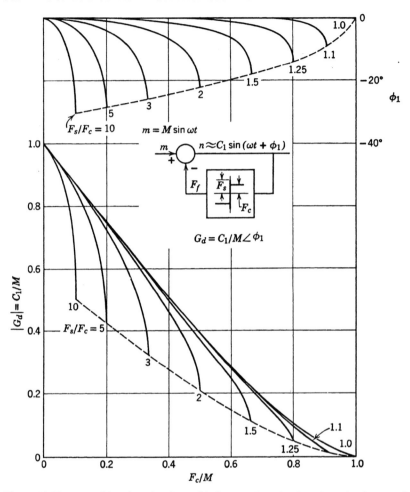

Figure 10-29 Describing function for a friction nonlinearity based on dominance of viscous forces.

It is important to note that Figs. 10-28 and 10-30*b* are fully equivalent representations of the nonlinearity under discussion and may be used interchangeably.

As an example of this nonlinearity, consider the two-stage hydromechanical servovalve in Fig. 10-31. Let us assume that the servo loop involved has been properly designed and, consequently, has a crossover frequency well below the hydraulic natural frequency. A Bode diagram similar to Fig. 8-6 would result. We would like to know what effect friction between the main spool and its sleeve would have on servovalve stability and closed

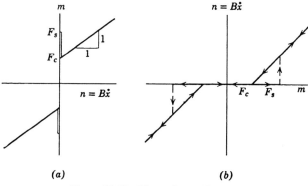

Figure 10-30   Plots of $m = B\dot{x} + F_f$.

loop performance. In this regard, let us assume that any self-sustained oscillations will occur at frequencies well below the hydraulic natural frequency so that fluid compressibility and load inertia can be neglected. The validity of this assumption can be checked once possible limit cycle frequencies have been established.

Referring to Fig. 10-31, the pilot spool position is given by (9-2) and the load flow is described by (6-3). Neglecting compressibility, the

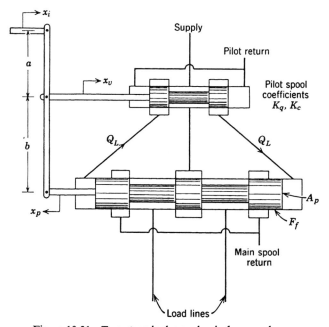

Figure 10-31   Two-stage hydromechanical servovalve.

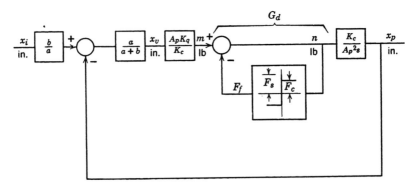

Figure 10-32 Block diagram of two-stage hydromechanical servovalve with friction on main spool.

continuity equation is

$$Q_L = A_p s x_p \qquad (10\text{-}11)$$

Leakage across the pilot valve ports can be included in the pilot spool coefficient K_c. Because friction load on the main spool is dominant,

$$P_L A_p = F_f \qquad (10\text{-}12)$$

where F_f is described by Fig. 10-24. Combining (6-3), (10-11), and (10-12), we obtain

$$\frac{K_q A_p}{K_c} x_v = F_f + \frac{A_p^2}{K_c} s x_p \qquad (10\text{-}13)$$

Equations 9-2 and 10-13 can be used to construct the block diagram in Fig. 10-32 which bears a close resemblance to Fig. 10-25b. Note that

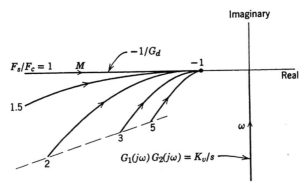

Figure 10-33 Describing function analysis of a servo loop with a friction nonlinearity based on dominance of viscous forces.

$B = A_p^2/K_c$ is the viscous damping coefficient of the valve and piston (actually main spool) combination, and it is usually a very large number. The linear portion of this loop is given by K_v/s, where

$$K_v = \frac{K_q}{(a + b)A_p} \sec^{-1}$$

is the velocity constant. The describing function locus $-1/G_d$ is computed from Fig. 10-29. Hence, the describing function analysis for this case is shown in Fig. 10-33 and indicates no particular stability problem. It is

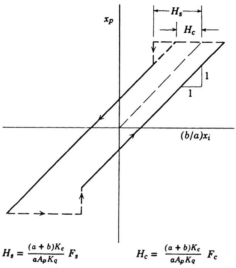

$$H_s = \frac{(a + b)K_c}{aA_pK_q} F_s \qquad\qquad H_c = \frac{(a + b)K_c}{aA_pK_q} F_c$$

Figure 10-34   Low frequency closed loop input-output characteristic of a two-stage hydromechanical servovalve.

clear that limit cycle frequencies, if such would exist, are lower than the crossover frequency, and this justifies the original assumption. If the loop were a lag-lead compensated type 1 servo, intersections might occur if phase margin were very small. The stability of conditionally stable loops would also be adversely affected. Therefore, it is reasonable to conclude that this friction nonlinearity is not a serious destabilizing influence. However, because it is comparable to deadband, it does result in a static error equal to the deadband zone.

Let us now look at the closed loop response of this two-stage servovalve. If a low frequency sine wave was used as the input to the system in Fig. 10-32, the input-output characteristic in Fig. 10-34 would result. This result is best visualized using the deadband representation of the nonlinearity in Fig. 10-30b. The action is similar to that of a loop with deadband followed by an integration which is a common analog computer circuit for generating backlash. In fact, if $F_s = F_c$, the input-output characteristic is exactly backlash. Thus, the two-stage servovalve introduces backlash, which is well known to be destabilizing, into the system because of friction. As a matter of fact, the nonlinearity in Fig. 10-34 will cause any type 1 servo to limit cycle. This nonlinearity will be considered in detail in the next case. Before leaving, however, it should be apparent that the analysis just given is applicable to any hydromechanical servo.

Case c—Spring Force Dominant

Situations arise where mass and viscous forces are small and may be neglected entirely. For example, at frequencies far below the natural frequency, the response of a second order system is determined by the spring

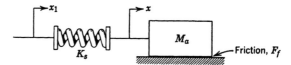

Figure 10-35   Spring and friction system.

constant and by friction if it is present. A model of this case is illustrated in Fig. 10-35.* Neglecting mass, the equation of motion is

$$K_s(x_1 - x) = F_f \tag{10-14}$$

where F_f is given in Fig. 10-24. The block diagram of this model is shown in Fig. 10-36 and is identical to the nonlinearity indicated in Fig. 10-25c. With some reasoning of the physics in Fig. 10-35 or of the signals involved in Fig. 10-36, it is possible to conclude that the input-output characteristic in Fig. 10-37 is an equivalent representation of this nonlinearity if $H_s = F_s$ and $H_c = F_c$.

Analysis of the waveforms in Fig. 10-37 yields the following describing function for this nonlinearity [8]

$$G_d = \left[\left(\frac{a_1}{M}\right)^2 + \left(\frac{b_1}{M}\right)^2\right]^{1/2} \underline{/\tan^{-1}\left(\frac{a_1}{b_1}\right)} \tag{10-15}$$

* This model is also used to illustrate stick-slip. At low velocities a body propelled by a compliant source may not move smoothly along a surface but may exhibit a jumpy motion due to alternate sticking and slipping. This phenomenon is known "stick-slip" and is especially noticeable in the motion of machine tool slides at low feeds. The chattering motion of chalk on a blackboard or of a fingernail being pulled across a metal surface are other examples of stick-slip. Stick-slip occurs below a certain critical velocity [9] and the necessary ingredients for this limit cycle are a velocity source, a spring, a mass, and static and coulomb friction between the sliding surfaces.

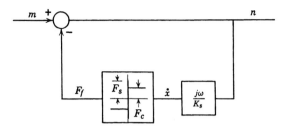

Figure 10-36 Friction nonlinearity based on dominance of spring forces. $m = K_s x_1$ and $n = K_s x$.

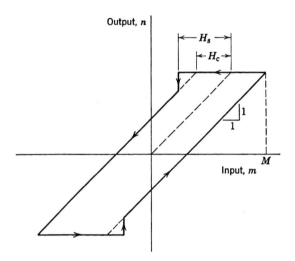

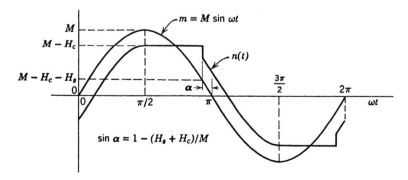

Figure 10-37 Input-output characteristic of spring-friction nonlinearity and waveforms for a sinusoidal input.

305

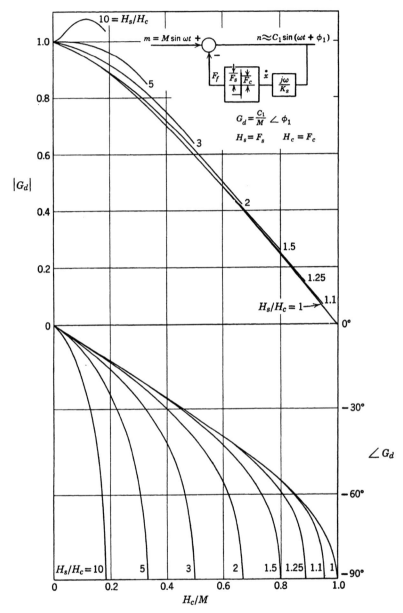

Figure 10-38   Describing function for a friction nonlinearity based on dominance of spring forces.

where $\dfrac{\pi a_1}{M} = 4\left(\dfrac{H_c}{M}\right)^2\left(1 + \dfrac{H_s}{H_c}\right) - 4\dfrac{H_c}{M} - \left(\dfrac{H_c}{M}\right)^2\left(1 + \dfrac{H_s}{H_c}\right)^2$

$$\dfrac{\pi b_1}{M} = \dfrac{\pi}{2} + \sin^{-1}\left[1 - \dfrac{H_c}{M}\left(1 + \dfrac{H_s}{H_c}\right)\right]$$

$$+ \left[1 - \dfrac{H_c}{M}\left(3 - \dfrac{H_s}{H_c}\right)\right]\left[\dfrac{2H_c}{M}\left(1 + \dfrac{H_s}{H_c}\right) - \left(\dfrac{H_c}{M}\right)^2\left(1 + \dfrac{H_s}{H_c}\right)^2\right]^{1/4}$$

which is plotted in Fig. 10-38 for various values of H_s/H_c. Because output
motion ceases when $M \le (H_s + H_c)/2$, this relation in the form

$$\left.\dfrac{H_c}{M}\right|_{max} = \dfrac{2}{1 + H_s/H_c} \tag{10-16}$$

defines the terminal points of the loci. By substituting (10-16) into (10-15)
the describing function at the end points is

$$G_d\big|_{end} = \dfrac{4}{\pi}\dfrac{(H_s - H_c)}{(H_s + H_c)}\underline{/-90^\circ} \tag{10-17}$$

Hence, we note that the $-1/G_d$ loci terminate on the -90° line of a Nyquist
diagram, and we can conclude that any type 1 (and higher) system having
this nonlinearity will limit cycle. Type 0 servo loops will also limit cycle
if the linear dynamics are such that intersections with the $-1/G_d$ loci occur.
Hence, this friction nonlinearity is much more critical regarding stability
compared with the previous two cases.

An important special case, representative of the majority of servo loops,
is where the loop dynamics consists simply of an integration, that is,
$G_1(j\omega)G_2(j\omega) = K_v/j\omega$ where K_v sec^{-1} is the velocity constant. Because
the amplitude is small during the limit cycle, the nonlinearity in Fig.
10-37 reduces to a rectangle of width $H_s + H_c$ and height $H_s - H_c$.
Hence the block diagram of this loop will be as shown in Fig. 10-39. The
describing function analysis for this case is shown in Fig. 10-40. Because
intersections occur at the end points of the $-1/G_d$ loci, this loop will limit
cycle. Equating the linear locus to the value of $-1/G_d$ at the end points,
obtained from (10-17), yields the limit cycle frequency as

$$\omega_{lc} = \dfrac{4}{\pi}\dfrac{K_v(H_s - H_c)}{(H_s + H_c)} \quad \text{rad/sec} \tag{10-18}$$

The amplitude of the limit cycle at the input to the nonlinearity in Fig.
10-39 is given by (10-16). Therefore,

$$M_{lc} = \dfrac{H_s + H_c}{2} \tag{10-19}$$

The amplitude at the output of the nonlinearity is

$$C_{lc} = \dfrac{H_s - H_c}{2} \tag{10-20}$$

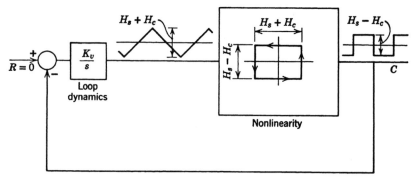

Figure 10-39   Simple Type 1 servo loop containing spring-friction nonlinearity.

Since the nonlinearity reduces to a rectangle for small inputs, the output of the nonlinearity must be a square wave. And this square wave passed through the integration in the loop results in a triangular waveform at the input to the nonlinearity. Therefore (10-19) and (10-20) are the amplitudes (zero to peak) of a triangular and a square wave, respectively.

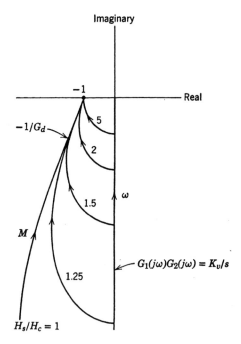

Figure 10-40   Describing function analysis of a simple Type 1 servo loop containing a spring-friction nonlinearity.

For this particular case it is possible to make an exact analysis of the limit cycle if triangular and square waveforms are assumed. The amplitude of the limit cycle remains that given by (10-19) and (10-20) but the frequency of oscillation becomes

$$\omega_{lc} = \frac{\pi}{2} \frac{K_v(H_s - H_c)}{(H_s + H_c)} \text{ rad/sec}$$

which is $(\pi/2)/(4/\pi) = \pi^2/8 = 1.23$ times that predicted by the describing function analysis. This frequency deviation is not large but can be attributed to the fact that the input to the nonlinearity is triangular rather than sinusoidal, as assumed in the describing function technique.

A common example of this nonlinearity occurs when there is a friction load on a compliant drive system, as illustrated by the machine tool slide in Fig. 10-41. The motor shaft must be rotated enough to build up the necessary break-away torque through the compliant drive before motion

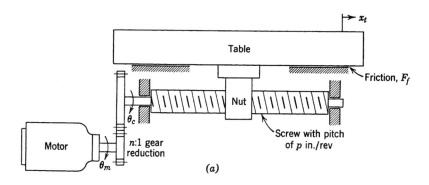

(a)

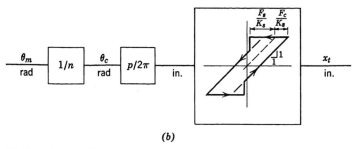

(b)

Figure 10-41 (a) Compliant machine tool drive with table on plain (friction) ways; (b) Input-output characteristic of system. K_s is the drive spring rate at the table and is obtained by locking the motor shaft.

takes place. If position feedback is taken from the table, then this friction nonlinearity will be in the loop and will cause a limit cycle. If feedback is taken from motor shaft position, then the loop will be stable because the nonlinearity is outside of the loop; however, this nonlinearity would cause an error in table position of up to $H_s + H_c$ in.

There are many physical examples of this particular friction nonlinearity [8]. All that is needed is a friction load on a compliant mechanical member. This nonlinearity causes even very simple servo loops to limit cycle and is responsible for many of the low frequency oscillations often observed. It cannot be compensated effectively except by use of a dither. However, dither is not useful in all cases because it may be difficult to add in the loop, it may cause wear, and it may appear in the output.

Summary

The equation of motion including friction (10-8) is of such complexity that it is not possible to obtain a describing function that includes all terms. However, such functions can be derived for cases where friction is combined with each of the usual forces, that is, mass, viscous, and spring forces, to oppose the driving force. Thus, three cases were distinguished. The first two cases were not a serious threat to stability but the third case was. Familiarity with examples and experience are the main factors in determining which case is applicable to a given problem.

10-9 USE OF DESCRIBING FUNCTION CONCEPT IN SINUSOIDAL TESTING

Quite often a suitable mathematical description of a component must be obtained to facilitate analysis and design of the system as a whole. This leads to the question as to what tests are necessary to obtain a satisfactory description. Familiarity with describing function analyses makes one aware that attenuation and phase shift can result from the amplitude as well as the frequency of the input sinusoid. Attenuation and phase shift due to input amplitude identifies the component as being nonlinear.* Because we have seen that an input-output characteristic is a very good description of common nonlinearities, this fact suggests a test procedure.

---

* It is interesting to note that phase lead as well as lag is possible from a nonlinearity. For example, backlash can give up to 90° phase lag. However, if the backlash is in the feedback path of a loop, then the closed loop response will have the characteristic shown in Fig. 10-1h at low frequencies which will yield up to 90° of phase lead, depending on the input amplitude. Case *a* in Section 10-8 is another example of a nonlinearity which gives phase lead.

Once an input and an output have been identified, a component can be represented by a "black box," as illustrated in Fig. 10-42a. An approximate mathematical description can be synthesized as a cascade arrangement of the input-output characteristic (to describe nonlinearities) and the transfer function (to describe the linear dynamics) of the component, as shown in Fig. 10-42b. It might be that the nonlinear and linear natures of the component are so interspersed that such a lumped representation is not possible. However, crude as this approximation might be, such a representation should be assumed unless a more acceptable model is known.

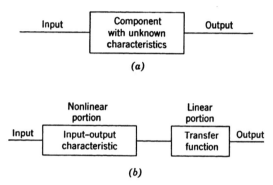

(a)

(b)

Figure 10-42   (a) Block representation of a component; (b) approximate mathematical description of the component.

Which lump is placed first in Fig. 10-42b can be deduced from the output waveform for high frequency sinusoidal inputs. If the output waveform is well filtered, one can conclude that the linear transfer occurs after the nonlinear characteristic. If the output waveform shows flat spots, discontinuities, cusps, and other features typical of the output of a nonlinearity, then one can conclude that the linear transfer function should precede the input-output characteristic. If neither extreme exists, the assumption that the nonlinear and linear natures of the component can be lumped into discrete blocks becomes questionable.

Care must be used in measuring the input-output characteristic of a component. Continuous plotting of data on an x-y recorder is recommended for best results. A low frequency input sinusoid is used so that linear dynamics do not affect the output. It is important to use a small input amplitude because a large value can mask any backlash and/or deadband which is present. However, a large input is required to establish saturation limits.

Measurement of the linear transfer function is difficult because it is not possible to isolate the effect of nonlinearities on the output. This is the reason many components, such as electrohydraulic servovalves, show a phase lag at low frequencies where the linear dynamics indicate no phase lag should exist. Fairly good results can be obtained by using a sinusoidal input amplitude large enough to mask nonlinearities which dominate at small inputs (such as deadband and backlash) but not so large as to encounter nonlinearities that become effective at large inputs (such as saturation). Sweeping the frequency range with an input between these limits will usually provide a frequency response from which the transfer function can be approximated using asymptotic slopes. Again, continuous data plotting of the frequency response, both gain and phase, is recommended.

In both these tests some juggling of input amplitude and frequency, as well as many trial runs, are usually necessary to achieve results having a high confidence level.

10-10   TROUBLESHOOTING LIMIT CYCLE OSCILLATIONS

In this chapter we have discussed whether certain known nonlinearities will cause instability problems. However, in spite of careful design, systems often oscillate and the problem is to find out why. Finding the cause of such oscillations can be time consuming and costly unless some systematic approach is used.

Because every signal in the loop is oscillating at the limit cycle frequency, it is impossible to trace the oscillation to its source. (We are assuming self-excited rather than forced oscillations.) At the outset it is quite important and fundamental to recognize that no such simple source exists. Self-excited oscillations spring into being because the dynamics of components in the loop combine to produce an unstable loop. If the loop is broken at any point the oscillation will cease. Modification of the static or dynamic characteristics of one or more components in the loop may cause the loop to become stable. Because limit cycle oscillation is a characteristic of the loop as a whole, there is no such thing as a "bad" component. However, it is quite possible for a component to have certain characteristics, such as backlash, which make a loop more prone to limit cycle. In this sense the component may be declared "bad." However, the oscillation is due to the combination of components and not to the individual component. Hence, limit cycle oscillation must be treated as a system problem. The purpose of this section is to describe techniques which have proved useful in troubleshooting limit cycle oscillations.

The first step in troubleshooting a limit cycle is to obtain a mathematical block diagram of the servo loop and use Nyquist's or Routh's criterion to establish stability. Stability is a basic performance limitation to any feedback control system and designs are often pushed to the unstable regions. An elementary analysis using gain constants and dominant time constants is often sufficient. Needless to say, analysis must show the loop to be stable before a discrepancy with test results can be declared. A simple gain reduction will often make the loop stable and is all that can be done in difficult cases. In complicated systems stability may be impossible to analyze. Testing and troubleshooting are then the only recourses, and systematic, rational techniques must be used.

The second step in a limit cycle problem is to identify the loop that is oscillating. Two facts aid this task. Because a loop ceases to oscillate when opened, one can start at the outermost loop and work toward inner

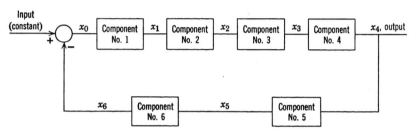

Figure 10-43   An unstable control loop.

loops, breaking each in turn, to determine the unstable loop. Another clue to the unstable loop is the frequency of oscillation. Loops tend to oscillate at frequencies below the crossover frequency (relay type servos are an exception). Hence, a high frequency oscillation would be indicative of an inner loop or component instability while a low frequency suggests instability of the major control loops. A comparison of the frequency of oscillation with known resonant frequencies, especially structural modes of vibration, may also be helpful. Once the oscillating loop is determined, then the sequence of events around the loop should be traced and all physical variables identified. A block diagram of the loop should be drawn and, if at all possible, a stability analysis should be made.

Suppose the block diagram of the unstable loop is shown in Fig. 10-43. All physical variables, x_0 through x_6, are oscillating in some periodic manner at the limit cycle frequency. Some waveforms are sinusoidal, but others are not which indicates the presence of nonlinearities. The problem now is to determine why this particular loop is oscillating.

The third step in this troubleshooting process is based on the fact that *the total phase lag around a loop that is limit cycling is exactly* 360° *and the loop gain is exactly unity.* Because the sign inversion of the negative feedback gives −180°, the components around the loop, that is, from x_0 to x_6 in Fig. 10-43, must contribute exactly −180° of phase shift. If the phase lag were greater or less than $-180°$, the limit cycle amplitude would be increasing or decreasing. A steady limit cycle amplitude is reached only when the phase lag is precisely $180°$ (see Section 10-2). This fact suggests a systematic test procedure. Instrument and record several signals in the loop, say x_2, x_4, and x_6 in Fig. 10-43, *while the loop is oscillating*. The phase lag across various components in the loop can then be obtained from this test record of the limit cycle. These values are then compared with theoretical values computed at the limit cycle frequency. Hence, the component that contributes an unusually large phase lag can be immediately identified.

For example, suppose a $50°$ phase lag is expected from x_4 to x_2, excluding the sign inversion, and this is closely verified in the test. Because the loop is oscillating, the phase lag from x_2 to x_4 must be $130°$, which should also correlate with the test record. Let us assume that component No. 3 is an electrohydraulic servovalve and component No. 4 is a hydraulic motor. Let us further assume the limit cycle frequency is low enough that servovalve and motor-load dynamics contribute negligible phase lag. Under these conditions a phase lag of only $90°$, due to the hydraulic motor integration, would be expected from x_2 to x_4. Because the test record shows a lag of $130°$, these two components are identified as causing the limit cycle. We must now instrument and record x_3, which is perhaps servovalve spool position, to determine whether the excess lag is due to the servovalve or to hydraulic motor load. Assuming this test shows $40°$ lag across the servovalve, this component is then undesirable. The question now is why this component contributes excessive phase lag. The answer is usually a nonlinearity, such as backlash, and not the linear dynamics of the component. The waveforms of the limit cycle signals provide a valuable clue to the particular nonlinearity involved. The output waveform of a component with deadband, saturation, backlash, or hysteresis is readily distinguished.

This technique of recording the limit cycle at several points in the loop and comparing the phase lag between these points with that expected works very well in complex systems with many components because physical variables can be easily identified and measured. It cannot be used effectively in troubleshooting component instabilities because of difficulties in identifying the oscillating loop and measuring the quantities involved.

In many cases attempts are made to troubleshoot a limit cycle by breaking the loop and injecting a sinusoidal signal from an oscillator to obtain component transfer functions. These measured transfer functions are then compared with theoretical results and discrepancies are noted. Inadvertently, however, the injected signal amplitude is much larger than that of the limit cycle and masks nonlinearities, such as backlash, important in limit cycle generation. It is far better to record the limit cycle signals directly and to deduce conclusions. Although these signal levels are low and must be amplified for recording, no amplitude calibration is necessary because only phase information is required.

The troubleshooting technique described is useful for another purpose. After a servo system is designed, built, and tested, it is desirable to anticipate limit cycle problems and determine how and why the system would oscillate if the gain were increased beyond the stability limit. Raising the gain beyond the stable value and recording several points in the loop, the test record of the resulting limit cycle can be analyzed to determine the critical nonlinearity and/or component. In this manner significant data concerning system stability can be obtained.

It has always been the author's experience that limit cycle oscillations or other transient problems can be traced to a few fundamental physical characteristics of the system and have relatively simple explanations. A "more thorough" analysis of the linear counterpart of the system taking into account more sources of dynamics might increase the system order from, say, third to perhaps tenth order, but usually sheds no new light on these problems. A simple argument, usually involving a nonlinearity, is sufficient to explain most system difficulties.

Another factor to consider is the system type. Type 1 systems, that is, systems with an integration in the loop, inherently have very good stability characteristics because they can be made stable by a sufficient reduction in only one parameter—the velocity constant. The −1 slope of the Bode diagram at low frequencies forces the gain down so that the crossover frequency can be established on a −1 slope, which is the ideal situation. In effect, the integration acts as a large low frequency lag which dominates the loop dynamics and, therefore, dominates the stability picture. This attribute of a Type 1 servo, that is, of having a sound crossover frequency independent of other system dynamics providing the velocity constant is low enough, is as important, if not more so, as the usually stated advantage of zero position error and accounts in large measure for the dependable operation of these systems.

Type 0 servo loops are quite another matter. The crossover frequency results from the combination of leads, lags, and quadratics encountered in the loop gain. If a quadratic or two simple lags of similar value dominate (i.e., are lowest in frequency) the loop gain, then crossover occurs on a −2 slope with practically zero phase margin. If the lags and/or quadratics are close together in value and are subject to variation with operating point and environmental conditions, it is virtually impossible to control the crossover frequency so that it occurs on a −1 slope. Hence, Type 0 loops are quite undependable because instability can be produced with many combinations of values for dynamic parameters. If the loop dynamics are dominated by a simple lag, then good stability can be achieved. In many Type 0 loops it is desirable to insert a simple lag to compensate or control stability (see example in Section 8-4).

It is quite important to realize that Type 0 and 1 systems differ greatly in their inherent stability characteristics, and this fact can be useful in diagnosing difficulties. Type 2 systems, from a practical point of view, are basically unstable (see comments on conditionally stable loops in Section 8-3) and require considerable effort to stabilize.

In closing this section it might prove useful to list some of the more usual cases of oscillation in hydraulic control systems.

1. Instability due to linear dynamics—The linear dynamics in a loop may be such that Nyquist's and/or Routh's criteria are not met. Three particular cases are excessive loop gain, attempting bandwidths greater than the hydraulic natural frequency, and appearance of structural resonances thought to be negligible.

2. Instabilities due to backlash—this nonlinearity, possessed to some degree by most components, is capable of inserting up to 90° of phase lag, depending on input amplitude, and can soak up the available phase margin in a loop and cause instability. It is especially severe in lag-lead compensated servos where the phase margin is usually less than 60°.

3. Instability due to static and coulomb friction load on a compliant drive—this nonlinearity, described in case c of Section 10-8, can cause any Type 1 servo loop to limit cycle. The output waveform is triangular in shape and the frequency is low.

4. Instability due to silting—valve silting can cause periodic jerks in the output (see comments in Section 5-8). As the servovalve silts up, the loop gain decreases and the servo loses control. The actuator drifts from position, causing an error. When the error opens the valve sufficiently, the dam of silt is broken and the particles are flushed out. Loop gain is reestablished and the output jumps back to its controlled position. The servovalve silts again and the cycle repeats, but with varying periods. Dither and/or clean oil are correction techniques.

5. Instability due to upsweep gain characteristic—overlapped servovalves and servovalves with round ports give an increasing flow gain characteristic which can cause servo instability above certain output velocities.

6. Instability due to overdamped hydraulic resonance—at higher flows the servovalve increases damping on the hydraulic resonance, which can become overdamped. The break frequency of one of the linear factors of the quadratic is lowered in value and can result in instability of lag-lead compensated positioning control system, where phase margin is low, above certain output velocities.

7. Instability due to stick-slip phenomenon—static and coulomb friction load can cause alternate sticking and slipping of the output actuator if the drive is compliant (see footnote on page 304). This instability occurs only while moving below a certain critical velocity and the oscillation waveform is saw-toothed in shape.

8. Instability due to air entrainment—air in the hydraulic fluid can lower the bulk modulus to such an extent that the hydraulic resonance is low enough to cause instability of the linear system. Trapped air can be avoided by proper passage design (see Section 2-5).

9. Instability due to hydraulic lines—the quarter wave length frequency of hydraulic lines* may coincide with structural resonances in the torque motor or with hydraulic natural frequencies in the servovalve and may cause oscillations (see comments in Section 7-6). Lines are often implicated in oscillation problems and can be identified by their very high frequency. Physical changes in line length or piecing together lines of different diameters can make changes in the line natural frequency and are useful correction techniques.

10. Instability due to flow forces—the steady-state flow force on a spool valve can cause instability if the driving source the valve is not sufficiently stiff. This phenomenon is particularly apparent in single-stage electro-hydraulic servovalves (see Section 7-3). Transient flow forces can detract from spool valve damping, if the damping length is destabilizing, and has been known to cause oscillation problems.

11. Instability due to pressure control valves—pressure relief, pressure reducing, and pressure-compensated flow control valves often exhibit a higher frequency oscillation which may appear in hydraulic control circuits and servos.

12. Instability due to compliant mountings—it should be apparent that the torque or force of an actuator is produced by reacting against the structure on which it is mounted. If the adjacent supporting structure is compliant, it is possible for the dynamic characteristics of the structure to appear in the control loop and cause instability, usually at a frequency corresponding to the mode of vibration of the structure. This kind of oscillation will sometimes occur in servovalves if they are not firmly mounted to a solid structure.

13. Instability of the hydraulic power supply—constant pressure power supplies may use a constant delivery pump and bypass oil through a relief valve to control pressure or may regulate the stroke of a variable delivery pump to control pressure. In either case instability, usually at the natural

---

* The fundamental natural frequency of a hydraulic line open at one end and closed at the other (quarter wavelength frequency) is $f = c/4L$ cps, where c is the velocity of sound in the fluid in in./sec and L is the line length in in. For petroleum base fluids, $c = 113.5\sqrt{\beta_e}$, where β_e is the effective bulk modulus in lb/in^2.

frequency of the relief valve or the stroke regulating mechanism, may occur, depending on the load attached to the supply. The volume of oil in the supply line is sometimes especially critical. Because accumulators add to this volume, they may cause the supply pressure to oscillate. Also, the supply line length can be such that conduit resonances may coincide with that of the relied valve or stroke control mechanism and cause sustained oscillations.

14. Forced oscillations—pumps, motors, and other devices produce pulsations at vane or piston frequencies which may appear in the output of hydraulic actuators. Accumulators and/or restrictors are sometimes used to reduce these undesirable pulsations.

REFERENCES

[1] Linebarger, R., and R. Brennan, "Digital Simulation for Control System Design," *Instruments & Control Systems*, October 1965, 147–152.
[2] Grabbe, E. M., S. Ramo, and D. E. Wooldridge, *Handbook of Automation, Computation and Control*, vol. 1. New York: Wiley, 1958, Chapter 25.
[3] Gibson, J. E., *Nonlinear Automatic Control*. New York: McGraw-Hill, 1963.
[4] Neal, C. B., and D. B. Bunn, "The Describing Function for Hysteresis," 1964 *Joint Automatic Control Conf.* (JACC).
[5] Tou, J., and P. M. Schultheiss, "Static and Sliding Friction in Feedback Systems," *J. Appl. Phys.*, 24, No. 9, September 1953, 1210–1217.
[6] Silberberg, M. Y., "A Note on the Describing Function of an Element with Coulomb, Static, and Viscous Friction," *Trans. AIEE*, 75, Pt. 2, 1956, 423–425.
[7] Merritt, H. E., and J. T. Gavin, "Friction Load on Hydraulic Servos," *Proc. Natl. Conf. Indl Hydraulics*, 16, 1962, 174–184.
[8] Merritt, H. E., and G. L. Stocking, "How to Find When Friction Causes Instability," *Control Engineering Magazine*, December, 1966.
[9] Singh, B. R., "Study of Critical Velocity of Stick-Slip Sliding," *ASME Trans., J. Eng. Ind.*, 1960.

11

Pressure and Flow Control Valves

Servovalves have been discussed as the principal component in controlling flow and/or pressure in hydraulic servo systems. However, other valves are also necessary in the generation and utilization of hydraulic power. These valves serve auxiliary functions in closed loop hydraulic controls. They may serve principal functions in the open loop type of control of actuators encountered in the hydraulic circuits of mobile and industrial machinery. The purpose of this chapter is to discuss some of these valves —chiefly those which employ feedback in their operation and, therefore, constitute a servo loop.

11-1 FUNCTIONAL CLASSIFICATION OF VALVES

Valves may be classified by their function, for example, relief valve, or by a feature of their construction (see Section 5-1) for example, spool valve. The many types of valves available are best classified according to their function. Three broad functional types can be distinguished: *directional control valves, pressure control valves*, and *flow control valves*.

Directional control valves direct or prevent flow through certain selected passages. They do not regulate flow or pressure, at least as a primary function, but simply perform a switching function. As such they have a discrete number, usually two or three, of valve positions. These valves usually employ spool type construction and may be actuated manually, mechanically, hydraulically, pneumatically, electrically, or pilot operated from a smaller valve. Two-way, three-way, and four-way valves are common. Examples of directional control valves are check valves and solenoid operated four-way spool valves.

Because directional valves are basically either open or closed, no feedback loop is involved and, therefore, stability problems are not encountered in their design and operation. These valves are usually rated by the somewhat antiquated "pipe size," but most manufacturers also give the pressure drop at rated flow capacity. Major application factors are rated flow capacity, pressure drop, cyclic life, maximum pressure rating, leakage, and actuation time.

Pressure control valves act to regulate pressure in a circuit and may be subdivided into *pressure relief valves* and *pressure reducing valves*. These devices may be single-stage (direct-operated) (Fig. 11-1) or two-stage (compound) (Fig. 11-7). Pressure relief valves, which are normally closed, open up to establish a maximum pressure and bypass excess flow to maintain the set pressure. Pressure reducing valves, which are normally open, close to maintain a minimum pressure by restricting flow in the line.

There are many other types of pressure control valves that are usually classified by function and are similar in construction to relief valves.

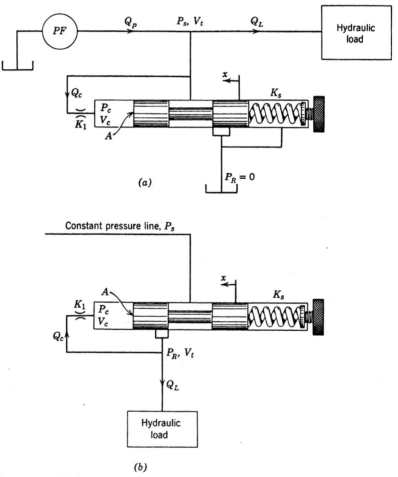

(a)

(b)

Figure 11-1   Single-stage pressure control valves; (a) relief valve; (b) reducing valve.

These are:

Safety or overload relief valves—usually a spring loaded ball or poppet type valve whose primary function is to provide a pressure limit in event of system malfunction.

Back pressure or counterbalance valve—a relief valve that maintains a constant back pressure on an actuator, usually to prevent a load from falling.

Unloading valve—a relief type valve which controls pump pressure and opens under the application of a pilot pressure to unload the pump.

Load dividing valve—a pressure control valve which proportions pressure between two pumps in series.

Sequence valve—basically a relief valve which allows flow to pass to another circuit once the set pressure level is exceeded. They are used to control the sequence of operation of two or more cylinders.

Decompression valve—a pressure control valve which controls the rate at which the pressure of a compressed fluid is released.

Shock suppression valve—a pressure control valve which bypasses flow such that the rate of pressure rise is limited to a safe value.

Flow control valves regulate flow in a circuit and vary in arrangement from simple two-way needle, globe, and gate valves to fixed and adjustable pressure-compensated flow control valves of considerable complexity. Pressure-compensated flow control valves employ a pressure control valve to maintain constant pressure drop across an orifice. With constant pressure drop, the flow through the orifice is then directly proportional to the orifice area and is not influenced by upstream and downstream pressure changes.

Some special types of flow control valves are the following:

Flow divider valve—a valve that divides a single flow into two or more prescribed flows.

Deceleration valve—usually a cam operated valve which controls flow to achieve a smooth deceleration of an actuator.

Pressure control valves and pressure-compensated flow control valves employ feedback and may be properly regarded as servo control loops. Therefore, proper dynamic design is necessary to achieve stability. A knowledge of dynamic characteristics is also helpful in the application of these devices.

11-2 SINGLE-STAGE PRESSURE CONTROL VALVES

Schematics of a single-stage pressure relief and a pressure reducing valve are shown in Figures 11-1a and 11-1b, respectively. In both cases the pressure to be controlled, pump output pressure P_s for the relief valve and reduced pressure P_R for the reducing valve, is sensed on the spool end area and compared with a spring force setting. The difference in force is used to actuate the spool valve which controls flow to maintain pressure at the set value. The particular construction illustrated in Fig. 11-1 may differ from practical constructions, but the principles of operation and dynamic analysis would be the same. The restriction in the sensing line is necessary for stability reasons and, as we shall see, plays a major role in dynamic performance. It may be a capillary, but it is usually a fixed orifice.

The following analysis is applicable to both valves in Fig. 11-1. The equation describing spool motion is

$$F_0 - AP_c = M_v s^2 x + K_e x \qquad (11\text{-}1)$$

where F_0 = spring preload force, lb

$K_e = K_s + 0.43w(P_{s_0} - P_{R_0})$ = equivalent spring rate (i.e., mechanical spring rate plus flow force spring rate), lb/in.

w = area gradient of main orifice, in²/in..

P_s = supply pressure, psi

P_R = reduced or low pressure, psi

P_c = sensed pressure, psi

M_v = spool mass plus $\frac{1}{3}$ mass of spring, lb-sec²/in.

A = spool end area, in.²

x = spool displacement, in.

$\omega_m = \sqrt{\dfrac{K_e}{M_v}}$ = mechanical natural frequency, rad/sec

s = Laplace operator, sec⁻¹

The linearized continuity equations at the sensed pressure chambers are

$$Q_c = K_1(P_s - P_c) = \frac{V_c}{\beta_e} sP_c - Asx \quad \text{for relief valve} \qquad (11\text{-}2)$$

$$Q_c = K_1(P_R - P_c) = \frac{V_c}{\beta_e} sP_c - Asx \quad \text{for reducing valve} \qquad (11\text{-}3)$$

where $K_1 = \dfrac{\partial Q}{\partial P}$ = flow-pressure coefficient of restrictor, in.³/sec/psi

V_c = sensing chamber volume, in³.

β_e = effective bulk modulus, lb/in.²

The linearized continuity equations at the chamber of the pressure being controlled are:

$$Q_p - Q_L - K_l P_s - K_c P_s - K_1(P_s - P_c) + K_q x = \frac{V_t}{\beta_e} s P_s$$

for relief valve   (11-4)

$$K_c(P_s - P_R) - Q_L - K_l P_R - K_1(P_R - P_c) + K_q x = \frac{V_t}{\beta_e} s P_R$$

for reducing valve   (11-5)

where $K_c = \dfrac{\partial Q}{\partial P} = \dfrac{C_d w x_0 \sqrt{2/\rho}}{\sqrt{P_{so} - P_{Ro}}} = \dfrac{Q_0}{2(P_{so} - P_{Ro})} =$ flow-pressure coefficient of main orifice, in.³/sec/psi

$$K_q = \frac{\partial Q}{\partial x} = C_d w \sqrt{\frac{2}{\rho}(P_{so} - P_{Ro})} = \text{flow gain of main orifice,}$$
in.³/sec/in.

$V_t =$ total volume of chamber where pressure is controlled, in.³
$Q_L =$ arbitrary load flow, in.³/sec
$Q_p =$ ideal pump flow (constant), in.³/sec
$K_l =$ leakage coefficient from chamber where pressure is controlled to drain (includes load leakage and, for the relief valve case, pump leakage), in.³/sec/psi

These equations define the valves. Let us combine some of them into a more useful form. Solving (11-2) for P_c and substituting into (11-4) yields after some manipulation

$$(Q_p - Q_L)\left(1 + \frac{s}{\omega_1}\right) + K_q\left[1 + \left(\frac{1}{\omega_1} + \frac{A}{K_q}\right)s\right]x$$

$$= \left[(K_{ce} + K_1)\left(1 + \frac{s}{\omega_3 + (V_c/V_t)\omega_1}\right)\left(1 + \frac{s}{\omega_1}\right) - K_1\right]P_s \quad (11\text{-}6)$$

where $\omega_1 = \dfrac{\beta_e K_1}{V_c} =$ break frequency of sensing chamber, rad/sec

$\omega_3 = \dfrac{\beta_e K_{ce}}{V_t} =$ break frequency of main volume, rad/sec

$K_{ce} = K_c + K_l =$ equivalent flow-pressure coefficient, in.³/sec/psi

Expanding the right side and using the fact that $\omega_3/\omega_1 = K_{ce}V_c/K_1 V_t$, (11-6) becomes

$$(Q_p - Q_L)\left(1 + \frac{s}{\omega_1}\right) + K_q\left[1 + \left(\frac{1}{\omega_1} + \frac{A}{K_q}\right)s\right]x$$

$$= K_{ce}\left[1 + \frac{1}{\omega_3}\left(1 + \frac{\omega_3}{\omega_1} + \frac{V_c}{V_t}\right)s + \frac{s^2}{\omega_1\omega_3}\right]P_s \quad (11\text{-}7)$$

Because $V_c/V_t \ll 1$, the right side can be factored to give the final form

$$(Q_p - Q_L)\left(1 + \frac{s}{\omega_1}\right) + K_q\left[1 + \left(\frac{1}{\omega_1} + \frac{A}{K_q}\right)s\right]x$$

$$= K_{ce}\left(1 + \frac{s}{\omega_1}\right)\left(1 + \frac{s}{\omega_3}\right)P_s \quad (11\text{-}8)$$

for the relief valve case. In an entirely analogous manner, again making

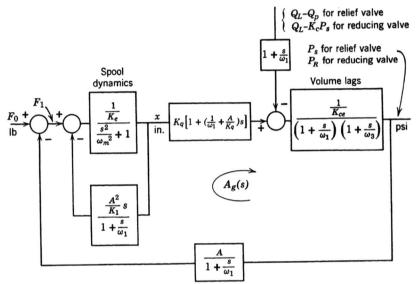

Figure 11-2   Block diagram of single-stage pressure control valves (relief and reducing).

the assumption that $V_c \ll V_t$, (11-3) and (11-5) can be combined to yield the following result for the pressure reducing valve:

$$(K_c P_s - Q_L)\left(1 + \frac{s}{\omega_1}\right) + K_q\left[1 + \left(\frac{1}{\omega_1} + \frac{A}{K_q}\right)s\right]x$$

$$= K_{ce}\left(1 + \frac{s}{\omega_1}\right)\left(1 + \frac{s}{\omega_3}\right)P_R \quad (11\text{-}9)$$

Equations 11-1, 11-2, and 11-8 define the relief valve and can be used to construct the block diagram in Fig. 11-2. Equations 11-1, 11-3, and 11-9 define the reducing valve and can be represented by the same block diagram.

If there were no restrictor in the sensing line, K_1 would be infinite (i.e., very large). The quantities $(A/K_1)s$ and s/ω_1 would be zero and the loop gain function for this case can be written directly from Fig. 11-2 as

$$A_g(s)\bigg|_{\substack{\text{without}\\\text{restrictor}}} = \frac{\dfrac{K_q A}{K_e K_{ce}}\left(1 + \dfrac{A}{K_q}s\right)}{\left(\dfrac{s^2}{\omega_m^2} + 1\right)\left(1 + \dfrac{s}{\omega_3}\right)} \tag{11-10}$$

Assuming the lead at K_q/A effectively cancels the lag at ω_3, we can see that the loop gain is dominated by the lowly damped mechanical resonance at ω_m and has a Bode diagram similar to Fig. 8-20. Variation in the parameters K_q, K_e, and K_{ce} with the operating point, low damping on the quadratic factor, and the crossover frequency occurring on a -2 slope (see Section 8-4) makes the stability of direct-operated (i.e., without restrictor) pressure control valves impossible to control in design. This fact is borne out by experience as these valves often oscillate and each application must be checked for stability. If ω_3 is small enough (by having a large V_t) it can dominate the loop dynamics and make the valve stable. However, cut-and-try application techniques are necessary.

The most effective way to stabilize these valves is to use a restrictor in the sensing line. These restrictors have the disadvantage that a large transient pressure overshoot may occur while the valve is synchronizing to a sudden flow change because of the lag in buildup of pressure P_c. However, some sacrifice in the accuracy and/or response of most control devices is usually necessary to achieve stability, and pressure control valves are no exception.

Referring to Fig. 11-2, we note that the restrictor cause a rate type feedback around the spool dynamics which, at least intuitively, should increase damping. However, action of the restrictor is best seen from the closed loop response of this inner loop. The response of this loop is

$$\frac{x}{F_1} = \frac{\dfrac{1}{K_e}\left(1 + \dfrac{s}{\omega_1}\right)}{\dfrac{s^3}{\omega_m^2\omega_1} + \dfrac{s^2}{\omega_m^2} + \left(\dfrac{1}{\omega_1} + \dfrac{1}{\omega_2}\right)s + 1} \tag{11-11}$$

where $\omega_2 = \dfrac{K_1 K_e}{A^2}$ = break frequency due to restrictor, rad/sec

$\dfrac{K_h}{K_e} = \dfrac{\omega_1}{\omega_2}$ = ratio of hydraulic to mechanical spring rates, dimensionless

$K_h = \dfrac{\beta_e A^2}{V_c}$ = spring rate of trapped fluid in sensing chamber, lb/in.

$\omega_h = \sqrt{\dfrac{K_h}{M_v}}$ = hydraulic natural frequency due to trapped oil spring and spool mass, rad/sec

The cubic denominator can be factored into a linear and a quadratic term. Hence, (11-11) becomes

$$\frac{x}{F_1} = \frac{\dfrac{1}{K_e}\left(1 + \dfrac{s}{\omega_1}\right)}{\left(1 + \dfrac{s}{\omega_r}\right)\left(\dfrac{s^2}{\omega_0^2} + \dfrac{2\delta_0}{\omega_0}s + 1\right)} \tag{11-12}$$

Normally, it is not easy to factor a cubic. However, in this case relatively simple charts can be derived, Figs. 11-3, 11-4, and 11-5, from which ω_r, ω_0, and δ_0 can be obtained for given values of ω_2/ω_m and K_h/K_e. For practical designs the trapped oil spring rate is very large so that $K_h/K_e \gg 1$. With

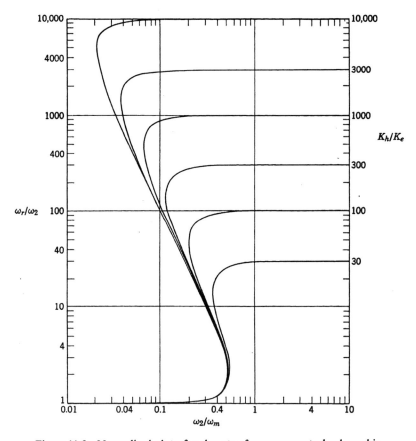

Figure 11-3   Normalized plot of real roots of pressure control valve cubic.

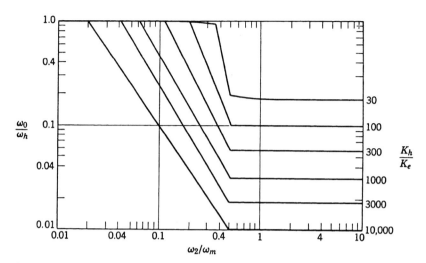

Figure 11-4 Normalized plot of natural frequency of quadratic factor of pressure control valve cubic.

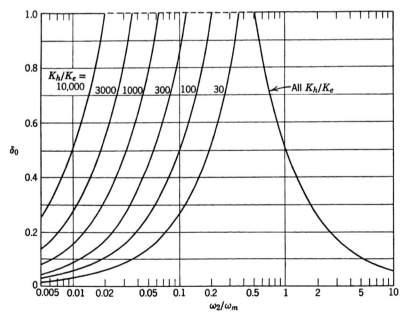

Figure 11-5 Damping ratio of quadratic factor of pressure control valve cubic.

327

this condition all three cubic roots are real for ratios of ω_2/ω_m in the range

$$\frac{2}{\sqrt{K_h/K_e}} \leq \frac{\omega_2}{\omega_m} \leq 0.5 \tag{11-13}$$

and the three break frequencies can be obtained from Fig. 11-3. An underdamped quadratic and a linear factor occur for values of ω_2/ω_m outside of this range. It is this region, which is of most interest, we are fortunate because simple literal expressions for the factors are possible.

For $K_h/K_e \gg 1$ and $\omega_2/\omega_m \geq 0.5$, $\omega_r = \omega_1$ and $\omega_0 = \omega_m$ and $\delta_0 = \frac{1}{2}(\omega_m/\omega_2)$. Hence, (11-12) becomes

$$\frac{x}{F_1} = \frac{\dfrac{1}{K_e}}{\dfrac{s^2}{\omega_m{}^2} + \dfrac{2\left(\dfrac{\omega_m}{2\omega_2}\right)}{\omega_m} s + 1} \tag{11-14}$$

And the loop gain function from Fig. 11-2 is

$$A_g(s) = \frac{\dfrac{K_q A}{K_e K_{ce}}\left[1 + \left(\dfrac{1}{\omega_1} + \dfrac{A}{K_q}\right)s\right]}{\left[\dfrac{s^2}{\omega_m{}^2} + \dfrac{2\left(\dfrac{\omega_m}{2\omega_2}\right)}{\omega_m} s + 1\right]\left(1 + \dfrac{s}{\omega_1}\right)^2\left(1 + \dfrac{s}{\omega_3}\right)} \tag{11-15}$$

Because ω_1 is large, this expression is not greatly different from (11-10) for the case with no restrictor. Hence, stability would be difficult to control because the loop dynamics is dominated by a resonance. However, the resonance can be well damped by selecting ω_2/ω_m to be near 0.5, and this would help stability. A large load volume making ω_3 near ω_m in value might cause instability. But if the volume were increased further so that $\omega_3 \ll \omega_m$, ω_3 would dominate and make the valve stable. Therefore, each application would require testing to determine stability.

We have seen that $\omega_2/\omega_m \geq 0.5$ does not represent a satisfactory design objective because it only results in increased damping on the mechanical resonance. For $K_h/K_e \gg 1$ and $\omega_2/\omega_m \leq 2\sqrt{K_e/K_h}$, $\omega_r = \omega_2$ and $\omega_0 = \omega_h$ and $\delta_0 = \frac{1}{2}(\omega_2/\omega_m)\sqrt{K_h/K_e} = \frac{1}{2}(\omega_1/\omega_h)$ are very good approximations. Hence, (11-12) becomes

$$\frac{x}{F_1} = \frac{\dfrac{1}{K_e}\left(1 + \dfrac{s}{\omega_1}\right)}{\left(1 + \dfrac{s}{\omega_2}\right)\left[\dfrac{s^2}{\omega_h{}^2} + \dfrac{2\left(\dfrac{\omega_1}{2\omega_h}\right)}{\omega_h} s + 1\right]} \tag{11-16}$$

and the loop gain function can be written from Fig. 11-2 as

$$A_g(s) = \frac{\dfrac{K_q A}{K_e K_{ce}}\left[1 + \left(\dfrac{1}{\omega_1} + \dfrac{A}{K_q}\right)s\right]}{\left(1 + \dfrac{s}{\omega_2}\right)\left(1 + \dfrac{s}{\omega_3}\right)\left(1 + \dfrac{s}{\omega_1}\right)\left[\dfrac{s^2}{\omega_h{}^2} + \dfrac{2\left(\dfrac{\omega_1}{2\omega_h}\right)}{\omega_h}s + 1\right]} \tag{11-17}$$

Some fundamental differences compared with (11-10) can now be noted.

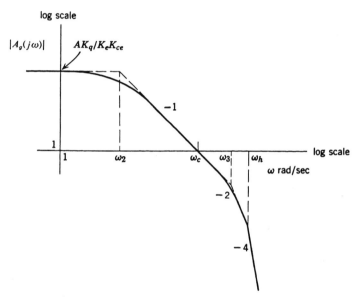

Figure 11-6   Bode diagram of a pressure control valve using a restrictor to achieve stability.

The mechanical resonance at ω_m has been replaced with a much higher hydraulic resonance at ω_h. Hence, the restrictor serves to trap an oil volume making a dynamic fluid spring which essentially replaces the mechanical spring in determining the resonance. Furthermore, a low frequency lag occurs at ω_2 and, because it dominates the loop dynamics, can be used to control the crossover frequency to stabilize the valve. The Bode diagram of (11-17) is shown in Fig. 11-6, assuming the lead effectively cancels the lag at ω_1. It is quite apparent that the loop is stable for all values of V_t and is, therefore, fairly independent of the load. Stability is also retained (keep in mind that ω_h is very large), even with substantial changes in the parameters K_q, K_e, and K_{ce}. Hence, this valve should be well behaved and applications should not be particularly critical.

The steady state compliance of pressure control valves, that is, the change from the set pressure for an increase in load flow, is often of interest and is a measure of the accuracy of control. Referring to Fig. 11-2, the steady-state compliance is

$$\frac{\Delta P_s}{\Delta Q_L} = - \frac{1/K_{ce}}{1 + AK_q/K_eK_{ce}} \tag{11-18}$$

for the relief valve (replace P_s with P_R for the reducing valve). Because the loop gain constant AK_q/K_eK_{ce} is much greater than unity for a good design, the following approximation is sufficient.

$$\frac{\Delta P_s}{\Delta Q_L} \approx - \frac{K_e}{AK_q} \tag{11-19}$$

Changes in K_q and K_e (due to flow forces) with the operating point alters the compliance to some extent.

Because ω_h and ω_1 are very large, (11-17) can be approximated by

$$A_g(s) \approx \frac{AK_q/K_eK_{ce}}{(1 + s/\omega_2)(1 + s/\omega_3)} \tag{11-20}$$

These two lags dominate the valve dynamics and require design consideration. The design process might proceed as follows:

1. Make a rough layout of the valve which meets static requirements of pressure and flow. Stabilizing damping lengths such as those in Fig. 11-1 should be used. Preliminary values for K_h/K_e, ω_m, and ω_h can be computed from the layout.

2. The restrictor should be sized such that

$$\omega_2 \leq 2\omega_m \left(\frac{K_e}{K_h}\right)^{1/2}$$

is satisfied. A practical restrictor size might require that K_h/K_e be reduced. This has the attendent disadvantage of lowering ω_h.

3. Compute the static compliance and compare with specifications.

4. ω_1 can now be computed and values for ω_3 can be obtained using application data. A Bode diagram of (11-17) is then sketched. Stability must be checked at all expected operating points and load conditions.

Some repeats of this procedure along with revised layouts will no doubt be required to arrive at a final design. Beveling or notching of the main orifice is often required to reduce the flow gain K_q.

Direct-operated (i.e., without stabilizing restrictor) pressure control valves respond rapidly to sudden flow demands but have poor stability and pressure control accuracy. The poor stability characteristics make these valves prone to generate noise in the form of whistles, screams, or chatter. Use of a restrictor improves stability and accuracy but with penalties of slower response and large pressure overshoots to rapid changes in load flow. Two-stage devices have the best stability and accuracy characteristics and are discussed in the next section.

11-3 TWO-STAGE PRESSURE CONTROL VALVES

Accurate pressure control, that is, low steady-state compliance, over a wide flow range usually requires a two-stage control valve (Fig. 11-7).

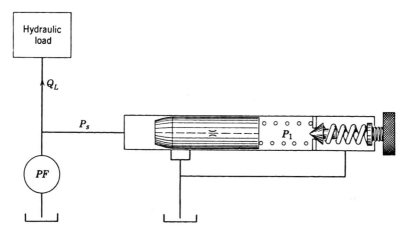

Figure 11-7   Two-stage pressure relief valve.

These valves are also referred to as *pilot-operated* or *compound*. The pilot section is usually a spring loaded poppet while the main valve is a spool type. The pilot valve can be located remotely from the main stage. Although these valves give accurate pressure control, they have slower response and are more susceptible to clogging with dirty fluids.

Referring to Fig. 11-7, the main orifice is held closed by a light seating spring on the main spool until flow through the fixed orifice builds up pressure P_1 to a value sufficient to open the pilot poppet. When the pilot valve opens, permitting flow to the return line, a pressure drop occurs across the fixed orifice. The resulting unbalance in force across the main spool opens the main orifice. Flow is then bypassed to the return line in such a manner that the supply pressure is held at the value set by the pilot spring. There are other configurations for two-stage pressure control valves, and the attributes of each have not been established and is an area deserving of study.

These two-stage valves are designed using experience and cut-and-try techniques because practical and useful analyses of their performance are lacking. However, this is the situation for many other hydraulic control components as well. Such studies should be done and offer a challenge to the universities having fluid power laboratories.

11-4 FLOW CONTROL VALVES

In many hydraulic circuits, control of flow is accomplished using simple adjustable needle, globe, or gate type valves. In some cases these valves

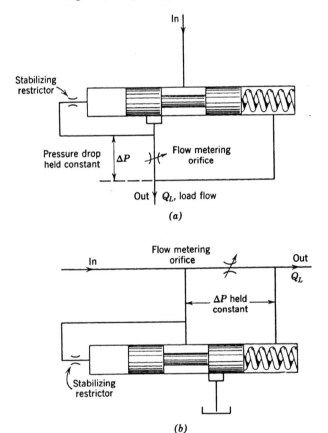

Figure 11-8 Pressure-compensated flow control valves: (a) restrictor type; (b) bypass type.

are used to regulate the speed of hydraulic actuators. However, varying loads on the actuator will cause changes in the pressure drop across the valve which in turn will alter the flow and, therefore, the speed of the actuator. In many cases these speed changes are tolerable. Because the flow through an orifice varies with the square root of pressure (rather than directly), an orifice is inherently a fairly good flow regulator.

When precise flow control is required it is necessary to hold constant pressure drop across the flow metering orifice using a pressure control valve. Two types of such *pressure-compensated flow control valves* can be distinguished (Fig. 11-8). In the usual type (Fig. 11-8a) the pressure control valve spool restricts the in-line flow to maintain constant pressure drop across the metering orifice. The other type, Fig. 11-8b, bypasses flow to maintain constant pressure drop.

With constant pressure drop (usually 40 to 100 psi) across the flow metering orifice, the flow is directly proportional to the orifice area. The area gradient of the metering orifice can be made linear, logarithmic, or specially contoured to obtain the desired relation between flow and position of the mechanical input which varies the orifice area. Because the pressure drop is held constant, design of the pressure control valve is made easier using the outline in Section 11-2.

Some pressure-compensated flow control valves have special features such as built-in check valves to allow free flow in reversed flow direction, built-in overload relief valves, or temperature compensation.

REFERENCES

[1] Leskiewicz, H. J., "Approach to the Theory of Hydraulic Pressure-Regulating Relief Valves," *ASME Paper No. 63-AHGT*-47.
[2] Ma, C. Y., "Analysis and Design of Hydraulic Pressure Reducing Valves," *ASME Trans., J. Eng. Ind., Paper No. 66-WA/MD*-4, 1967.

12

Hydraulic Power Supplies

In open loop type hydraulic control circuits the pump is an integral part of the control scheme. However, in hydraulic servo controls the pump is part of an auxiliary unit that can be considered simply as a source of hydraulic power. Hydraulic servovalves may be fed from either constant flow or constant pressure sources. If constant flow is used, the servovalve must have an open center so that it can pass full pump flow. The large standby power loss and the low pressure sensitivity of open center valves

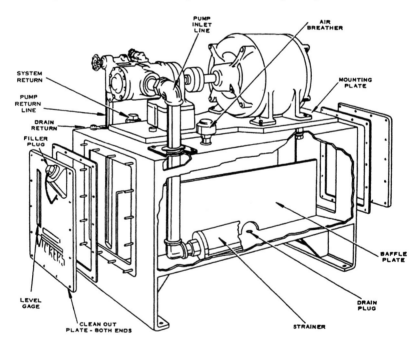

Figure 12-1 Constant pressure hydraulic power supply using a bypass relief valve pressure regulator (courtesy of Vickers, Inc., Machinery Hydraulics Division, Ferndale, Michigan).

usually make it more desirable to operate servovalves from constant pressure sources. These constant pressure power supplies are the subject of the present chapter. A typical industrial power supply is illustrated in Fig. 12-1.

12-1 BASIC CONFIGURATIONS OF HYDRAULIC POWER SUPPLIES

A constant pressure hydraulic power supply usually consists of a pump, a prime mover (an electric motor in stationary supplies) to drive the pump,

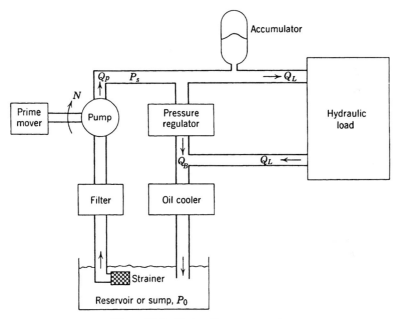

Figure 12-2 Hydraulic power supply.

a reservoir of fluid, a pressure regulator, a relief valve for safety purposes, filters, and, perhaps, an oil cooler and an accumulator. Such a supply is shown schematically in Fig. 12-2. The accumulator functions to filter pressure pulsations from the pump and to provide additional fluid under pressure to accommodate peak flow demands. Accumulators are often not necessary.

The reason for a pressure regulator should be thoroughly understood. A pump is essentially a *constant flow device*. The ideal pump flow is

$$Q_p = D_p N \qquad (12\text{-}1)$$

where D_p = displacement of the pump, in.^{3}/rad

N = pump speed, rad/sec

Thus, the pump simply moves an amount of fluid, and it does not determine the output or supply pressure. The pressure is determined primarily by the load to which the pump is connected. This is similar to an electric generator, an essentially constant voltage device, where the current drawn is determined by the load impedance. Actually, a pump has some internal leakage which is usually proportional to the pressure across the pump, that is, $Q_l = C_l P_s$. Thus, the effective pump output flow is $Q_p - Q_l$. If V_t is the total volume at the high pressure side of the pump and β_e is the

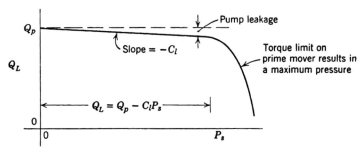

Figure 12-3   Load flow versus pressure for a pump with no pressure control.

effective bulk modulus, then the continuity equation gives

$$Q_p - C_l P_s - Q_L = \frac{V_t}{\beta_e} s P_s \tag{12-2}$$

where Q_L is the load flow. Rearranging yields

$$P_s = \frac{Q_p - Q_L}{C_l[1 + (V_t/C_l \beta_e)s]} \tag{12-3}$$

where P_s is the pump outlet or supply pressure. The steady-state portion of this equation, plotted in Fig. 12-3, shows that the load flow is relatively independent (except for leakage) of P_s. Also note that this pressure would become very large if Q_L approached zero. In fact, if there is no load flow or leakage, then

$$P_s = \frac{Q_p}{(V_t/\beta_e)s} \tag{12-4}$$

Pump output pressure increases rapidly as the integral of flow and the pump casing or lines would eventually rupture. For this reason, simple relief valves are used on pumps as a safety measure and often in addition to the pressure regulator.

The purpose of the pressure regulator is to limit P_s to a predetermined value and to modify pump stroke or pump flow to maintain it constant and independent of Q_L. Thus, an ideal pressure regulator is such that $\Delta P_s/\Delta Q_L = 0$, that is, there is no change in supply pressure as a result of flow changes.

There are two types of pressure regulators: *bypass regulators* and *stroke regulators*. Bypass regulators are used with fixed displacement pumps and their action is to sense pump output pressure, compare it with a reference, and bypass the flow not being used by the load in such a manner as to hold pump pressure at the reference value. The common relief valve is of this variety. These systems are quite fast but have the disadvantage of generating large quantities of heat in circulating full flow through the relief valve when there is no load flow. Stroke regulators are used with a variable displacement pump and their action is to compare pump output pressure with a reference and to vary the stroke arm of the pump so as to hold pump pressure at the reference value. This type regulator has the advantage of lower heat generation but has slow response and requires a more elaborate and expensive pump. These two types of pressure regulated supplies are schematically shown in Figs. 12-4 and 12-6. The features of each type of regulator tend to complement the other, and the application requirements would settle the choice to be made.

12-2 BYPASS REGULATED HYDRAULIC POWER SUPPLIES

Bypass regulated power supplies almost always use a two-stage relief valve because of their superior accuracy and stability over a wide range of flows. The analysis of this type of supply is identical to the analysis of its relief valve. Because two-stage relief valves are quite complex when all factors are included, an approximate analysis will be given.

Consider the bypass regulated constant pressure supply in Fig. 12-4. The continuity equation at the supply pressure chamber is

$$Q_p - C_l P_s - Q_B - Q_L = \frac{V_t}{\beta_e} s P_s \qquad (12\text{-}5)$$

The linearized orifice equation describing the bypassed flow is

$$Q_B = K_{qb} x_p + K_{cb} P_s \qquad (12\text{-}6)$$

where K_{qb} = flow gain of bypass valve, in.^{3}/sec/in.

K_{cb} = flow-pressure coefficient of bypass valve, in.^{3}/sec/psi

The force equation of the pilot spool is

$$P_s A_v - F_0 = \frac{1}{K_s}\left(\frac{s^2}{\omega_{nv}^2} + 1\right) x_v \qquad (12\text{-}7)$$

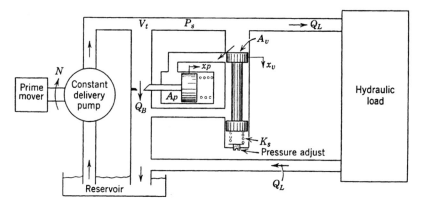

Figure 12-4   Bypass regulated constant pressure hydraulic power supply.

where F_0 = spring preload reference force, lb

   ω_{nv} = natural frequency of pilot spool mass and spring, rad/sec

For an elementary, and perhaps crude, analysis, bypass valve position is related to pilot valve position by

$$A_p s x_p = K_q x_v \tag{12-8}$$

where K_q = flow gain of pilot spool valve, in.^{3}/sec/in.

These four equations can be combined to form the block diagram in Fig. 12-5.

The loop gain function for this loop is

$$A(s) = \frac{K_v}{s\left(1 + \dfrac{s}{\omega_v}\right)\left(\dfrac{s^2}{\omega_{nv}^2} + 1\right)} \tag{12-9}$$

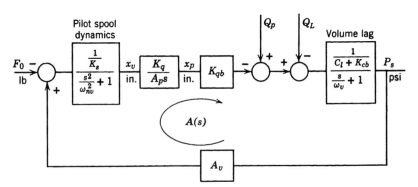

Figure 12-5   Block diagram of bypass type pressure regulator.

where $K_v = \dfrac{K_q A_v K_{qb}}{K_s A_p (C_l + K_{cb})}$ = velocity constant, sec^{-1}

$\omega_v = \dfrac{\beta_e (C_l + K_{cb})}{V_t}$ = volume lag, rad/sec

Because the loop contains an integration it can be made stable by selecting parameter values such that K_v is sufficiently small. However, stability may be difficult to control because K_v is dependent on many parameters, especially the leakage coefficients C_l and K_{cb}, which vary with the operating point.

Referring to Fig. 12-5 and neglecting the break at ω_{nv}, the dynamic compliance of the regulator, which is a measure of its control accuracy, is

$$\frac{\Delta P_s}{\Delta Q_L} = -\frac{\dfrac{1}{K_v (C_l + K_{cb})}\, s}{\dfrac{s^2}{K_v \omega_v} + \dfrac{s}{K_v} + 1} \qquad (12\text{-}10)$$

At steady-state the compliance is zero, indicating perfect regulation. As frequency is increased the compliance increases to a maximum value and then decreases as frequency is increased further. The compliance is zero at zero frequency and at infinite frequency. In practice, the compliance is not actually zero at steady-state because of flow forces and the spring force on the bypass valve.

Bypass pressure regulators (i.e., relief valves) modulate the bypass flow to hold the required output pressure. If the load suddenly takes all the flow the pump can deliver, the bypass valve closes and the regulator is inoperative in controlling pressure. Hence, the regulator should be supplied with enough flow, perhaps 5 to 10% of pump flow, so that it remains operative for all expected load flows. When load flow is zero, full pump flow circulates through the relief valve, and the heat generated and the resulting power loss become prohibitive. For these reasons, bypass regulators are used mainly with low power supplies or as safety devices in large power supplies. They are relatively fast and inexpensive.

12-3 STROKE REGULATED HYDRAULIC POWER SUPPLIES

Variable delivery pumps are used in constant pressure power supplies when efficiency is desired. These pumps employ a built-in hydraulically operated mechanism, essentially a three-way valve controlled piston with spring return, which varies the tilt angle of the swash plate and, therefore, the pump delivery to maintain constant supply pressure. Because pressure is regulated by varying the hydraulic horsepower generated, this type of power supply is much more efficient than the bypass type where the generated horsepower is constant. A stroke regulated power supply is illustrated in Fig. 12-6 and a much simplified analysis will be given.

The ideal pump flow is proportional to pump speed and to stroke control position. Therefore,

$$Q_p = D_p N = -k_p N x_p \qquad (12\text{-}11)$$

where K_p = displacement gradient of pump stroke control, in.^{3}/rad/in.

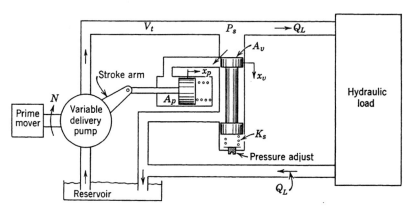

Figure 12-6   Stroke regulated constant pressure hydraulic power supply.

The negative sign simply denotes that pump flow is reduced when the stroke position is increased and is necessary to give feedback in the stroke control loop. For an elementary analysis, the pilot spool position and stroke control piston position can be approximately related by (6-49).

$$\frac{x_p}{x_v} = \frac{\dfrac{K_q}{A_p}}{s\left(\dfrac{s^2}{\omega_h^2} + \dfrac{2\delta_h}{\omega_h} s + 1\right)} \qquad (12\text{-}12)$$

where K_q = flow gain of pilot spool valve, in.^{3}/sec/in.
     ω_h = hydraulic natural frequency due to trapped oil spring and swash plate mass, rad/sec
     δ_h = hydraulic damping ratio, dimensionless
This equation neglects the effect of the piston return spring (as indicated in Section 6-3, this would cause the integration to become a large lag) and friction and pressure induced forces on the swash plate members. Neglecting flow to the pilot spool, the continuity equation for flows in the

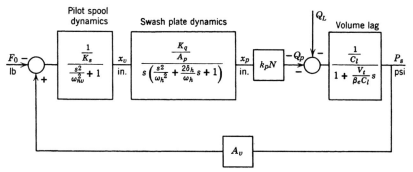

Figure 12-7   Block diagram of stroke control pressure regulator.

supply pressure chamber can be written

$$Q_p - C_l P_s - Q_L = \frac{V_t}{\beta_e} s P_s \qquad (12\text{-}13)$$

Equations 12-7, 12-11, 12-12, and 12-13 can be used to construct the block diagram in Fig. 12-7. The loop gain function and dynamic compliance can be obtained and are similar to that in Section 12-2. However, the swash plate dynamics cannot be neglected and account in large measure for the slower response of stroke regulated pumps.

12-4 INTERACTION OF HYDRAULIC POWER SUPPLY AND SERVO LOOP

A typical system consists of a variable delivery pump with a stroke regulator as a power supply for a servovalve and actuator, as shown schematically in Fig. 12-8. Any change in supply pressure will result in a change in load flow, which will initiate corrective regulator action. Thus

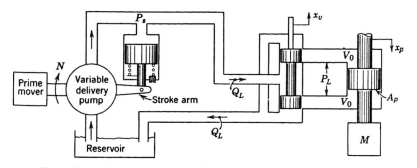

Figure 12-8   Constant pressure hydraulic power supply with servo load.

the hydraulic supply and the load share a control loop, and it is the stability of this loop that is to be investigated.

The loop gain of the interacting loop is simply the negative of the product of the regulator compliance and the transfer function $\Delta Q_L/\Delta P_s$ of the servo load. Therefore,

$$A(s) = - \left.\frac{\Delta P_s}{\Delta Q_L}\right|_{\substack{\text{press} \\ \text{reg}}} \times \left.\frac{\Delta Q_L}{\Delta P_s}\right|_{\substack{\text{servo} \\ \text{load}}} \tag{12-14}$$

Because the regulator compliance is always negative, this function is always positive. The interacting loop always has negative feedback, and stability can be determined from a Bode diagram of this equation.

As an example, the linear flow equation for the servovalve in Fig. 12-8 is

$$Q_L = K_q x_v - K_c P_L + K_c P_s \tag{12-15}$$

where K_q = servovalve flow gain, in.^{3}/sec/in.
       K_c = servovalve flow-pressure coefficient, in.^{3}/sec/psi
The continuity equation for the actuator chamber is (see Section 6-2)

$$Q_L = A_p s x_p + \frac{V_0}{2\beta_e} s P_L \tag{12-16}$$

neglecting actuator leakage. Assuming simple inertia load on the actuator the equation of motion is

$$P_L A_p = M s^2 x_p \tag{12-17}$$

These three equations can be combined to yield the following transfer function.

$$\frac{\Delta Q_L}{\Delta P_s} = \frac{K_c \left(\dfrac{s^2}{\omega_h^2} + 1 \right)}{\dfrac{s^2}{\omega_h^2} + \dfrac{2\delta_h}{\omega_h} s + 1} \tag{12-18}$$

where $\omega_h = \sqrt{\dfrac{2\beta_e A_p^2}{V_0 M}}$ = hydraulic natural frequency, rad/sec

$\delta_h = \dfrac{K_c}{A_p} \sqrt{\dfrac{\beta_e M}{2V_0}}$ = hydraulic damping ratio, dimensionless

This relation can be substituted into (12-14) along with the pressure regulator compliance, and a Bode plot can be made to determine stability.

Some general comments concerning stability of the interacting loop can be made. The quantity $\Delta Q_L/\Delta P_s$ for the servo load is normally small. The compliance of the regulator is also small. Because the interacting loop gain is the product of two small quantities, it is usually much less than unity at all frequencies and stability is no problem. Thus, the assumption of constant supply pressure often made in analyzing hydraulic servo systems is justified from the stability viewpoint.

Although stability is not a problem, one should not conclude that there is no interaction between hydraulic supply and the servo load. Sudden velocity changes in the servo result in flow demands in the hydraulic supply which can cause substantial changes in supply pressure and impair transient performance. Because of the many variables and nonlinearities involved, prediction of these pressure transients is virtually impossible. Direct measurement in the evaluation program for the system is necessary.

12-5 RESERVOIRS OF HYDRAULIC SYSTEMS

The primary function of a reservoir is to maintain sufficient supply of working fluid. It also stores the fluid, allowing entrained air to rise, contaminants to settle, and heat to dissipate. In many cases it serves as a convenient place to mount pump, heat exchanger, filter, and other equipment associated with the hydraulic supply.

The reservoir may be built-in or integral with the machinery being controlled or it may be separate if space and appearance permit. Integral reservoirs are more difficult to maintain, do not dissipate heat well, and may cause undesirable thermal distortion of adjacent members. Separate reservoirs are usually favored for stationary industrial applications, as shown in Fig. 12-1.

The reservoir capacity in gallons should be two to three, and preferably more, times the pump capacity in gallons per minute. This rule of thumb usually provides sufficient capacity to fill the system at start-up, to maintain a reasonable oil level despite flow fluctuations during operation, to sustain the system during an emergency shut down if a hydraulic line breaks, to absorb heat for short periods, to have enough wall surface for some cooling by radiation, to allow air and dirt to separate out of the oil, to contain fluid which may drain back to the tank from the system, and to replenish lost oil.

Referring to Fig. 12-1, some other important considerations in reservoir design and construction are:

1. Pump inlet and main return lines should be *below* the minimum working fluid level to prevent cavitation and aeration. Main (active) return lines should be directed against tank walls to assure peripheral circulation for good heat dissipation. Atmospheric drain lines should be discharged onto a plate above the working fluid level to spread the fluid and facilitate the release of entrained air. All lines should enter the top of the reservoir so that the tank is not drained when a line is removed.

2. Baffles should be used between the return and suction lines to reduce fluid velocities and to prevent continuous recirculation of the same fluid. This allows more time for air bubbles to escape and for contaminants to settle and helps to cool the oil.

3. The reservoir must be completely enclosed to prevent entrance of any foreign matter. Pressured reservoirs are desirable because airborne contamination is eliminated and pump suction is improved. However, pressurized drain lines are then required.

4. The air breather must be of sufficient size to maintain near atmospheric pressure in the tank at all flows. Otherwise, the pump may cavitate.

5. A liquid level indicator showing high and low levels is very desirable.

6. The reservoir should have convenient access plates and plugs for cleaning, draining liquid, and removing strainers and filters.

7. The filler opening should be conveniently located and have an internal screen to keep out foreign particles.

8. The pump inlet should be provided with a strainer in addition to the system filter.

Care must be taken in the assembly of hydraulic power supplies to make sure the reservoirs are thoroughly cleaned and properly painted for the particular type of working fluid. Certain types of synthetic fluids are very excellent paint removers.

12-6 HEAT GENERATION AND DISSIPATION IN HYDRAULIC SYSTEMS

Because hydraulic systems generate heat, they must operate above ambient temperature to dissipate the heat to the environment or to reject it to the coolant of a heat exchanger. Operation at excessive temperatures can break down the oil, causing sludges, varnishes, etc., which can clog orifices. Decrease in viscosity and lubricity at elevated temperatures may drastically shorten service life of components such as pumps. Seals, packings, hoses, filters, etc. have a definite temperature range for satisfactory operation. Thermal distortion in hydraulic components and adjacent structures may be undesirable. Therefore, it is quite important that the hydraulic system be designed so that a heat balance is achieved at a satisfactory operating temperature.

There are several sources of heat generation in hydraulic systems.

1. The major heat generators are the orifices and valves in the system used to throttle and control the flow. As pointed out in Section 3-6, the hydraulic power consumed by these devices is dissipated in heating the fluid and, to a much lesser extent, local heating of a valve itself. It is

important to realize that valves are inherently heat generators, but this is the price that must be paid for the ability to control. Relief valves and servovalves are good examples of heat generators.

2. Another source of heat generation is the resistive pressure drops in hydraulic lines, fittings, filters, and passageways in components such as valves, motors, and heat exchangers. Undersized and/or dirty passages should be avoided to minimize heat losses.

3. Leakage flow losses in pumps, motors, and valves add to the heat generated by a system.

4. Seal friction, mechanical friction, windage losses, and viscous drag between surfaces in pumps and motors also generate heat.

5. The compression of oil and, especially, entrained air to higher pressures in pumps during the pumping portion of the stroke causes heat generation. Rapidly cycling of gas-charged accumulators can cause gas temperatures higher than that of the oil which results in heat flow to the oil.

Each of the latter four sources of heat are usually much smaller than that produced by the metering orifices of control valves but collectively represent a significant contribution. The hydraulic system can absorb heat from external sources such as prime movers which also must be taken into account.

The operating temperature of the oil in a system should be determined to see whether it is satisfactory. This requires a computation of the heat generated and the natural heat dissipation capability of the system. One method of determining the heat generated is to add up the losses of each component. This technique is tedious and requires estimation of many quantities, such as leakage rates, efficiencies, and pressure drops. Because the hydraulic horsepower initially generated must be used in mechanical power at the output or converted into heat, a far better technique is to compute the total hydraulic horsepower generated by the pump (or pumps) and subtract the mechanical power delivered to the load by actuation devices (pistons or motors) to yield the heat that must be dissipated.

The mechanical horsepower and heat power equivalents of hydraulic pressure and flow will prove useful. Manipulation of appropriate conversion constants yields the horsepower developed as

$$\text{hp} = \frac{\Delta PQ}{1714.3} \tag{12-19}$$

and the equivalent heat power as

$$q = 2540\,\text{hp} = 1.485\,\Delta PQ \tag{12-20}$$

where ΔP = pressure difference across device, psi

Q = flow through device, gal/min

q = heat power, Btu/hr

hp = horsepower (1 hp = 550 ft-lb/sec = 42.4 Btu/min)

The hydraulic horsepower initially generated depends on the type of power supply. For a constant pressure supply with a bypass type (i.e., relief valve) regulator, full power is generated at all times and the heat power is

$$q|_{PF} = 1.485\,P_sQ_p \text{ Btu/hr} \qquad (12\text{-}21)$$

where P_s = constant supply pressure, psi

$Q_p = (60/231)D_pN$ = ideal pump flow, gal/min

Because the total power generated is desired, the ideal rather than actual pump flow (actual is the ideal flow minus leakage) is used because the difference represents a heat loss. If the constant pressure supply used a stroke regulated variable delivery pump, the heat power will be less and depends on the load flow, Q_L. Therefore

$$q|_{PV} = 1.485P_sQ_p\left(\frac{Q_L}{Q_p}\right) \qquad (12\text{-}22)$$

where Q_p now denotes the pump flow at maximum stroke and the ratio Q_L/Q_p is between zero and one. An analysis of the load duty cycle is necessary to establish Q_L. If the duty cycle is not repetitive, then Q_L/Q_p might be estimated at say 0.5 or the most conservative value of unity used. It should be clear that the two equations given do not represent the heat generated by the pump but rather the heat power equivalent of the generated hydraulic power.

The load duty cycle must now be analyzed to determine the average mechanical power delivered to the load. If the output actuator is holding a given position during most of the cycle, no power is consumed. This is usually the case in servo controlled systems where nearly all the generated power is eventually dissipated as heat. Systems using a variable delivery pump have much less heat dissipation when the actuator is holding position because the pump flow is reduced to only that necessary to supply leakage losses. However, the heat losses would be comparable to that of a bypass supply if the actuator moved at high velocities and required low pressure differences, but such loads are exceptional.

The heat power to be dissipated is the equivalent heat power generated at the pump minus the mechanical power at the actuator. Heat is dissipated in hydraulic systems by the three basic methods of conduction, radiation, and convection. Heat conduction to adjacent structures is the principal cooling means when the reservoir is built-in. The heat conducted is given by

$$q = kA \frac{dT}{dx} \text{ Btu/hr} \qquad (12\text{-}23)$$

where k = thermal conductivity of the structure material, Btu/ft-hr-°F

$\frac{dT}{dx}$ = thermal gradient in the direction of heat flow, °F/ft

A = area normal to heat flow path, ft²

For complex and irregular structures, as they all are, it is obviously difficult to identify the thermal gradients, and specific tests are required to determine the oil temperature.

Separate reservoirs are self-cooled by radiation and, to a lesser degree, by free convection of heat from the heat sink formed by the mass of the oil, reservoir, housings, and tubing. The heat transferred from separate reservoirs is usually written

$$q = UA \, \Delta T \qquad (12\text{-}24)$$

where U = over-all heat transfer coefficient, Btu/ft²-hr-°F

ΔT = temperature difference between oil and ambient, °F

q = heat dissipation rate, Btu/hr

A = surface area of reservoir (usually taken as the area of the sides and bottom, if above the floor, but not the top), ft²

U man depends on many variables (including paint color, which should be dark for good radiation) but mainly on the air circulation around the tank. A value in the range 1.5 to 3 Btu/ft²-hr-°F is applicable for normal installations where air flow is somewhat inhibited and the tank is not particularly free of dirt. If the air circulation is improved, as on a moving vehicle or aided with a fan, then values of 10 and higher are possible. Much higher heat transfer coefficients are, of course, possible with heat exchangers. Assuming $U = 2$ Btu/ft²-hr-°F for a reservoir in free air, from (12-20) and (12-24) we get that an area to horsepower ratio of about 25 ft²/hp is needed if the oil temperature is not to exceed 50°F above ambient. Thus, if 5 hp had to be dissipated, then 125 ft² of area is required. A 50 gal separate type reservoir in free air usually has enough surface area to dissipate heat at a rate of about 1 hp (42.4 Btu/min) with a 50°F rise above ambient.

If the heat to be dissipated exceeds the natural dissipation capability of the system for reasonable oil temperatures, a heat exchanger is necessary. Most heat exchangers are air cooled or water cooled. However, refrigeration units are used where water is too expensive or not available and air coolers are not applicable. The air cooled units consist of a fan or blower forcing ambient air across a core section of tubes and fins through which the hot oil flows. Water cooled heat exchangers usually employ shell-and-tube construction with water flowing through the tubes and oil flowing across the tubes. The water flow rate can be automatically controlled to maintain a given oil temperature.

Manufacturers recommendations should be followed in sizing of the heat exchanger. Heat exchangers should be installed in the return or low pressure line of the system to eliminate the need for expensive high pressure units and should be protected against pressure surges.

12-7   CONTAMINATION AND FILTRATION

Contamination of the working fluid, that is, the presence of foreign materials, is responsible for the vast majority of system troubles. An essential part of system design is the control of contamination with easily serviced filters so that particle sizes are restricted to a level satisfactory for operation.

Contamination in servovalves causes friction between the spool and sleeve which increases hysteresis, erosion of the metering edges which increases center flow, silting, sticking of the spool and, in extreme cases, complete failure due to clogging of internal orifices. Abrasive contaminants such as sand, metallic particles, and lapping compound residues promote wear and may cause rapid failure of some pumps by scoring the valving plate. Contamination can also result in a slow deterioration of the over-all system performance. When intermittent and irratic performance and/or troubles of undefined origin occur, it is sound to suspect dirty oil.

Contaminants may be self-generated within the system, such as the metallic particles from the normal wear of components such as pumps and motors, elastometer particles from seal wear, shedding of the filter media and flexible lines. Improperly cleaned components during their manufacture is the major source of contaminants and causes cloth fibers, grits and chips from grinding and machining, pipe sealing compounds, and lapping compound residues to appear in the working fluid. Airborne particles may enter the system as a contaminant. Carelessness in handling components during maintenance may contribute dirt.

One micron (1 μ), which is one millionth of a meter, is used as the unit of length in dealing with contamination and filtration. Useful conversion constants are 1 μ = 0.0004 in. and 25 μ = 0.001 in.

In recent years there has been much progress in the art of contamination analysis and specification. Several methods of determining the contamination level in a hydraulic fluid are in use.

1. Visual inspection—because the lower limit of human visibility is 40 μ an oil that appears dirty is very dirty because particle sizes greater than 40 μ are being observed. This test, usually the first to be made, is satisfactory and conclusive in noncritical applications.

2. Patch test—a sample of fluid is filtered through a filter paper which collects the contaminant. The color or shade of the paper is then used as a rough measure of the contamination level. This technique requires experience with a particular type of hydraulic system to develop standards for comparison.

3. Silting index—a sample of fluid is discharged through a small pore filter under constant pressure. As the filter clogs, the flow rate decays with the latter half of the sample taking more time to pass through than the first half. The difference in these times is taken as a measure of contamination and can be used to compute a number called Silting Index. Because large particles do not clog the filter, this technique is a measure of the finer particles of about 5 μ or less.

4. Gravimetric analysis—a volume of contaminated fluid is passed through a dry preweighed filter paper which retains the contaminant. The filter is flushed with a solvent to remove the oil retained, dried, and weighed to yield the contaminant weight. The weight of the contaminant (usually in milligrams) per unit of fluid volume (usually in gallons or liters) is used as a measure of the contamination level. This technique is relatively simple but does not take into consideration particle size.

5. Electronic particle counter—the fluid sample is passed through a transparent tube. A light beam is projected through the tube and sensed with a photocell. Electronic counters record the number of light interruptions to obtain a measure of the number of particles in certain micron ranges. These instruments are expensive but give a direct and rapid particle count. The result is expressed as the number of particles per 100 ml sample in various size ranges.

6. Optical particle count—a 100 ml fluid sample is passed through an analysis filter with ruled squares where the contaminants are deposited. The filter is placed under a microscope and the particles in a statistically significant number of squares are counted in several different size ranges. The results are expressed as the number of particles per 100 ml sample in the various size ranges. This technique is the most commonly used. It is most sensitive and gives particular information concerning the nature and shape of the contaminant. Relatively simple equipment is necessary, but a skilled technician is required because of the many sources of error, such as quality of optical instruments and lighting, distribution of contaminant, and operator fatigue.

Particle counts, gravimetric analysis, and silting index are all quantitative measures of contamination level. Particle count is the best measure, but it is awkward to simply state. To avoid this difficulty, various classes of contamination levels have been proposed (Table 12-1). The particle count of a sample would be compared with the values in Table 12-1 to select the contamination class descriptive of the system. In this manner a single number is used to give particle count information.

Table 12-1 SAE, ASTM, AIA Tentative Hydraulic Contamination Standards Particles per 100 ml by Class of System (tentative)

Size Range Micron	Contamination Class							
	0	1	2	3	4	5	6	7–10
2.5–5	Pending							
5–10	2,700	4,600	9,700	24,000	32,000	87,000	128,000	P e n d i n g
10–25	670	1,340	2,680	5,360	10,700	21,400	42,000	
25–50	93	210	380	780	1,510	3,130	6,500	
50–100	16	28	56	110	225	430	1,000	
>100	1	3	5	11	21	41	92	

Typically and Approximately   Classes 3 and 4—critical systems, in general
Class 0—rarely attained   Class 5—poor missile system
Class 1—MIL-H-5606B   Class 6—fluid as received
Class 2—good missile system   Class 7—industrial service

It is the nature of random contamination in a system that a plot of the cumulative number of particles versus particle size is a straight line on semilog paper (Fig. 12-9). Hence only a few points are required to establish the number of particles of any size. The effect of filtration is to lower this curve. The ability to measure and specify contamination levels makes it possible to evaluate filters, intelligently apply them, and have a meaningful preventative maintenance program.

Fluids must be filtered to remove contaminants. There are two basic types of filter media: *surface* and *depth*. The comparative retentivity characteristics of each are shown in Fig. 12-10. However, some filter media may have properties of both these basic types.

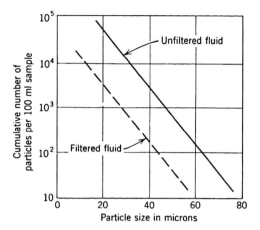

Figure 12-9   Typical plot of cumulative number of particles versus particle size.

The distinguishing feature of surface type, also known as absolute or screen type, filter media is their uniform and specific pore size. This type of filter has absolute retention of all particles, except perhaps long fibers, larger than its pore size. The contaminant collects on the surface, hence the name, and, because it loads up readily, this filter has low dirt-holding capacity. However, these filter media often have good mechanical strength, low shedding, and are cleanable. Examples of surface type filter media are pierced metal, wound wire, woven wire cloth using square, dutch, twill, and dutch twill weaves, and the Millipore filter.*

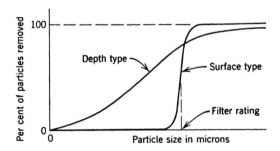

Figure 12-10   Retention characteristics of surface and depth type filter media.

* The Millipore filter, made by Millipore Filter Corporation, Bedford, Massachussetts, is composed of a thin porous membrane of pure cellulose esters. Porosity grades range from 0.01 to 8μ. Because the pore size is extremely uniform, this filter has become a standard in most contamination analysis procedures. It is used where fine filtration is desired.

A depth type filter medium is composed of a relatively deep matrix of randomly distributed windings, fibers, or particles. The contaminated fluid must flow through numerous, long, and tortuous passages of differing cross sections. Both particles and fiber types of contaminants are absorbed and/or entrapped in the interstices of the filter matrix. Because this type of filter medium has no specific pore size, there is no definite limit to the particle size which may pass. However, the density of the filter barrier is

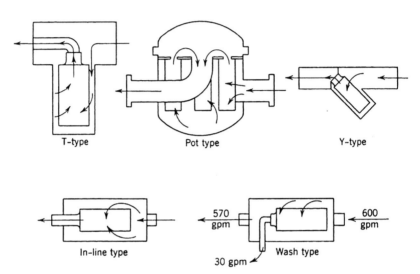

Figure 12-11 Basic configurations of filter assembles (from *Filtration in Modern Fluid Systems* by H. L. Wheeler, Jr., 1964, and courtesy of the Bendix Corp., Madison Heights, Michigan).

such that it is penetrated by only a few particles above the rated size of the filter. In contrast to surface type media, depth filters remove substantial particles below the rated size of the filter. Hence, this type of filter is often preferred for its ability to remove fine particles. Depth type filter media have large dirt-holding capacity, can trap fibers in the long mazelike passages, have low pressure drop, and are inexpensive. However, they are not cleanable, vibration and pressure pulsations may force contaminants through the filter, they may collapse or burst under high pressure differences, the filter size is often large and bulky, and the filter medium tends to shed, which contributes to contamination. Examples of depth type filter media are sintered metals, resin-impregnated papers, matted fibers such as felt, cellulose, and fiberglass, ceramics, sand, fullers earth, etc.

The filter medium must be suitably housed for installation in a system. The five basic filter configurations are shown in Fig. 12-11. The T-type configuration is the most popular as a system filter because of its compactness and ease of servicing. The in-line type filter is often built into components to protect critical orifices such as those in the pilot stage of servovalves. The filter housing may contain pressure difference indicators which show when filter is clogged and should be serviced. Relief valves which open and bypass the flow around the filter when it becomes clogged are often built into the filter housing. This improves system reliability because operation is possible if the filter is completely clogged.

Systems always require filtration; it is simply the type and size of filter that must be determined. Safe contamination levels usually evolve from past experiences with similar systems and the recommendations of component manufacturers. The micron rating of the filter is selected accordingly. Additional factors in filter selection are dirt-holding capacity (oversize the filter if possible), ability to pass required flow with a minimum of pressure drop, ability to withstand the pressure levels at the place where it is installed, shedding of the filter medium, and the number of active devices in the system which generate contaminants.

Ideally, both types of filter media should be used especially in critical systems. A depth type prefilter is used to remove large quantities of contaminants, especially fine particles, and is followed by a surface type filter to retain all particles of a harmful size. The surface filter would clog rapidly without the depth prefilter. However, cost and space usually dictate that only one filter can be used, and a depth type is generally chosen.

Ideally, the filter should be installed at the last possible place before the critical components. Several filters might be required because several critical elements, such as servovalves and pumps, may be involved. However, very often only one filter is used, and it is placed in the pump outlet line. The pump is then not directly protected except for a coarse surface type filter (strainer) which is always placed in the pump inlet line as a guard against large particles entering and damaging the pump. If the filter is placed in the pump inlet, then wear particles generated by the pump can pass to the servovalve, and cavitation of the pump might occur. Sometimes the filter is placed in the return line. Because fluid velocities are low at this point, good filtration can be achieved with an inexpensive low pressure unit. However, the reservoir must be maintained free of dirt rather than considered a place for contaminants to settle and to possibly enter the pump. This can be accomplished by placing the pump inlet at a low point in the reservoir. Sometimes a bypass type filter arrangement is used in which a separate pump forces fluid through only a filter. The filter and its pressure drop are eliminated from the main hydraulic circuit, but there is only a statistical certainty of clean oil entering critical components.

Experience has shown that a system cleaned at the time of manufacture can be kept clean with the system filter. Therefore, system flushing with several filter changes is recommended to clean the new hydraulic fluid, tubing, manifolds, and passageways in components of a newly constructed system. This flushing should be done before the installation of critical components, such as servovalves, to prevent premature wear by abrasive contaminants.

Because filters physically retain the contaminants, they are inexorably doomed to clog and eventually to fail to function. Hence, very fine filtration requires frequent filter inspections and a sound preventative maintenance program.

REFERENCES

[1] Lustig, R., "Hydraulic System Reservoirs," *Machine Design*, June 6, 1963, 146–150.

[2] Dodge, L., "Oil-system Cooling," *Product Eng.*, June 25, 1962, 92–96.

[3] Wheeler, Jr., H. L., *Filtration in Modern Fluid Systems*. Bendix Corporation, 434 W. 12 Mile Road, Madison Heights, Michigan, 1964.

[4] *Detection and Analysis of Contamination*. Millipore Filter Corporation, Bedford, Massachussetts, ADM-30, 1964.

[5] *Ultracleaning of Fluids and Systems*. Millipore Filter Corporation, Bedford, Massachussetts, ADM-60, 1963.

Index

355